Fundamentals of Industrial Instrumentation and Process Control

About the Author

WILLIAM DUNN has degree in physics from the University of London and electrical engineering degree from the University of Ottawa. He has also worked for over 40 years in product research, design, and management in the electronics industry. More recently he has worked as an adjunct professor teaching digital electronics, industrial instrumentation and process control, logic controllers, and industrial motor control. He holds more than 25 patents and has presented over 30 papers at industrial conferences.

Fundamentals of Industrial Instrumentation and Process Control

William C. Dunn

Second Edition

Mc
Graw
Hill
Education

New York Chicago San Francisco Athens
London Madrid Mexico City Milan
New Delhi Singapore Sydney Toronto

Library of Congress Control Number: 2018933117

Fundamentals of Industrial Instrumentation and Process Control, Second Edition

1 2 3 4 5 6 7 8 9 QVS 23 22 21 20 19 18

ISBN 978-1-265-79365-4
MHID 1-26-579365-4

Sponsoring Editor
Mike McCabe

Editorial Supervisor
Donna M. Martone

Acquisitions Coordinator
Elizabeth Houde

Project Manager
Sonam Arora,
Cenveo® Publisher Services

Copy Editor
Nupur Mehra

Proofreader
Diksha Rai

Production Supervisor
Pamela A. Pelton

Composition
Cenveo Publisher Services

Art Director, Cover
Jeff Weeks

Contents

Preface

The evolution of industrial processing and control has gone through several industrial revolutions to the complexities of modern day microprocessor controlled processing. Today's technology evolution has made it possible to measure parameters deemed impossible only a few years ago, as well as improvements in accuracy, tighter control, and waste reduction.

This text was specifically written as an introduction to modern day industrial instrumentation and process control for the two-year technical, vocational, or degree student and as a reference manual for managers, engineers, and technicians working in the field of instrumentation and process control. It is anticipated that the perspective student will have a basic understanding of mathematics, electricity, and physics. This course should adequately prepare a perspective candidate wishing to become a technician, or for the perspective engineer wishing to get a solid basic understanding of instrumentation and process control in this ever changing field.

Instrumentation and process control involves a wide range of technologies and sciences, and is used in an unprecedented number of industrial applications from applications in the home and office such as heating and cooling control and hot water systems, to chemical and automotive instrumentation and process control. This text is designed to cover all aspects of industrial instrumentation, such as sensing a wide range of variables, the transmission and recording of the sensed signal, controllers for signal evaluation, and the control of the manufacturing process for a quality and uniform product.

Chapter 1 gives an introduction to industrial instrumentation. Chapters 2–8 describe the wide range of sensors and their use in the measurement of a wide variety of physical variables, such as level, pressure, flow, temperature, humidity, density, position, force, and safety. Chapter 9 discusses signal conditioning, amplification, bridge circuits, and temperature compensation. The use of regulators and actuators for controlling pressure, flow, and the control of the input variables to a process are discussed in Chap. 10. System configurations are given in Chap. 11 which cover ON/OFF systems and the types of feedback used in closed-loop systems. In Chap. 12, documentation as applied to instrumentation and control is introduced, together with standard symbols recommended by the Instrument Society of America (ISA) for use in instrumentation control diagrams. Electronics and pneumatics are the mediums for signal transmission, and control. The use of the HART protocol is considered together with analog-to-digital signal conversion in Chap. 13. Chapter 14 introduces number conversions, logic gates, and functional building blocks. Chapter 15 discusses the programmable logic controller and ladder diagrams. Finally, in Chap. 16, various types of electric motors and their use in process control are discussed.

The primary reason for writing this book was that the author felt there was not a clear, concise, and up-to-date text for understanding the basics of instrumentation and process control from the perspective of the technician and engineer. The Second Edition has been expanded to try and cover as many of the basics as possible that are used in process control. Every effort has been made to ensure that the text is accurate, easily readable, and understandable.

Both engineering and scientific units are discussed in the text. Each chapter contains worked examples for clarification, with exercise problems at the end of each chapter. A glossary is given at the end of the text. There are answers to the odd numbered questions in the Appendix. An instructor's manual with answers to the problems is available online.

Acknowledgment

I would like to thank my wife Nadine for her patience, understanding, and many helpful suggestions during the writing of this text.

Fundamentals of Industrial Instrumentation and Process Control

CHAPTER 1

Introduction

Chapter Objectives

This chapter is to introduce you to instrumentation, the various measurement units used, and why process control relies extensively on instrumentation. This chapter will help you become familiar with instrument terminology and standards.

Topics discussed in this chapter are as follows:

- The basics of a process-control loop
- Definition of the elements in a control loop
- The difference between the various types of variables
- Considerations in a process facility
- Units, standards, and prefixes used in parameter measurements
- Comparison of the English and the SI units of measurement
- Instrument accuracy and the parameters that affect an instrument's performance

1.1 Introduction

Instrumentation is the basis used in industry for process control. However, it comes in many forms, from domestic water heaters and HVAC (heating, ventilation, and air-conditioning), where the variable temperature is measured and used to control gas, oil, or electricity flow to the water heater, or heating system, or electricity to the compressor for refrigeration, to complex industrial process-control applications such as the ones used in the petroleum or chemical industry.

In industrial control a wide number of variables can be sensed simultaneously from temperature, flow, and pressure to time and distance, all of which can be interdependent variables in a single process requiring complex microprocessor systems for total control. Due to the rapid advances in technology, instruments in use today may be obsolete tomorrow; new and more efficient measurement techniques are constantly being introduced. These changes are being driven by the need for higher accuracy, quality, precision, and performance. To measure parameters accurately, techniques have been developed that were thought impossible only a few years ago.

1.1.1 History

Manufacturing can be dated back to when humans first made tools, clay pots, bowls, plates, and the like in the early stone age or before. It may have been crude manufacturing, but nonetheless it was a manufacturing process, with control in the hands of the maker. Limited process control came with the mining and smelting of copper and then tin to make bronze for the manufacturing of tools, weapons, and so on some two millennia BCE. Iron was then discovered and used to replace bronze for tools, weapons, and suits of armor and chain mail appeared as manufacturing processes became better understood and a degree of quality control was initiated. About this time the water wheel came into being and was used for pumping water, crushing ore, and grinding wheat, corn, and the like. The next big step in manufacturing came with Watt's steam engine in the 1770s. The steam engine provided the power to drive manufacturing machines, and hence the start of the industrial revolution. To support and control the manufacturing process, came the need for sensors, instruments, and process control. Pneumatics then came into being to give the feedback signals for control and automation. The development of the electric motor gave greater flexibility in power generation and control until today with the development of the electronics industry we have process controllers that can simultaneously change many variables and control many actuators and valves for temperature and flow. The assembly and production lines give tight control and high-quality processing. We must also mention the many organizations that have come into being for developing process rules, regulations, and standardization.

1.2   Process Control

In order to produce a product with consistent high quality, tight process control is necessary. A simple-to-understand example of process control would be the supply of water to a number of cleaning stations, where the water temperature needs to be kept constant in spite of the demand. A simple control block is shown in Fig. 1.1a: steam and cold water are fed into a heat exchanger, where heat from the steam is used to bring the cold water to the required working temperature. A thermometer is used to measure the temperature of the water (the measured variable) from the process or exchanger. The temperature is observed by an operator who adjusts the flow of steam (the manipulated variable) into the heat exchanger to keep the water flowing from the heat

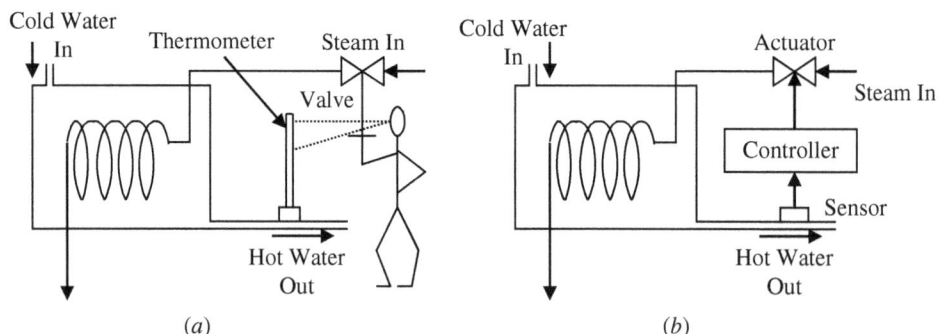

FIGURE 1.1   Process control showing (a) the manual control of a simple heat exchanger process loop and (b) automatic control of a heat exchanger process loop.

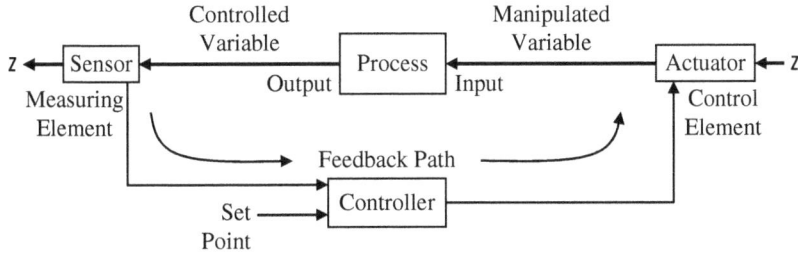

FIGURE 1.2   Block diagram of a process-control loop.

exchanger at the constant set temperature. This operation is referred to as process control, and in practice would be automated as shown in Fig. 1.1*b*.

Process control is the automatic control of an output variable by sensing the amplitude of the output parameter from the process and comparing it to the desired or set level and feeding an error signal back to control an input variable in this case steam. See Fig. 1.1*b*. A temperature sensor attached to the outlet pipe senses the temperature of the water flowing. As the demand for hot water increases or decreases, a change in the water temperature is sensed and converted to an electrical signal which is then amplified and sent to a controller that evaluates the signal and sends a correction signal to an actuator, which adjusts the flow of steam to the heat exchanger to keep the temperature of the water at its predetermined value.

The diagram in Fig. 1.1*b* is an oversimplified feedback loop and is expanded in Fig. 1.2. In any process there are a number of inputs, i.e., from chemicals to solid goods. These are manipulated in the process and a new chemical or component emerges at the output. The controlled inputs to the process and the measured output parameters from the process are called variables.

In a process-control facility the controller is not necessarily limited to one variable, but it can measure and control many variables. A good example of the measurement and control of multivariables that we encounter on a daily basis is performed by the processor in the automobile engine.

Figure 1.3 lists some of the functions performed by the engine processor. Most of the controlled variables are six or eight devices depending on the number of cylinders in the engine. The engine processor has to perform all of these functions in approximately 5 ms. This example of engine control can be related to the operations carried out in a process-control operation.

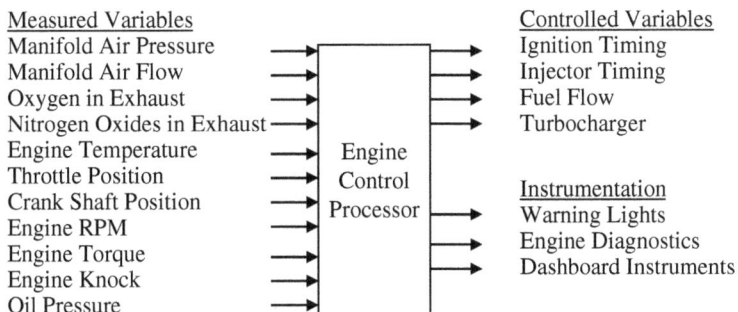

FIGURE 1.3   Automotive engine showing some of the measured and controlled variables.

1.3   Definition of the Elements in a Control Loop

Figure 1.4 breaks down the individual elements of the blocks in a process-control loop. The measuring element consists of a sensor, a transducer, and a transmitter with its own regulated power supply. The control element has an actuator, a power control circuit, and its own power supply. The controller has a processor with a memory and a summing circuit to compare the set point to the sensed signal so that it can generate an error signal. The processor then uses the error signal to generate a correction signal to control the actuator and the input variable. The function and operation of the blocks in different types of applications will be discussed in Chaps. 9, 10, and 11. The definition of these blocks is given as follows:

Feedback loop is the signal path from the output back to the input to correct for any variation between the output level from the set level. In other words, the output of a process is being continually monitored. The error between the set point and the output parameter is determined, and a correction signal is then sent back to one of the process inputs to correct for changes in the measured output parameter.

Controlled or measured variable is the monitored output variable from a process, and the value of the monitored output parameter is normally held within tight given limits.

Manipulated variable is the input variable or parameter to a process that is varied by a control signal from the processor to an actuator. By changing the input variable, the value of the measured variable can be controlled.

Set point is the desired value of the output parameter or variable being monitored by a sensor. Any deviation from this value will generate an error signal.

Instrument is the name of any various device types for indicating or measuring physical quantities or conditions, performance, position, or direction, and the like.

Sensors are devices that can detect physical variables, such as temperature, light intensity, or motion, and have the ability to give a measurable output that varies in relation to the amplitude of the physical variable. The human body has sensors in

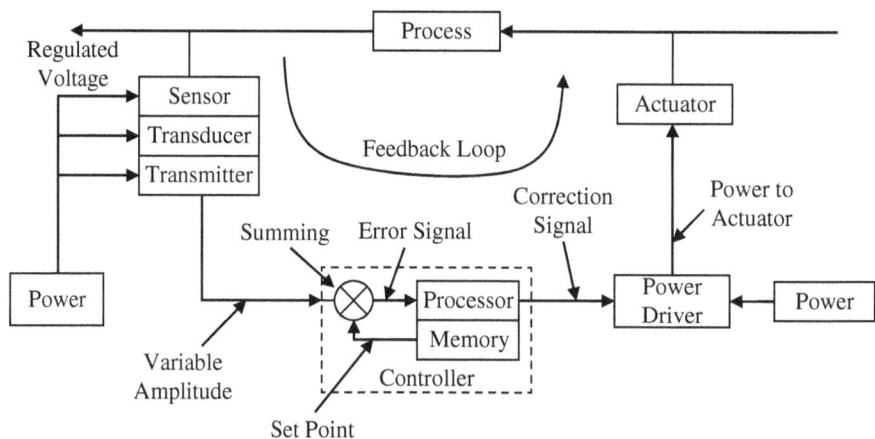

Figure 1.4 Block diagram of the elements that make up the feedback path in a process-control loop.

the fingers that can detect surface roughness, temperature, and force. A thermometer is a good example of a line-of-sight sensor, in that it will give an accurate visual indication of temperature. In other sensors such as a diaphragm pressure sensor, a strain transducer may be required to convert the deformation of the diaphragm into an electrical or pneumatic signal, before it can be measured.

Transducers are devices that can change one form of energy to another, e.g., a resistance thermometer converts temperature into electrical resistance, or a thermocouple converts temperature into a voltage. Both of these devices give an output that is proportional to the temperature. Many transducers are grouped under the heading of sensors.

Converters are devices that are used to change the format of a signal without changing the energy form, i.e., a change from a voltage to a current signal.

Actuators are devices that are used to control an input variable in response to a signal from a controller. A typical actuator will be a flow control valve, which can control the rate of flow of a fluid in proportion to the amplitude of an electrical signal from the controller. Other types of actuators are magnetic relays that turn on and off electrical power, such as power to the fans and compressor in an air-conditioning system in response to signals from the room temperature sensors.

Controllers are devices that monitor signals from transducers and take the necessary action to keep the process within specified limits according to a predefined program by activating and controlling the necessary actuators.

Programmable logic controllers (PLCs) are used in process-control applications, and are microprocessor-based systems. Small systems have the ability to monitor several variables and control several actuators, with the capability of being expanded to monitor 60 or 70 variables and control a corresponding number of actuators, as may be required in a petrochemical refinery. PLCs have the ability to use analog or digital input information and output analog or digital control signals. They can also communicate globally with other controllers, are easily programmed on line or off line, and supply an unprecedented amount of data and information to the operator. Ladder networks are normally used to program the controllers.

An error signal is the difference between the set point and the amplitude of the measured variable.

A correction signal is the signal used to control power to the actuator to set the level of the input variable.

Transmitters are devices used to amplify and format signals so that they can transmit data over long distances without loss of accuracy. The transmitted signal can be in one of several formats, i.e., pneumatic, digital, analog voltage, analog current, or as a radio frequency (RF) modulated signal. Digital transmission is preferred in newer systems since the controller is microprocessor based. Analog data transmission is still used over short distances, and in some systems both analog data and digital data are combined (see Chap. 13). The controller compares the amplitude of the signal from the sensor to a predetermined set point, which in Fig. 1.1*b* is the amplitude of the signal of the hot water sensor. The controller will then send a signal that is proportional to the difference between the reference and the transmitted signal to the actuator telling the actuator to open or close the valve controlling the flow of steam to adjust the temperature of the water to its set value.

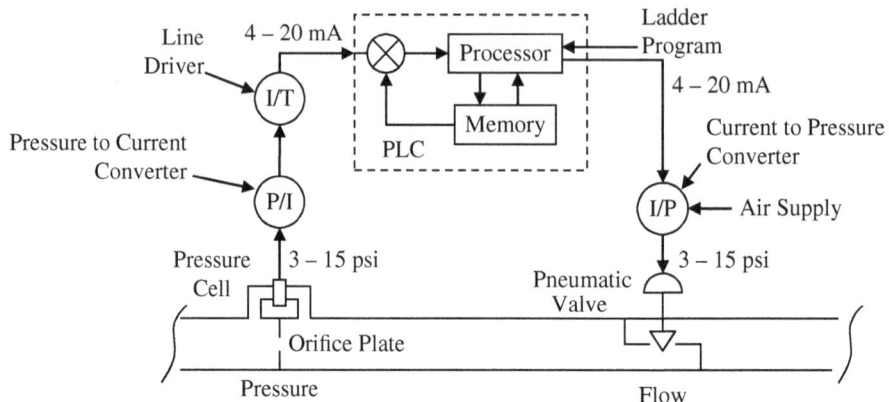

FIGURE **1.5** Process control with a flow regulator for use in Example 1.1.

Example 1.1 Figure 1.5 shows the block diagram of a closed-loop flow control system. Identify the following elements: (*a*) the sensor, (*b*) the transducer, (*c*) the actuator, (*d*) the transmitter, (*e*) the controller, (*f*) the manipulated variable, and (*g*) the measured variable.

Solution (*a*) The sensor is labeled pressure cell in the diagram. (*b*) The transducer is labeled converter. There are two transducers—one for converting pressure to current and the other for converting current to pressure to operate the actuator. (*c*) The actuator in this case is the pneumatic valve. (*d*) The transmitter is the line driver. (*e*) The controller is labeled as a PLC. (*f*) The manipulated variable is the differential pressure developed by the fluid flowing through the orifice plate constriction. (*g*) The controlled variable is the flow rate of the liquid.

Simple and ideal process-control systems have been discussed. In practical process control, the scenarios are much more complex with many scenarios and variables to be considered, such as stability, reaction time, and accuracy. Many of the basic problems are discussed in the following chapters.

1.4 Process Facility Considerations

The process facility has a number of basic requirements including safety precautions and well regulated, reliable electrical, water, and air supplies.

An *electrical supply* is required for all control systems and must meet all standards in force at the plant. The integrity of the electrical supply is most important. Many facilities have backup systems to provide an uninterruptible power supply (UPS) to take over in case of loss of external power. Power failure can mean plant shutdown and the loss of complete production runs. Isolating transformers should be used in the power supply lines to prevent electromagnetic interference (EMI) generated by motors, contactors, relays, and so on from traveling through the power lines and affecting sensitive electronic control instruments.

Grounding is a very important consideration in a facility for safety reasons. Any variations in the ground potential between electronic equipment can cause large errors in signal levels. Each piece of equipment should be connected to a heavy copper bus that is properly grounded. Ground loops should also be avoided by grounding cable screens and signal return lines at one end only. In some cases, it may be necessary to use signal isolators to alleviate grounding problems in electronic devices and equipment.

An *air supply* is required to drive pneumatic actuators in most facilities. Instrument air in pneumatic equipment must meet quality standards. The air must be dirt, oil, contaminant, and moisture free. Frozen moisture, dirt, and the like can block or partially block restrictions and nozzles giving false readings or complete equipment failure. Air compressors are fitted with air dryers and filters, and have a reservoir tank with a capacity large enough for several minutes' supply in case of system failure. Dry clean air is supplied at a pressure of 90 psig (630 kPa(g)) and with a dew point of 20°F (10°C) below the minimum winter operating temperature at atmospheric pressure. Additional information on the quality of instrument air can be found in ANSI/ISA-7.0.01-1996, *Quality Standard for Instrument Air.*

Water supply is required for many cleaning and cooling operations, and for steam generation. Domestic water supplies contain large quantities of particulates and impurities, and may be satisfactory for cooling, but are not suitable for most cleaning operations. Filtering and other similar processes can remove some of contaminates making the water suitable for some cleaning operations. But in case of ultrapure water a reverse osmosis system may be required.

Installation and maintenance must be considered when locating instruments, valves, and so on. Each device must be easily accessible for maintenance and inspection. It may also be necessary to install hand-operated valves so that equipment can be replaced or serviced without complete plant shutdown. It may be necessary to contract out maintenance of certain equipment or have the vendor install equipment, if the necessary skills are not available in-house.

Safety is a top priority in a facility. The correct material must be used in container construction, plumbing, seals, and gaskets to prevent corrosion and failure leading to leakage and spills of hazardous materials. All electrical equipment must be properly installed to code with breakers. Electrical systems must have the correct fire retardant for use in case of electrical fires. More information can be found in ANSI/ISA-12.01.01-1999, *Definitions and Information Pertaining to Electrical Apparatus in Hazardous Locations.*

Environment is also a priority in a facility. The comments on safety also apply here. Spills, escape of hazard gases, dumping of waste, and emissions are all extremely detrimental to the environment. The Environmental Protection Agency (EPA) has published rule governing dumping hazardous and nuclear waste, pollution, and emissions, and its effect on the air, water, land, endangered species, and so on.

1.5   Units and Standards

As with all disciplines a set of standards has evolved over the years to ensure consistency and avoid confusion. The Instrument Society of America (ISA) has developed a complete list of symbols for instruments, instrument identification, and process-control drawings, which will be discussed in Chap. 12.

The units of measurement fall into two distinct systems: first, the English system and second, the International system, SI (Systéme International D'Unités) based on the metric system. The English system has been the standard used in the United States, but the SI system is slowly making inroads, so that students need to be aware of both systems of units and able to convert units from one system to the other. Confusion can arise over some units such as the pound mass and pound weight. The pound mass is the slug

Quantity	English			SI		
Base Units	Units	Symbol	Units	Symbol		Conversion to SI
Length	Foot	ft	Meter	m		1 ft = 0.305 m
Mass	Pound (slug)	lb (slug)	Kilogram	kg		1 lb (slug) = 14.59 kg
Time	Second	s	Second	s		
Temperature	Rankine	R	Kelvin	K		1°R = 5/9 K
Liquid measure	Gallon (US)	gal	Liter	L		1 gal = 3.78 L

TABLE **1.1**   Basic Units

(no longer in common use) which is the equivalent of the kilogram in the SI system of units, whereas the pound weight is a force similar to the Newton which is the unit of force in the SI system. The conversion factor of 1 lb = 0.454 kg which is used to convert mass (weight) between the two systems, is in effect equating 1 lb force to 0.454 kg mass; this being the mass that will produce a force of 4.448 N or a force of 1 lb. Care must be taken not to mix units from the two systems. For consistency some units may have to be converted before they can be used in an equation.

Table 1.1 gives a list of the base units used in instrumentation and measurement in the English and SI systems and also the conversion factors. Other units are derived from these base units.

Example 1.2   How many meters are there in 110 yard?

Solution
110 yard = 330 ft = (330 × 0.305) m = 100.65 m

Example 1.3   What is the equivalent length in inches of 2.5 m?

Solution
2.5 m = (2.5/0.305) ft = 8.2 ft = 98.4 in

Example 1.4   The weight of an object is 2.5 lb. What is the equivalent force and mass in the SI system of units?

Solution
2.5 lb = (2.5 × 4.448) N = 11.12 N
2.5 lb = (2.5 × 0.454) kg = 1.135 kg

Table 1.2 gives a list of some commonly used units in the English and SI systems, conversion between units, and also their relation to the base units. As explained earlier pound is used as both the unit of mass and force. Hence, the unit for pound in energy and power is mass, whereas the unit for pound in pressure is force, where the lb (force) = lb (mass) × g (acceleration due to gravity).

Example 1.5   What is the pressure equivalent of 18 psi in SI units?

Solution
1 psi = 6.897 kPa
18 psi = (18 x 6.897) kPa = 124 kPa

Quantity	English				SI		
	Name	Symbol	Units		Name	Symbol	Units
Frequency	Hertz	Hz	s^{-1}		Hertz	Hz	s^{-1}
Energy	Foot-pound	ft·lb	$lb·ft^2/s^2$		Joule	J	$kg·m^2/s^2$
Force	Pound(f)	lb	$lb·ft/s^2$		Newton	N	$kg·m/s^2$
Resistance	Ohm	Ω			Ohm	Ω	$kg·m^2$ per (s^3A^2)
Electric potential	Volt	V			Volt	V	$A\Omega$
Pressure	Pounds/in^2	psi	lb/in^2		Pascal	Pa	N/m^2
Charge	Coulomb	C			Coulomb	C	$A·s$
Inductance	Henry	H			Henry	H	$kg·m^2$ per $(s^3·A^2)$
Capacitance	Farad	F			Farad	F	$s^4·A^2$ per $(kg·m^2)$
Magnetic flux	Weber	Wb			Weber	Wb	$V·s$
Power	Horse power	hp	$lb·ft^2/s^3$		Watt	W	J/s

Conversion to SI:
1 ft·lb =1.356 J.
1 lb (F) = 4.448 N.
1 psi = 6897 Pa.
1 hp = 746 W.

TABLE 1.2 Units in Common Use in the English and SI Systems

Multiple	Prefix	Symbol	Multiple	Prefix	Symbol
10^{12}	tera	T	10^{-2}	centi	c
10^9	giga	G	10^{-3}	milli	m
10^6	mega	M	10^{-6}	micro	μ
10^3	kilo	k	10^{-9}	nano	n
10^2	hecto	h	10^{-12}	pico	p
10	deca	da	10^{-15}	femto	f
10^{-1}	deci	d	10^{-18}	atto	a

TABLE 1.3 Standard Prefixes

Standard prefixes are commonly used for multiple and submultiple quantities to cover the wide range of values used in measurement units. These are given in Table 1.3.

1.6 Instrument Accuracy

The *accuracy* of an instrument or device is the difference between the indicated value and the actual value. Accuracy depends on linearity, hysteresis, offset, drift, and sensitivity. The resulting discrepancy is stated as a ± deviation from true, and is specified as a percentage of full-scale deflection (% FSD), percent of span, or percent of reading.

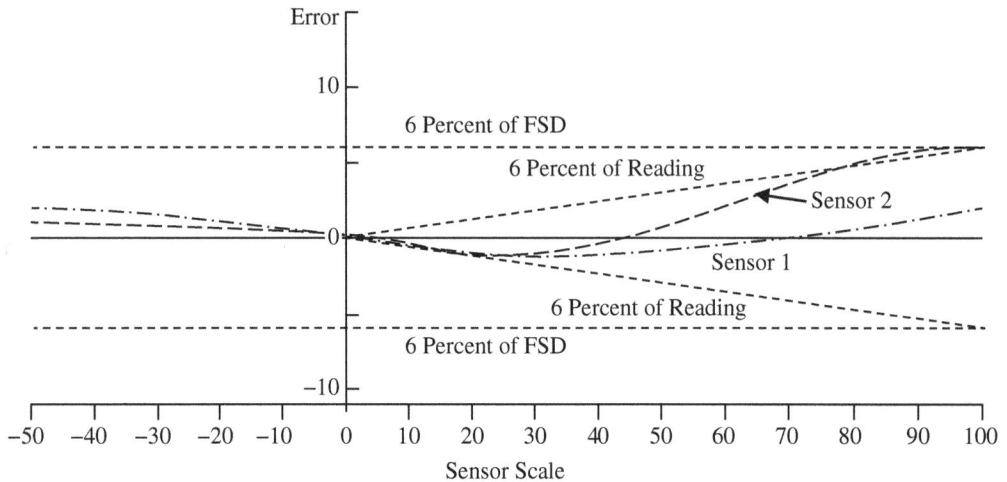

FIGURE **1.6**   Sensor scale and effect of FSD and span error on readings.

Figure 1.6 shows the scale of an instrument that goes from −50 to 100 units. The span of this instrument is 150 units and the FSD is 100 units. One set of dashed lines shows deviation from true of 6 percent of reading and the other deviation of 6 percent of FSD. Also shown is the output from two sensors and their deviation from true. Although sensor 1 is showing about ±6 percent accuracy if it is read as percent of reading, it has an accuracy of better than ±2 percent over its full range if its accuracy is specified as FSD. Sensor 2 is showing an accuracy of ±7 percent of reading or ±6 percent of FSD (±4 percent span) at the high end, but could be selected to read from −50 to 50 with an accuracy of 1 percent of span. This shows the care that is needed when interpreting instrument accuracy.

Summary

• This chapter introduces the concept of process control, and simple process loops, which will be expanded in later chapters.

The main points described in this chapter were as follows:

1. A description of the operation of a basic process loop with a definition of the terms used in process control.

2. Some of the basic considerations for electrical, air, and water requirements in a process facility. Consideration needs for safety.

3. A comparison of the units used for parameter measurement and their relation to basic units.

4. The relation between English and SI units, which are based on metric units. The use of standard prefixes to define multiples.

5. The accuracy of sensors and instruments and the parameters such as linearity, used to evaluate accuracy.

Problems

1.1   What is the difference between controlled and manipulated variables?

1.2   What is the difference between set point, error signal, and correction signal?

1.3   How many pounds are equivalent to 63 kg?

1.4   How many micrometers are equivalent to 0.73 milli-in?

1.5   How many pounds per square inch are equivalent to 38.2 kPa?

1.6   How many foot pounds of energy are equivalent to 195 J?

1.7   What force in pounds is equivalent to 385 N?

1.8   How many amperes are required from 110 V supply to generate 1.2 hp? Assume 93 percent efficiency.

1.9   How many Joules are equivalent to 27 ft·lb of energy?

1.10   What is the sensitivity of an instrument whose output is 17.5 mV for an input change of 7°C?

1.11   A temperature sensor has a range of 0 to 120°C and an absolute accuracy of ±3°C. What is its FSD percent accuracy?

1.12   A flow sensor has a range of 0 to 25 m/s and a FSD accuracy of ±4.5 percent. What is the absolute accuracy?

1.13   A pressure sensor has a range of 30 to 125 kPa. The absolute accuracy is ±2 kPa. What is its percent full scale and span accuracy?

1.14   A temperature instrument has a range −20 to 500°F. What is the error at 220°F? Assume the accuracy is (a) ±7 percent of FSD and (b) ±7 percent of span.

1.15   What is the difference between the span and full scale of an instrument?

1.16   What is the difference between a transducer and a converter?

1.17   What are the environmental concerns in a process facility?

1.18   What is the difference if any between an imperial gallon and a US gallon?

1.19   Plot a graph of the readings in Table 1.4 for a pressure sensor to determine its accuracy as a percentage of reading and percentage of FSD from best-fit straight line?

1.20   Plot a graph of the readings in Table 1.5 for a temperature sensor to determine the deviation from a straight line, what is its accuracy as a percentage of FSD?

True Pressure (kPa)	0	20	40	60	80	100
Gauge Pressure (kPa)	0	15	32	49.5	69	92

TABLE 1.4   Pressure Readings

True Temperature (°C)	0	20	40	60	80	100
Temperature Reading (°C)	0	16	34	56	82	110

TABLE 1.5   Temperature Readings

CHAPTER 2

Pressure

Chapter Objectives

This chapter will help you understand the units used in static liquid and gas pressure measurements and become familiar with the most common methods of using the various pressure measurement standards.

Topics discussed in this chapter are as follows:

- The terms—pressure, specific weight, specific gravity (SG), and buoyancy
- Difference in static, dynamic, and impact pressures
- The difference between atmospheric, absolute, gauge, and differential pressure values
- Various pressure units in use, i.e., British units versus SI (metric) units
- Various types of pressure measuring devices
- Laws applied to pressure
- Application considerations

2.1 Introduction

Pressure is the force exerted by gases and liquids due to their weight, such as the pressure of the atmosphere on the surface of the earth and the pressure containerized liquids exert on the bottom and walls of a container. Here weight is defined as the force produced by the gravitational attraction on the mass of the material.

Pressure units are a measure of force acting over a specified area. It is most commonly expressed in pounds per square inch (psi), sometimes pounds per square foot (psf) in English units, or pascals (Pa or kPa) in metric units.

$$\text{Pressure} = \frac{\text{force}}{\text{area}} \tag{2.1}$$

Example 2.1   The liquid in a container has a total weight of 250 lb; the container has a 3.0 ft^2 base. What is the pressure in pounds per square inch?

Solution

$$\text{Pressure} = \frac{250}{3 \times 144} \text{ psi} = 0.58 \text{ psi}$$

13

2.2   Basic Terms

In the English system the pound is used for both mass and force; to distinguish the pound mass is sometimes called the slug or lbm and the pound force is pound (lb) or lbf. The weight of a substance is sometimes expressed in kilograms, i.e., 1 lb = 0.45 kg (a lb is a force and kg a mass); in this case 0.45 kg is the mass that will produce a force of 4.5 N.

2.2.1   Density

Density ρ is defined as the mass per unit volume of a material, i.e., pound (slug) per cubic foot (lb (slug)/ft³), or kilogram per cubic meter (kg/m³). Do not confuse pound mass and pound force.

Specific weight γ is defined as the weight per unit volume of a material, i.e., pound per cubic foot (lb/ft³) or newton per cubic meter (N/m³).

Specific gravity of a liquid or solid is a dimensionless value since it is a ratio of two measurements in the same units. It is defined as the density of a material divided by the density of water or the specific weight of the material divided by the specific weight of water at 4°C temperature. The specific weights and specific gravities of some common materials are given in Table 2.1. The specific gravity of a gas is its density/specific weight divided by the density/specific weight of air at 60°F and 1 atm (14.7 psia). In the SI system, the density in g/m³ or kg/m³ and SG have the same value.

2.2.2   Impact Pressure

Static pressure is the pressure of fluids or gases that are stationary or not in motion (see Fig. 2.1). The pressure at point A is considered as static pressure although the fluid above it is flowing.

Dynamic pressure is the pressure exerted by a fluid or gas when it impacts on a surface or an object due to its motion or flow. In Fig. 2.1, the dynamic pressure is the difference in pressure between points B and A.

Impact pressure (total pressure) is the sum of the static and dynamic pressures on a surface or object. The pressure at point B in Fig. 2.1 is the impact pressure.

	Temperature, °F	Specific Weight*		Specific Gravity
		lb/ft³	kN/m³	
Acetone	60	49.4	7.74	0.79
Alcohol (ethanol)	68	49.4	7.74	0.79
Glycerin	32	78.6	12.4	1.26
Mercury	60	846.3	133	13.55
Steel		400	76.93	7.85
Water	39.2	62.43	9.8	1.0

*Conversion factors: 1 ft³ = 0.028 m³, 1 lb = 4.448 N, g (gravitational constant) = 9.8 m/s² or 32.2 ft/s².

TABLE 2.1   Specific Weights and Specific Gravities of Some Common Materials

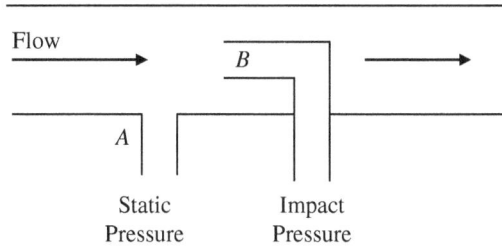

Figure 2.1 Illustration of static, dynamic, and impact pressures.

2.3 Pressure Measurements

There are six terms applied to pressure measurements. They are as follows:

Total vacuum which is zero pressure or lack of pressure, as would be experienced in outer space.

Vacuum is a pressure measurement made between total vacuum and normal atmospheric pressure (14.7 psi).

Atmospheric pressure is the pressure on the earth's surface due to the weight of the gases in the earth's atmosphere, and is normally expressed at sea level as 14.7 psi or 101.36 kPa. It is, however, dependant on atmospheric conditions. The pressure decreases above sea level, and at an elevation of 5000 ft drops to about 12.2 psi (84.122 kPa) due to the lower gravitational force and reduced atmospheric pressure.

Absolute pressure is the pressure measured with respect to a vacuum and is expressed in pounds per square inch absolute (psia) or kilopascals absolute kPa (a).

Gauge pressure is the pressure measured with respect to atmospheric pressure and is normally expressed in pounds per square inch gauge (psig) or kPa (g). Figure 2.2a shows graphically the relation between atmospheric, gauge, and absolute pressures.

NIST (National Institute of Standards and Technology) recommends using p_g or p_a = kPa instead of p = kPa (a) or (g) to avoid confusion.

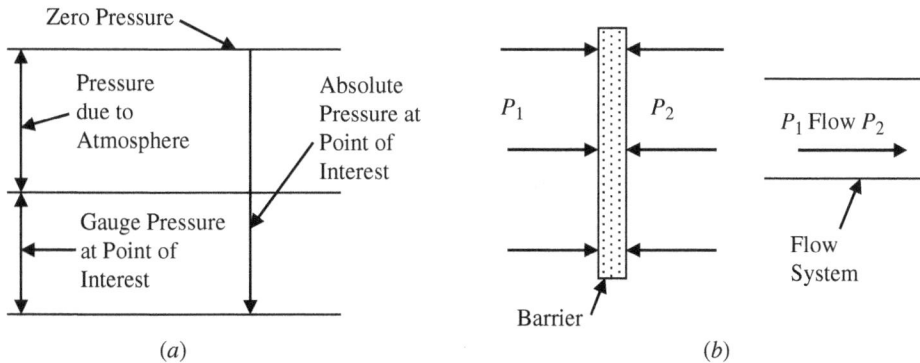

Figure 2.2 Illustration of (a) gauge pressure versus absolute pressure and (b) delta or differential pressure.

| | Water | | Mercury* | | | |
	in*	cm†	mm	ins	kPa	psi
1 psi	27.7	70.3	51.7	2.04	6.895	1
1 psf	0.19	0.488	0.359	0.014	0.048	0.007
1 kPa	4.015	10.2	7.5	0.259	1	0.145
Atmospheres	407.2	1034	761	29.96	101.3	14.7
Torr	0.535	1.36	1	0.04	0.133	0.019
Millibar	0.401	1.02	0.75	0.029	0.1	0.014

*At 39°F.
†At 4°C.
‡Mercury at 0°C.

TABLE 2.2   Pressure Conversions

Example 2.2   The atmospheric pressure is 14.5 psi. If a pressure gauge reads 1200 psf, what is the absolute pressure?

Solution

$$\text{Absolute pressure} = 14.5 \text{ psi} + \frac{1200 \text{ psf}}{144} = 14.5 \text{ psi} + 8.3 \text{ psi} = 22.8 \text{ psia}$$

Differential pressure is the pressure measured with respect to another pressure and is expressed as the difference between the two values. This would represent two points in a pressure or flow system, and is referred to as the *delta p*, or Δp. Figure 2.2b shows two situations where differential pressure exists across a barrier, and between two points in a flow system.

A number of *measurement units* are used to express atmospheric pressure. They are as follows:

1. Pounds per square foot (psf), or per square inch (psi)
2. Pascals (N/m^2)*
3. Atmospheres (atm)
4. Inches or centimeters of water
5. Inches or millimeters of mercury
6. Torr = 1 mm mercury
7. Bar (1.013 atm) = 100 kPa

Table 2.2 gives a table of conversions between various pressure measurement units.

Example 2.3   What pressure in pascals corresponds to 15 psi?

Solution

$$p = 15 \text{ psi} (6.897 \text{ kPa/psi}) = 102.9 \text{ kPa}$$

Note: Pressure can also be considered as potential energy in which case the pascal can be expressed as energy (Joule) per unit volume, i.e., $Pa = J/m^3$.

2.4 Pressure Formulas

Hydrostatic pressure is the pressure in a liquid. The pressure increases as the depth in a liquid increases. This increase is due to the weight of the fluid above the measurement point. The pressure is given by

$$p = \gamma h \qquad (2.2)$$

where p = pressure in pounds per unit area or pascals
γ = specific weight (lb/ft^3 in English units, or N/m^3 in SI units)
h = distance from the surface in compatible units (ft, in, cm, m, and so on)

Example 2.4 What is the gauge pressure in (a) kilopascals and (b) newtons per square centimeter at a distance 1 m below the surface in water?

Solution

(a) $p = 100 \ cm/m/10.2 \ cm/kPa = 9.8 \ kPa$

(b) $p = 9.8 \ N/m^2 = 9.8/10,000 \ N/cm^2 = 0.98 \times 10^{-3} \ N/cm^2$

The pressure in this case is the gauge pressure, i.e., kPa (g). To get the total pressure, the pressure of the atmosphere must be taken into account. The total pressure (absolute) in this case is $9.8 + 101.3 = 111.1$ kPa (a).

The g and a should be used in all cases to avoid confusion. In the case of pounds per square inch and pounds per square foot, this would become pounds per square inch gauge and pounds per square foot gauge, or pounds per square inch absolute and pounds per square foot absolute. Also it should be noted that if glycerin was used instead of water the pressure would be 1.26 times higher, as its specific gravity is 1.26.

Example 2.5 What is the specific gravity of mercury if the specific weight of mercury is 846.3 lb/ft^3?

Solution

$$SG = 846.3/62.4 = 13.56$$

Head is sometimes used as a measure of pressure. It is the pressure in terms of a column of a particular fluid; i.e., a head of 1 ft of water is the pressure that would be exerted by a 1-ft-tall column of water, i.e., 62.4 psfg, or the pressure exerted by 1 ft head of glycerin would be 78.6 psfg.

Example 2.6 What is the pressure at the base of a water tower, which has 50 ft of head?

Solution

$$p = 62.4 \ lb/ft^3 \times 50 \ ft = 3120 \ psfg = 3120 \ psf/144 \ ft^2/in^2 = 21.67 \ psig$$

The hydrostatic paradox states that the pressure at a given depth in a liquid is independent of the shape of the container or the volume of liquid contained. The pressure value is a result of the depth and density. Figure 2.3a shows various shapes of tanks. The total pressure or forces on the sides of the container depend on its shape, but at a specified depth. The pressure is given by Eq. (2.2).

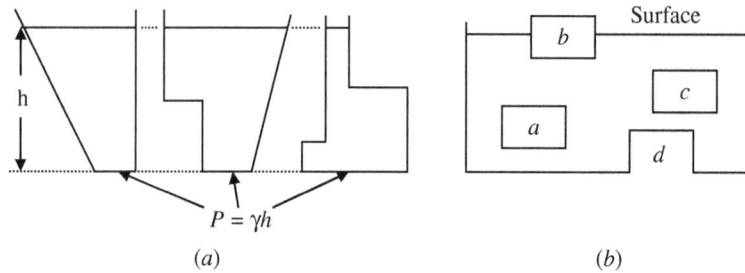

FIGURE 2.3   Diagrams demonstrating (a) hydrostatic paradox and (b) buoyancy.

Buoyancy is the upward force exerted on an object immersed or floating in a liquid. The weight is less than it is in air due to the weight of the displaced fluid. The upward force on the object causing the weight loss is called the buoyant force and is given by

$$B = \gamma V \tag{2.3}$$

where B = buoyant force in (lb)
γ = specific weight of liquid in (lb/ft³)
V = volume of the liquid displaced in (ft³)

If working in SI units B is in newtons, γ in newton per cubic meter, and V in cubic meters.

In Fig. 2.3b, columns a, b, c, and d are of the same size. The buoyancy forces on a and c are the same although their depth is different. There is no buoyancy force on d as the liquid cannot get under it to produce a force. The buoyancy force on b is half that of a and c, as only half of the object is submersed.

Example 2.7   What is the buoyant force on a wooden cube with 3-ft sides floating in water, if the block is half submerged?

Solution

$$B = 62.4 \text{ lb/ft}^3 \times 3 \text{ ft} \times 3 \text{ ft} \times 1.5 \text{ ft} = 842.4 \text{ lb}$$

Example 2.8   What is the apparent weight of a 3-m³ block of steel totally immersed in glycerin?

Solution

Weight of steel in air = 3×76.93 kN = 230.8 kN

Buoyancy force on steel = 3×12.4 kN = 37.2 kN

Apparent weight = 230.8 – 37.2 = 193.6 kN (19.75 Mg)

Pascal's law states that the pressure applied to an enclosed liquid (or gas) is transmitted to all parts of the fluid and to the walls of the container. This is demonstrated in the hydraulic press in Fig. 2.4. A force of F_s, exerted on the small piston (ignoring friction), will exert a pressure in the fluid given by

$$p = \frac{F_s}{A_s} \tag{2.4}$$

where A_s is the cross-sectional area of the smaller piston.

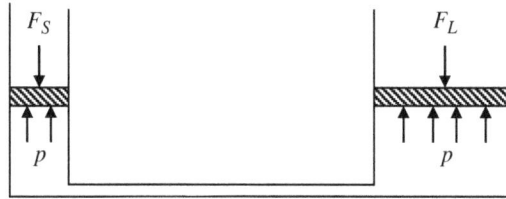

FIGURE 2.4 Diagram of a hydraulic press.

Since the pressure is transmitted through the liquid to the second cylinder (Pascal's law), the force on the larger piston (F_L) is given by

$$F_L = pA_L \tag{2.5}$$

where A_L is the cross-sectional area of the large piston (assuming the pistons are at the same level), from which

$$F_L = \frac{A_L F_S}{A_S} \tag{2.6}$$

It can be seen that the force F_L is magnified by the ratio of the piston areas. This principle is used extensively in hoists, hydraulic equipment, and the like.

Example 2.9 In Fig. 2.4, if the area of the small piston A_S is 0.3 m² and the area of the large piston A_L is 5 m², what is the force F_L on the large piston, if the force F_S on the small piston is 85 N?

Solution

$$\text{Force } F_L \text{ on piston} = \frac{5N}{0.3 \times 85} = 1416.7N$$

A *vacuum* is very difficult to achieve in practice. Vacuum pumps can only approach a true vacuum. Good small volume vacuums, such as in a barometer, can be achieved. Pressures less than atmospheric pressure are often referred to as "negative gauge" and are indicated by an amount below atmospheric pressure, e.g., negative 5 psig would correspond to 9.7 psia (assume atm = 14.7 psia).

2.5 Measuring Instruments

2.5.1 Manometers

Manometers are a good example of pressure measuring instruments, although, they are not as common as they used to be because of the development of new, smaller, more rugged, and easier to use pressure sensors.

U-tube manometers consist of U-shaped glass tubes partially filled with a liquid. When there are equal pressures on both sides, the liquid levels will correspond to the zero point on a scale as shown in Fig. 2.5a. The scale is graduated in pressure units.

When a pressure is applied to one side of the U-tube that is higher than on the other side as shown in Fig. 2.5b, the liquid rises higher in the lower pressure side, so that the

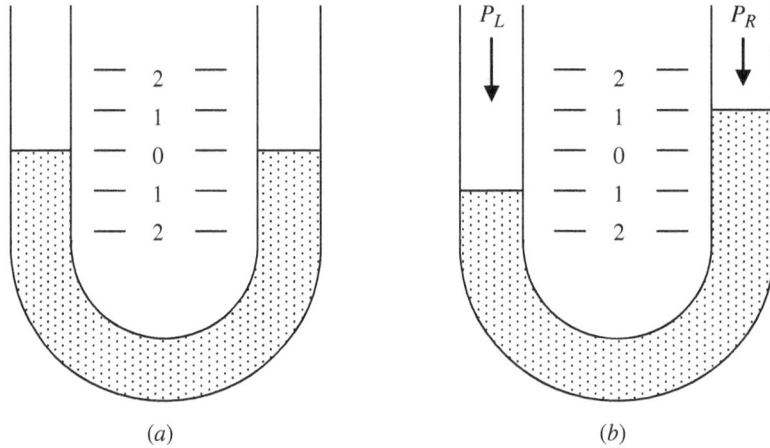

FIGURE 2.5 Simple U-tube manometers with (a) no differential pressure and (b) higher pressure on the left side.

difference in height of the two columns of liquid compensates for the difference in pressure as in Eq. (2.2). The pressure difference is given by

$$P_R - P_L = \gamma \times \text{difference in height of the liquid in the columns} \qquad (2.7)$$

where γ is the specific weight of the liquid in the manometer.

Inclined manometers were developed to measure low pressures. The low-pressure arm is inclined, so that the fluid has a longer distance to travel than in a vertical tube for the same pressure change. This gives a magnified scale as shown in Fig. 2.6a.

Well manometers are an alternative to inclined manometers for measuring low pressures using low-density liquids. In the well manometer, one leg has a much larger diameter than the other leg, as shown in Fig. 2.6b. When there is no pressure difference the liquid levels will be at the same height for a zero reading. An increase in pressure in the large leg will cause a larger change in the height of the liquid in the smaller leg.

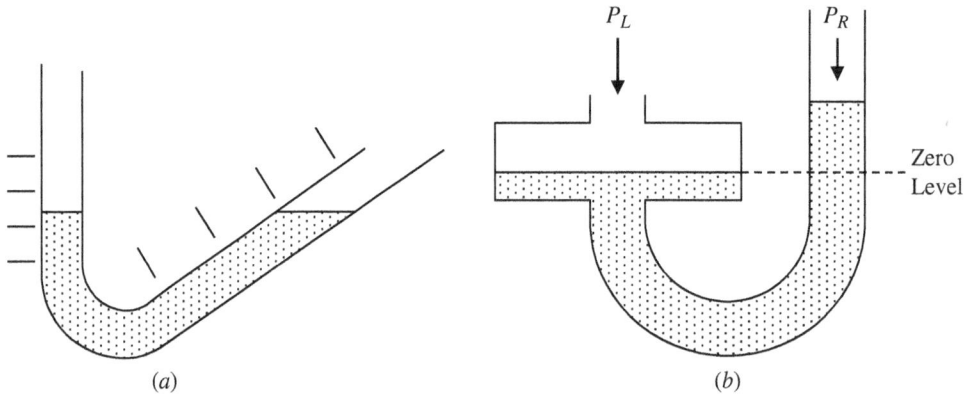

FIGURE 2.6 Other types of manometers are the (a) inclined-tube manometer and (b) well-type manometer.

The pressure across the larger area of the well must be balanced by the same volume of liquid rising in the smaller leg. The effect is similar to the balance of pressure and volume in hydraulic jacks.

Example 2.10 The liquid in a well manometer has a specific weight of 40 lb/ft³. How far will the liquid rise in the smaller leg, if the pressure in the larger leg is 1.5 lb/ft² higher than in the smaller leg?

Solution

$$h = \frac{p}{\gamma} = \frac{1.5 \text{ ft}}{40} = 0.45 \text{ in}$$

Example 2.11 The liquid in a manometer has a density of 850 kg/m³. What will be the difference in the liquid levels in the manometer tubes, if the differential pressure between the tubes is 5.2 kPa?

Solution

$$h = \frac{p}{\gamma} = \frac{(5.2 \text{ kPa})}{850 \text{ kg/m}^3} \times \frac{\text{N/m}^3}{9.8 \text{ Nm}^3} = 62 \text{ cm}$$

2.5.2 Diaphragms, Capsules, and Bellows

Gauges are a major group of pressure sensors that measure pressure with respect to atmospheric pressure. Gauge sensors are usually devices that change their shape when pressure is applied. These devices include diaphragms, capsules, bellows, and Bourdon tubes.

A *diaphragm* consists of a thin layer or film of a material supported on a rigid frame and is shown in Fig. 2.8a. Pressure can be applied to one side of the film for gauge sensing, or pressures can be applied to both sides of the film for differential or absolute pressure sensing. A wide range of materials can be used for the sensing film, from rubber to plastic for low-pressure devices, silicon for medium pressures, to stainless steel for high pressures. When pressure is applied to diaphragm, capsule, bellows, and Bourdon tube pressure sensors the change in shape of the sensor can be converted into an electrical signal using capacitive, LVDT (linear variable differential transformer), potentiometer, piezoelectric, or strain-gauge transducer (older techniques included carbon pile devices). Of all these devices, the micromachined silicon diaphragm is the most commonly used industrial pressure sensor for the generation of electrical signals.

A *silicon diaphragm* uses silicon, which is a semiconductor. This allows a strain gauge and signal amplifier electronics to be integrated into the top surface of the silicon structure after the back side of the silicon was etched to form a diaphragm. These devices have built-in temperature-compensated piezoelectric strain gauge and amplifiers that give a high-output voltage (5 V FSD [volt full-scale reading or deflection]). They are very small, accurate (commercial devices 2 percent FSD), reliable, have a good temperature operating range (−50 to 200°F), are low cost, can withstand high overloads, have good longevity, and are unaffected by many chemicals. Commercially made devices are available for gauge, differential, and absolute pressure sensing up to 200 psi (1.5 MPa). This range can be extended by the use of stainless steel diaphragms to 10,000 psi (70 MPa).

Figure 2.7a shows the cross sections of the three configurations of the silicon chips (sensor dies) used in microminiature pressure sensors, i.e., gauge, absolute, and differential. The given dimensions illustrate that the sensing elements are very small. The die is packaged into a plastic case (about 0.2-in thick × 0.6-in diameter).

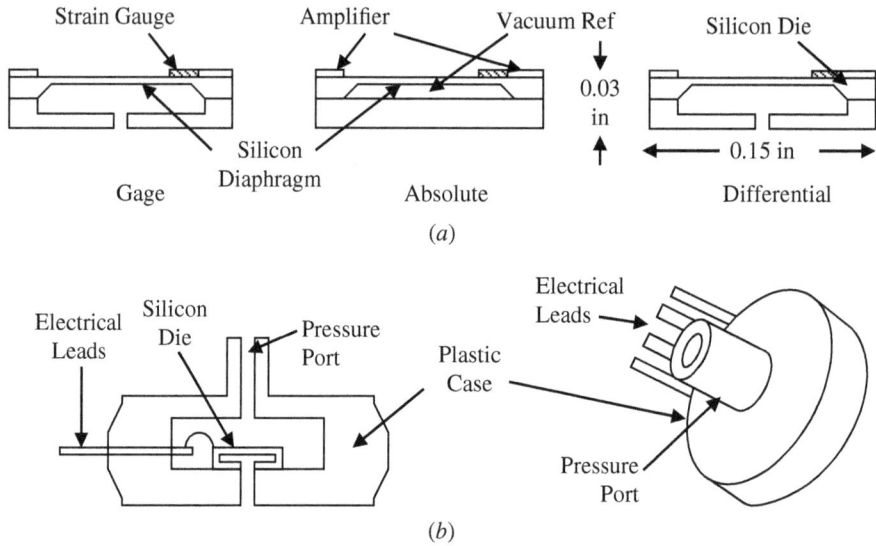

FIGURE 2.7 Cross section of (a) various types of microminiature silicon pressure sensor die and (b) a packaged microminiature gauge sensor.

A gauge assembly is shown in Fig. 2.7b. The sensor is used in blood pressure monitors and many industrial applications, and is extensively used in automotive pressure-sensing applications, i.e., manifold air pressure, barometric air pressure, oil, transmission fluid, break fluid, power steering, tire pressure, and the like.

Capsules are two diaphragms joined back to back, as shown in Fig. 2.8b. Pressure can be applied to the space between the diaphragms forcing them apart to measure gauge pressure. The expansion of the diaphragm can be mechanically coupled to an indicating device. The deflection in a capsule depends on its diameter, material thickness, and elasticity. Materials used are phosphor bronze, stainless steel, and iron nickel alloys. The pressure range of instruments using these materials is up to 50 psi (350 kPa). Capsules can be joined together to increase sensitivity and mechanical movement.

Bellows is similar to capsules except that the diaphragms instead of being joined directly together are separated by a corrugated tube or tube with convolutions, as shown in Fig. 2.8c. When pressure is applied to the bellows it elongates by stretching the convolutions, not the end diaphragms. The materials used for the bellows type of pressure sensor are similar to those used for the capsule giving a pressure range for the

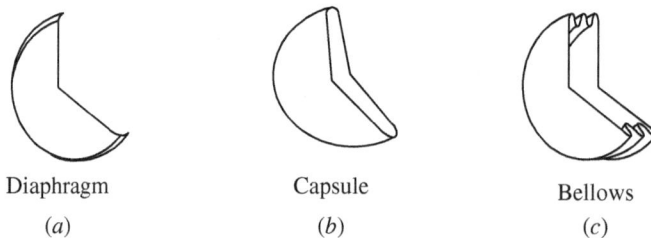

FIGURE 2.8 Various types of pressure-sensing elements: (a) diaphragm, (b) capsule, and (c) bellows.

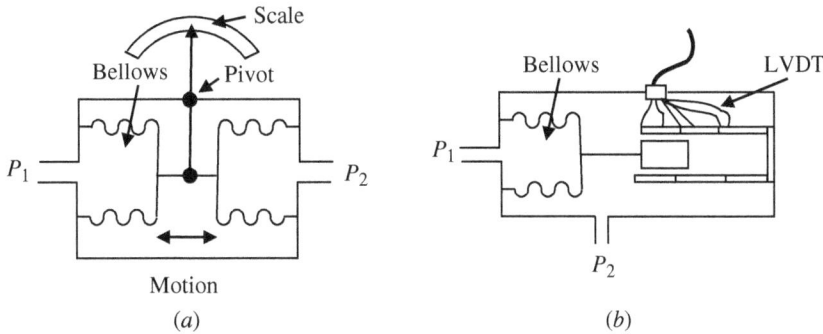

FIGURE 2.9 Differential bellows pressure gauges for (a) direct scale reading and (b) as a pressure transducer.

bellows of up to 800 psi (5 MPa). Bellows devices can be used for absolute and differential pressure measurements.

Differential measurements can be made by connecting two bellows mechanically to be opposing each other when pressure is applied to them, as shown in Fig. 2.9a. When pressures P_1 and P_2 are applied to the bellows a differential scale reading is obtained. Figure 2.9b shows a bellows configured as a differential pressure transducer driving an LVDT to obtain an electrical signal. P_2 could be atmospheric pressure for gauge measurements. The bellows is the most sensitive of the mechanical devices for low-pressure measurements, i.e., 0 to 210 kPa.

2.5.3 Bourdon Tubes

Bourdon tubes are hollow, cross-sectional beryllium, copper, or steel tubes shaped into a three-quarter circle, as shown in Fig. 2.10a. They may be rectangular or oval in cross section, but the operating principle is that the outer edge of the cross section has a larger surface than the inner portion. When pressure is applied, the outer edge has a proportionally larger total force applied because of its larger surface area, and the diameter of the circle increases. The walls of the tube are between 0.01- and 0.05-in thick. The tubes are anchored at one end so that when pressure is applied to the tube it tries to straighten,

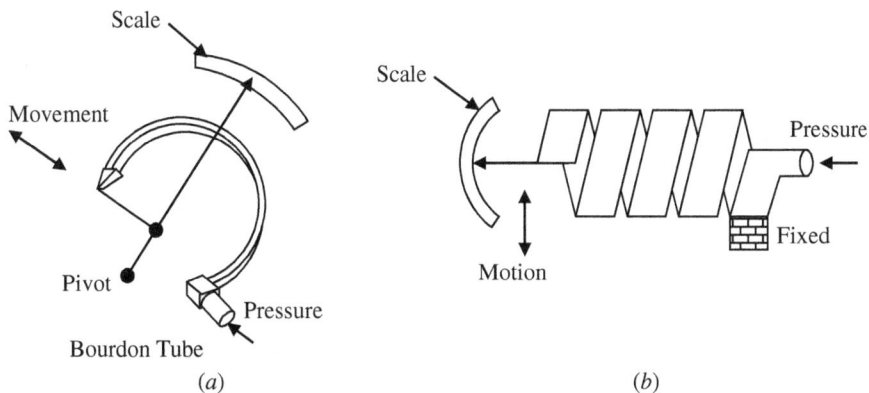

FIGURE 2.10 Pressure sensors shown are (a) the Bourdon tube and (b) the helical Bourdon tube.

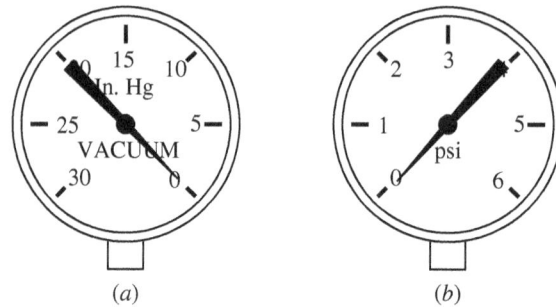

FIGURE **2.11** Bourdon tube-type pressure gauges for (a) negative and (b) positive pressures.

and in doing so the free end of the tube moves. This movement can be mechanically coupled to a pointer which when calibrated will indicate pressure as a line of sight indicator or it can be coupled to a potentiometer to give a resistance value proportional to pressure for electrical signals. The tubes can also be shaped into helical or spiral shape to increase their range. Figure 2.10b shows a helical pressure tube. This configuration is more sensitive than the circular Bourdon tube. The Bourdon tube dates from the 1840s. It is reliable, inexpensive, and one of the most common general purpose pressure gauges.

Bourdon tubes can withstand overloads of up to between 30 and 40 percent of their maximum rated load without damage, but if overloaded may require recalibration. The Bourdon tube is normally used for measuring positive gauge pressures, but can also be used to measure negative gauge pressures. If the pressure to the Bourdon tube is lowered then the diameter of the tube reduces. This movement can be coupled to a pointer to make a vacuum gauge. Bourdon tubes can have a pressure range of up to 100,000 psi (700 MPa). Figure 2.11 shows the Bourdon tube type of pressure gauge when used for measuring negative pressure (vacuum) (a) and positive pressure (b). Note the counter clockwise movement in (a) and the clockwise movement in (b).

2.5.4   Other Pressure Sensors

Barometers are used for measuring atmospheric pressures. A simple barometer is the mercury barometer shown in Fig. 2.12a. It is now little used due to its fragility and the toxicity of mercury. The aneroid (no fluid) barometer is favored for direct reading (bellows in

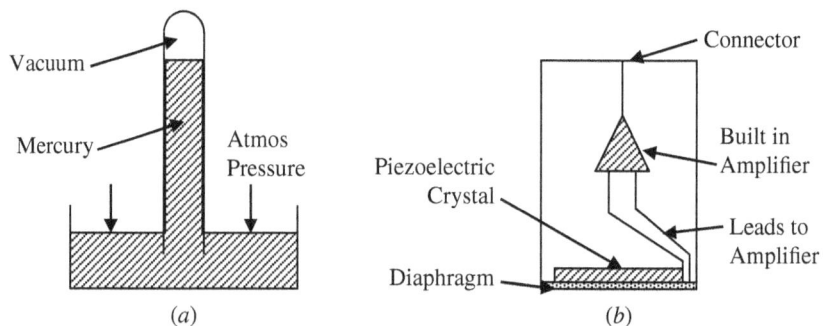

FIGURE **2.12**   Diagram of (a) barometer and (b) piezoelectric sensing element.

Fig. 2.9, or helical Bourdon tube in Fig. 2.10*b*), and the solid-state absolute pressure sensor for electrical outputs.

A *piezoelectric pressure gauge* is shown in Fig. 2.12*b*. Piezoelectric crystals produce a voltage between their opposite faces when a force or pressure is applied to the crystal. This voltage can be amplified and the device used as a pressure sensor.

Capacitive devices use the change in capacitance between the sensing diaphragm and a fixed plate to measure pressure. Some microminiature silicon pressure sensors use this technique in preference to a strain gauge. This technique is also used in a number of other devices to accurately measure small changes in diaphragm deformation.

2.5.5   Vacuum Instruments

Vacuum instruments are used to measure pressures less than atmospheric pressure. The Bourdon tube, diaphragms, and bellows can be used as vacuum gauges, but measure negative pressures with respect to atmospheric pressure. The silicon absolute pressure gauge has a built-in low-pressure reference, so it is calibrated to measure absolute pressures. Conventional devices can be used down to 20 torr (5 kPa). The range can be extended down to about 1 torr with special sensing devices.

The *Pirani gauge* and special setups using thermocouples can measure vacuums down to about 5 torr. These methods are based on the relation of heat conduction and radiation from a heating element to the number of gas molecules per unit volume in the low-pressure region, which determines the pressure.

Ionization gauges can be used to measure pressures down to about 2 torr. The gas is ionized with a beam of electrons and the current is measured between two electrodes in the gas. The current is proportional to the number of ions per unit volume, which is also proportional to the gas pressure.

McLeod gauge is a device set up to measure very low pressures, i.e., from 1 to 50 torr. The device compresses the low-pressure gas so that the increased pressure can be measured. The change in volume and pressure can then be used to calculate the original gas pressure, providing that the gas does not condense.

The American Society for Mechanical Engineers (ASME) has set standards for pressure sensors as follows.

B40.100-2013; Pressure Gauges and Gauge Attachments

PTC 19.2; Performance Test Code for Pressure Measurements

2.6   Application Considerations

When installing pressure sensors, care should be taken to select the correct pressure sensor for the application.

2.6.1   Selection

Pressure-sensing devices are chosen for pressure range, overload requirements, accuracy, temperature operating range, line-of-sight reading, or electrical signal, and response time. In some applications there are other special requirements. Parameters, such as hysteresis and stability should be obtained from the manufacturers' specifications. For most commercial applications, reading positive pressures, the Bourdon tube is a good choice for direct visual readings and the silicon pressure sensor for the generation of electrical signals.

Device	Maximum Pressure Range, lb/in²	Accuracy FSD, %	Response Time, s	Overload, %
Bourdon tube	10,000	2	1	40
Silicon sensor	10,000	2	1×10^{-3}	400

TABLE 2.3   Comparison of Commercial Bourdon Tube Sensor and Silicon Sensor

Both types of devices have commercially available sensors to measure from a few psi pressure FSD up to 10,000 psi (700 MPa) FSD. Table 2.3 gives a comparison of the two types of devices.

In industrial applications, pressure sensors are available with accuracies of 0.1 percent of span in both Bourbon tube devices and silicon devices. To achieve these accuracies, devices often have a limited temperature operating range and limited pressure range, so that great care must be taken in the selection of a device for a specific application.

2.6.2   Installation

The following should be taken into consideration when installing pressure-sensing devices:

1. Distance between sensor and source should be kept to a minimum.

2. Sensors should be connected via valves for ease of replacement.

3. Overrange protection devices should be included at the sensor.

4. To eliminate errors due to trapped gas in sensing liquid pressures, the sensor should be located below the source.

5. To eliminate errors due to trapped liquid in sensing gas pressures, the sensor should be located above the source.

6. When measuring pressures in corrosive fluids and gases, an inert medium is necessary between the sensor and source, or the sensor must be corrosion resistant.

7. The weight of liquid in the connection line of a liquid pressure-sensing device located above or below the source will cause errors in the zero, and a correction must be made by the zero adjustment, or otherwise compensated for in measurement systems.

8. Resistance and capacitance can be added to electronic circuits to reduce pressure fluctuations and unstable readings.

2.6.3   Calibration

Pressure-sensing devices are calibrated at the factory. In cases where a sensor is suspect and needs to be recalibrated the sensor can be returned to the factory for recalibration, or it can be compared to a known reference. Low-pressure devices can be calibrated against a liquid manometer. Higher pressure devices can be calibrated with a dead-weight tester. In a dead-weight tester, the pressure to the device under test is created by weights on a piston. High pressures can be accurately reproduced.

Summary

Pressure measurement standards in both English and SI units were discussed in this section. Pressure formulas and the types of instruments and sensors used in pressure measurements were given.

The main points described in this chapter were as follows:

1. Definitions of the terms and standards used in pressure measurements, both gauge and absolute pressures.

2. English and SI pressure measurement units and the relation between the two as well as atmospheric, torr, and millibar standards.

3. Pressure laws and formulae used in hydrostatic pressure measurements and buoyancy together with examples.

4. The various types of instruments including manometers, diaphragms, and microminiature pressure sensors. Various configurations for use in absolute and differential pressure sensing in both liquid and gas pressure measurements.

5. In the application section the characteristics of pressure sensors were compared, and the considerations that should be made when installing pressure sensors were given.

Problems

2.1   A tank is filled with pure water. If the pressure at the bottom of the tank is 17.63 psig, what is the depth of the water?

2.2   What is the pressure on an object on the bottom of a fresh water lake if the lake is 123-m deep?

2.3   An instrument reads 1038 psf. If the instrument was calibrated in kilopascals, what would it read?

2.4   What will be the reading of a mercury barometer in centimeters if the atmospheric pressure is 14.75 psi?

2.5   A tank 2.2 ft × 3.1 ft × 1.79 ft weighs 1003 lb when filled with a liquid. What is the specific gravity of the liquid if the empty tank weighs 173 lb?

2.6   An open tank 3.2-m wide by 4.7-m long is filled to a depth of 5.7 m with a liquid whose SG is 0.83. What is the absolute pressure on the bottom of the tank in kilopascals?

2.7   Two pistons connected by a pipe are filled with oil. The larger piston has 3.2-ft diameter and a force of 763 lb is applied to it. What is the diameter of the smaller piston if it can support a force of 27 lb?

2.8   A block of wood with a density of 35.3 lbs/ft³ floats in a liquid with three-fourths of its volume submersed. What is the specific gravity of the liquid?

2.9   A 15.5-kg mass of copper has an apparent mass of 8.7 kg in oil whose SG is 0.77. What is the volume of the copper and its specific weight?

2.10   A dam is 283-m high, when it is full of water. What is the pounds per square inch absolute at the bottom of the reservoir?

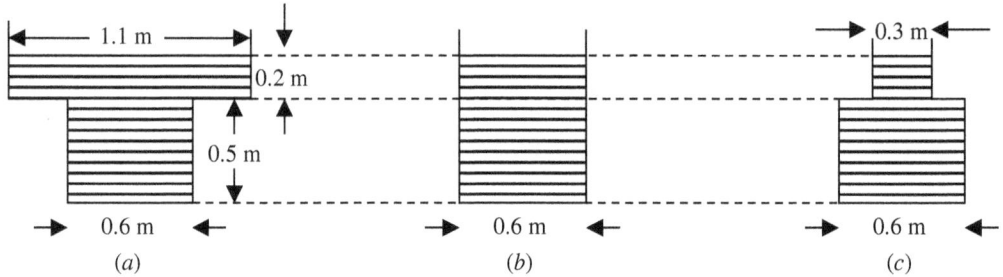

FIGURE **2.13** Figure for use with Problem 2.20.

2.11 A liquid has an SG of 7.38. What is its specific weight in pounds per cubic foot and kilograms per cubic meter?

2.12 What is the equivalent of 25, 49, and 83 kPa in pounds per square inch?

2.13 The cabin pressure in a spacecraft is maintained at 14.3 psia. What will be the force on a window 2.9-ft wide and 1.7-ft high when the craft is in outer space?

2.14 A U-tube manometer uses glycerin as the measuring fluid. What will be the differential pressure if the distance between the levels of glycerin is 103 in?

2.15 An open tank contains 1.9 m of water floating on 10.3 cm of mercury. What is the pressure in pounds per square foot absolute on the bottom of the tank?

2.16 Oil (SG = 0.93) is pumped from a well. If the pump is 11.7 ft above the surface of the oil, what pressure must the pump be able to generate to lift the oil up to the pump?

2.17 A piston 8.7-in diameter has a pressure of 3.7 kPa on its surface. What force in SI units is applied to the piston?

2.18 The water pressure at the base of a water tower is 107.5 psi. What is the head of water?

2.19 A U-tube manometer reads a pressure of 270 torr. What is the pressure in pounds per square inch absolute?

2.20 Each of the three circular containers in Fig. 2.13 contains a liquid with an SG of 1.37. What is the pressure in pascal gauge acting on the base of each container and the weight of liquid in each container?

CHAPTER 3

Level

Chapter Objectives

This chapter will help you understand the units used in level measurements, and become familiar with the most common methods of using the various level standards.

Topics discussed in this chapter are as follows:

- The formulas used in level measurements
- The difference between direct and indirect level measuring devices
- The difference between continuous and single-point measurements
- The various types of instruments available for level measurements
- Application of the various types of level sensing devices

Most industrial processes use liquids such as water, chemicals, fuel, and the like, as well as gases and free flowing solids. These materials are stored in containers ready for on-demand use. It is, however, imperative to know the levels and remaining volumes of these materials so that the containers can be replenished on an as-needed basis to avoid large volume storage and high pressures at the point of use.

3.1 Introduction

This chapter discusses the measurement of the level of liquids and free flowing solids in containers. The detector is normally sensing the interface between a liquid and a gas, a solid and a gas, a solid and a liquid, or possibly the interface between two liquids. Sensing liquid levels fall into two categories: first, single-point sensing and second, continuous level monitoring. In the case of single-point sensing, the actual level of the material is detected when it reaches a predetermined level, so that the appropriate action can be taken to prevent overflowing or to refill the container.

Continuous level monitoring measures the level of the liquid on an uninterrupted basis. In this case, the level of the material will be constantly monitored, and hence the volume known. An example of this is the fuel gauge in an automobile.

Level measurements can be direct or indirect; examples of these are using a float technique or measuring pressure and calculating the liquid level. Free flowing solids are dry powders, crystals, rice, or grain and so forth.

3.2   Level Formulas

Pressure is often used as an indirect method of measuring liquid levels. Pressure increases as the depth increases in a fluid. The pressure is given by

$$\Delta p = \gamma \Delta h \tag{3.1}$$

where Δp = change in pressure
γ = specific weight
Δh = depth

Note the units must be consistent, i.e., pounds and feet, or Newton and meters.

In this case, a pressure sensor is used near the base of the container.

Buoyancy is an indirect method used to measure liquid levels. The level is determined using the buoyancy of an object (denser than the liquid) partially immersed in a liquid. The buoyancy B or upward force on a body in a liquid can be calculated from the equation

$$B = \gamma \times \text{area} \times d \tag{3.2}$$

where area is the cross-sectional area of the object and d is the immersed depth of the object.

The liquid level is then calculated from the weight of a body in a liquid W_L, which is equal to its weight in air ($W_A - B$), from which we get

$$d = \frac{W_A - W_L}{\gamma \times \text{area}} \tag{3.3}$$

A force sensor or strain gauge can be used to monitor the effect of buoyancy on the object.

The weight of a container can be used to calculate the level of the material in the container. In Fig. 3.1a, the volume V of the material in the container is given by

$$V = \text{area} \times \text{depth} = \pi r^2 \times d \tag{3.4}$$

where r is the radius of the container and d is the depth of the material.

The weight of material W in a container is given by

$$W = \gamma V \tag{3.5}$$

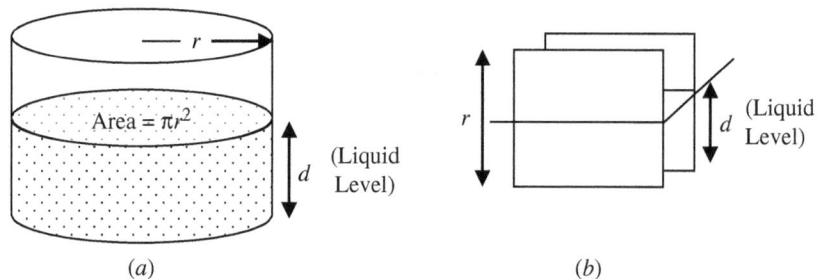

(a)   (b)

FIGURE 3.1   Illustration of the relation between (a) volume of liquid and the cross-sectional area and the liquid depth and (b) liquid level, plate capacitance, and a known dielectric constant in a nonconducting liquid.

Liquid	Dielectric Constant
Water	80 @ 20°C
	88 @ 0°C
Glycerol	42.5 @ 25°C
	47.2 @ 0°C
Acetone	20.7 @ 25°C
Alcohol (Ethyl)	24.7 @ 25°C
Gasoline	2.0 @ 20°C
Kerosene	1.8 @ 20°C

TABLE 3.1   Dielectric Constant of Some Common Liquids

A load cell is used to measure the weight of the container.

Capacitive probes can be used in nonconductive liquids and free flowing solids for level measurement. Many materials when placed between the plates of a capacitor increases the capacitance by a factor μ called the dielectric constant of the material. For instance, air has a dielectric constant of 1 and water 80. Figure 3.1*b* shows two capacitor plates partially immersed in a nonconductive liquid. The capacitance (Cd) is given by

$$Cd = Ca\mu \frac{d}{r} + Ca \qquad (3.6)$$

where Ca = capacitance with no liquid
$\quad \mu$ = dielectric constant of the liquid between the plates
$\quad r$ = height of the plates
$\quad d$ = depth or level of the liquid between the plates

The dielectric constants of some common liquids are given in Table 3.1; there are large variations in dielectric constant with temperature so that temperature correction may be needed. In Eq. (3.6), the liquid level is given by

$$d = \frac{Cd + Ca}{\mu Ca} r \qquad (3.7)$$

The capacitance can be measured using a capacitance bridge circuit.

3.3   Level Sensing Devices

There are two categories of level sensing devices. They are direct sensing, in which the actual level is monitored, and indirect sensing where a property of the liquid such as pressure is sensed to determine the liquid level.

3.3.1   Direct Level Sensing

Sight glass (open end/differential) or gauge is the simplest method for direct visual reading. As shown in Fig. 3.2, the sight glass is normally mounted vertically adjacent to the container. The liquid level can then be observed directly in the sight glass. The container in Fig. 3.2*a* is closed. In this case, the ends of the glass are connected to the top and bottom of the tank, as would be used with a pressurized container (boiler) or a container

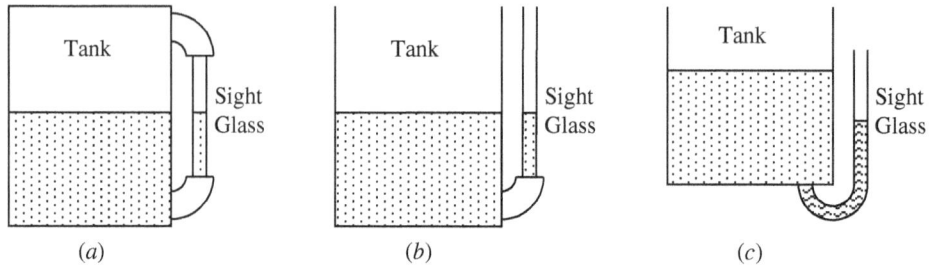

FIGURE 3.2 Various configurations of a sight glass to observe liquid levels: (a) pressurized or closed container, (b) open container, and (c) higher density sight glass liquid.

with volatile, flammable, hazardous, or pure liquids. In cases where the tank contains inert liquids such as water and pressurization is not required, the tank and sight glass can both be open to the atmosphere as shown in Fig. 3.2b. The top of the sight glass must have the same pressure conditions as the top of the liquid or the liquid levels in the tank and sight glass will be different. In cases where the sight glass is excessively long, a second inert liquid with higher density than the liquid in the container can be used in the sight glass (see Fig. 3.2c). Allowance must be made for the difference in the density of the liquids. If the glass is stained or reacts with the containerized liquid, the same approach can be taken using a different material for the sight glass. Magnetic floats can also be used in the sight glass so that the liquid level can be monitored with a magnetic sensor such as a Hall effect device.

Floats (angular arm or pulley) are shown in Fig. 3.3. The figure shows two types of simple float sensors. The float material is less dense than the density of the liquid and floats up and down on top of the material being measured. In Fig. 3.3a, a float with a pulley is used; this method can be used with either liquids or free flowing solids. With free flowing solids, agitation is sometimes used to level the solids. An advantage of the float sensor is that it is almost independent of the density of the liquid or solid being monitored. If the surface of the material being monitored is turbulent causing the float reading to vary excessively, some means of damping might be used in the system such as a larger float. In Fig. 3.3b, a ball float is attached to an arm; the angle of the arm is measured to indicate the level of the material (an example of the use of this type of sensor is the monitoring of the fuel level in the tank of an automobile).

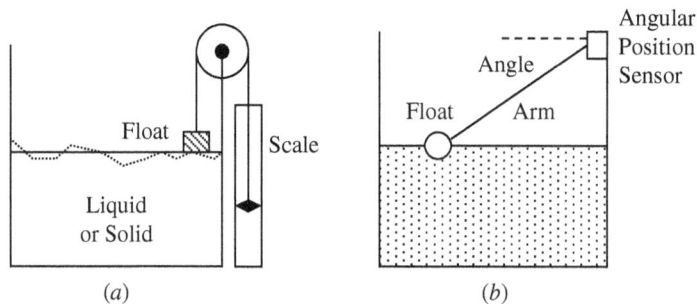

FIGURE 3.3 Methods of measuring liquid levels using (a) a simple float with level indicator on the outside of the tank and (b) an angular arm float.

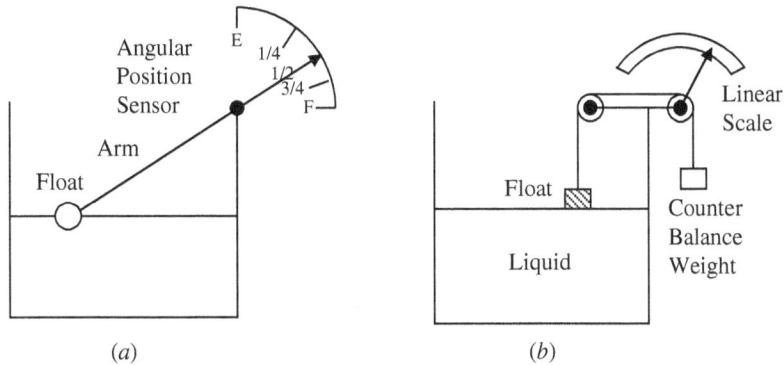

FIGURE 3.4   Scales used with float level sensors: (a) nonlinear scale with angular arm float and (b) linear scale with a pulley type of float.

Although very simple and cheap to manufacture, the disadvantage of this type of float is its nonlinearity as shown by the line of sight scale in Fig. 3.4a. The scale can be replaced with a potentiometer to obtain an electrical signal that can be linearized for industrial use.

Figure 3.4b shows an alternative method of using pulleys to obtain a direct visual scale that can be replaced by a potentiometer to obtain a linear electrical output with level.

3.3.2   Indirect Level Sensing

The most commonly used method of indirectly measuring a liquid level is to measure the hydrostatic pressure at the bottom of the container. The depth can then be extrapolated from the pressure and the specific weight of the liquid can be calculated using Eq. (3.1). The pressure can be measured by any of the methods given in the section on pressure. The dial on the pressure gauge can be calibrated directly in liquid depth. The depth of a liquid can also be measured using indirect methods, such as bubblers, radiation, resistive tapes, and by weight measurements.

> **Example 3.1**   A pressure gauge located at the base of an open tank containing a liquid with a specific weight of 54.5 lb/ft³ registers 11.7 psi. What is the depth of the fluid in the tank?
>
> **Solution**   From Eq. (3.1)
>
> $$h = \frac{p}{\gamma} = \frac{11.7 \text{ psi} \times 144}{54.5 \text{ lb/ft}^3} = 30.9 \text{ ft}$$

Bubbler devices require a supply of clean air or inert gas. The setup is shown in Fig. 3.5. Gas is forced through a tube whose open end is close to the bottom of the tank. The specific weight of the gas is negligible compared to the liquid and can be ignored. The pressure required to force the liquid out of the tube is equal to the pressure at the end of the tube due to the liquid, which is the depth of the liquid multiplied by the specific weight of the liquid. This method can be used with corrosive liquids as the material of the tube can be chosen to be corrosion resistant.

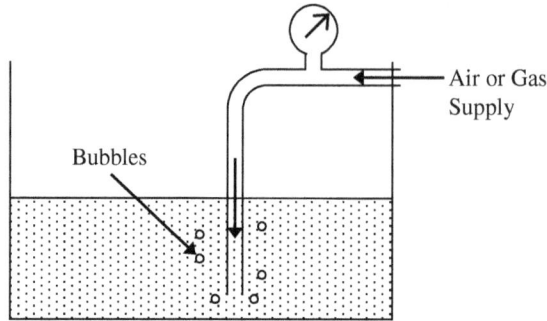

FIGURE 3.5 Liquid level measurements using a bubbler.

Example 3.2 How far below the surface of the water is the end of a bubbler tube, if bubbles start to emerge from the end of the tube when the air pressure in the bubbler is 148 kPa?

Solution From Eq. (3.1)

$$h = \frac{p}{\gamma} = \frac{148 \text{ kPa} \times 10^{-4}}{1 \text{ gm/cm}^3} = 14.8 \text{ cm}$$

A *displacer* with force sensing is shown in Fig. 3.6a. This device uses the change in buoyant force on an object to measure the changes in liquid level. The displacers must have a higher specific weight than that of the liquid level being measured and have to be calibrated for the specific weight of the liquid. A force or strain gauge measures the excess weight of the displacer. There is only a small movement in this type of sensor compared to a float sensor.

The buoyant force on a cylindrical displacer shown in Fig. 3.6b using Eq. (3.2) is given by

$$F = \frac{\gamma \pi d^2 L}{4} \tag{3.8}$$

where γ = specific weight of the liquid
$\quad\quad d$ = float diameter
$\quad\quad L$ = length of the displacer submerged in the liquid

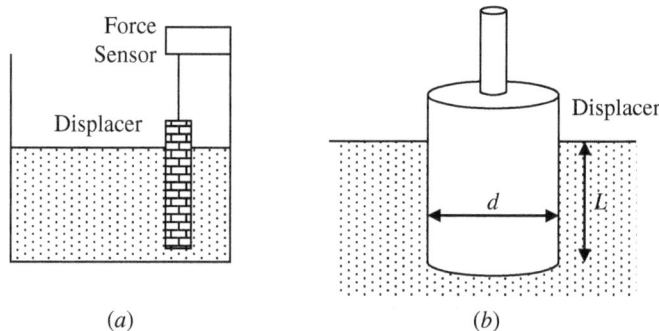

FIGURE 3.6 Displacer with a force sensor for measuring liquid level by (a) observing the loss of weight of the displacer due to the buoyancy forces of the displaced liquid and (b) dispenser dimensions.

The weight as seen by the force sensor is given by

$$\text{Weight on force sensor} = \text{Weight of displacer} - F \qquad (3.9)$$

It should be noted that the units must be in the same measurement system, and the liquid must not rise above the top of the displacer or the displacer must not touch the bottom or sides of the container.

Example 3.3 A displacer with a diameter of 8 in is used to measure changes in water level. If the water level changes by 1 ft what is the change in force sensed by the force sensor?

Solution From Eq. (3.9)

$$\text{Change in force} = (\text{weight of dispenser} - F_1) - (\text{weight of dispenser} - F_2) = F_2 - F_1$$

From Eq. (3.8)

$$F_2 - F_1 = \frac{62.4 \text{ lb/ft}^2 \times \pi(8 \text{ ft})^2 \times 12}{4} = 21.8 \text{ lb}$$

Example 3.4 A 3.5-cm diameter displacer is used to measure acetone levels. What is the change in force sensed if the liquid level changes by 52 cm?

Solution

$$F_2 - F_1 = \frac{7.74 \text{ kN/m}^3 \times \pi \times 3.5^2 \text{ cm}^2 \times 52 \text{ cm}}{4 \times 10^6 \text{ cm/m}} = 3.87 \text{ N } (395 \text{ g})$$

Probes for measuring liquid levels fall into three categories, i.e., conductive, capacitive, and ultrasonic.

Conductive probes are used for single-point measurements in liquids that are conductive and nonvolatile as a spark can occur. Conductive probes are shown in Fig. 3.7a. Two or more probes as shown can be used to indicate set levels. If the liquid is in a metal container, the container can be used as the common probe. When the liquid is in contact with two probes the voltage between the probes causes a current to flow indicating a set level has been reached. Thus, probes can be used to indicate when the liquid level is low and to operate a pump to fill the container. Another or third probe can be used to indicate when the tank is full and to turn-off the filling pump.

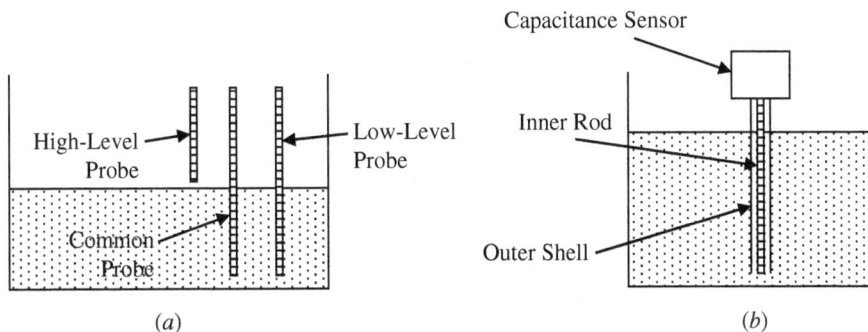

FIGURE 3.7 Methods of measuring liquid levels using (a) conductive probes for detecting set levels and (b) a capacitive probe for continuous monitoring.

Capacitive probes are used in liquids that are nonconductive and have a high μ and can be used for continuous level monitoring. The capacitive probe shown in Fig. 3.7b consists of an inner rod with an outer shell; the capacitance is measured between the two using a capacitance bridge. In the portion out of the liquid, air serves as the dielectric between the rod and outer shell. In the immersed section, the dielectric is that of the liquid that causes a large capacitive change, if the tank is made of metal it can serve as the outer shell. The capacitance change is directly proportional to the level of the liquid. The dielectric constant of the liquid must be known for this type of measurement. The dielectric constant can vary with temperature so that temperature correction may be required.

Example 3.5 A capacitive probe 30-in long has a capacitance of 22 pF in air. When partially immersed in water with a dielectric constant of 80 the capacitance is 1.1 nF. What is the length of the probe immersed in water?

Solution From Eq. (3.6)

$$d = \frac{(1.1 \times 10^3 \text{ pF} - 22 \text{ pF}) 30 \text{ in}}{80 \times 22 \text{ pF}} = 18.4 \text{ in}$$

Ultrasonics can be used for single-point or continuous level measurement of a liquid or a solid. A single ultrasonic transmitter and receiver can be arranged with a gap as shown in Fig. 3.8a to give single-point measurement. As soon as liquid fills the gap, ultrasonic waves from the transmitter reach the receiver. A setup for continuous measurement is shown in Fig. 3.8b. Ultrasonic waves from the transmitter are reflected by the surface of the liquid to the receiver; the time for the waves to reach the receiver is measured. The time delay gives the distance from the transmitter and receiver to the surface of the liquid, from which the liquid level can be calculated knowing the velocity of ultrasonic waves. As there is no contact with the liquid, this method can be used for solids, corrosive materials, and volatile liquids. In a liquid the transmitter and receiver can also be placed on the bottom of the container and the time measured for a signal to be reflected from the surface of the liquid to the receiver to measure of the depth of the liquid.

Radiation methods are sometimes used in cases where the liquid is corrosive, very hot, or detrimental to installing sensors. For single-point measurement only one transmitter and a detector are required. If several single-point levels are required, a detector

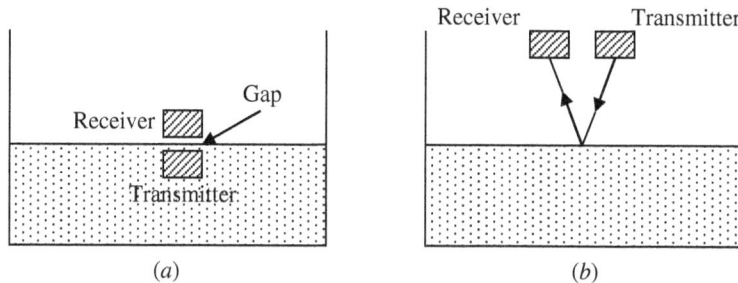

FIGURE 3.8 Use of ultrasonics for (a) single-point liquid level measurement and (b) continuous liquid level measurements made by timing reflections from the surface of the liquid.

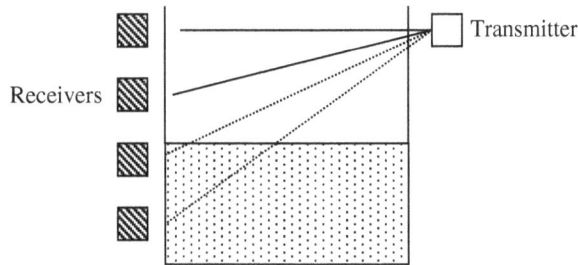

Figure 3.9 Liquid level measurements using a radiation technique.

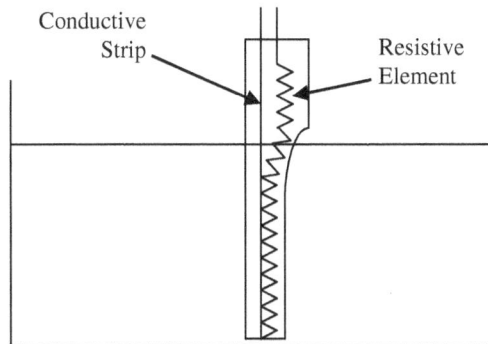

Figure 3.10 Demonstrating a resistive tape level sensor.

will be required for each level measurement as shown in Fig. 3.9. The disadvantages of this system are the cost and the need to handle radioactive material.

Resistive tapes can be used to measure liquid levels (see Fig. 3.10). A resistive element is placed in close proximity to a conductive strip in an easily compressible nonconductive sheath; the pressure of the liquid pushes the resistive element against the conductive strip shorting out a length of the resistive element proportional to the depth of the liquid. The sensor can be used in liquids or slurries. Though it is cheap, it is not rugged or accurate, is prone to humidity problems, and measurement accuracy depends on material density.

Load cells can be used to measure the weight of a tank and its contents. The weight of the container is subtracted from the reading, leaving the weight of the contents of the container. Knowing the cross-sectional area of the tank and the specific weight of the material, the volume and/or depth of the contents can be calculated. This method is well suited for continuous measurement and the material being weighed does not come into contact with the sensor. Figure 3.11 shows two elements that can be used in load sensors for measuring force. Figure 3.11a shows a cantilever beam used as a force or weight sensor.

The beam is rigidly attached at one end and a force is applied to the other end. A strain gauge attached to the beam is used to measure the strain in the beam; a second strain gauge is used for temperature compensation. Figure 3.11b shows a piezoelectric sensor used to measure force or weight. The sensor gives an output voltage proportional to the force applied.

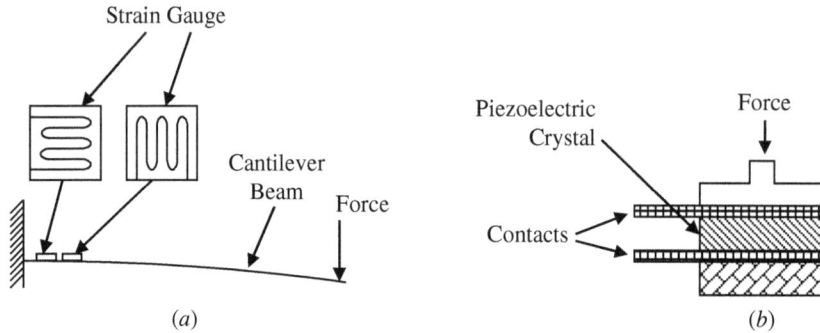

FIGURE 3.11 Force sensors can be used for measuring weight using (a) strain gauge technique or (b) a piezoelectric technique.

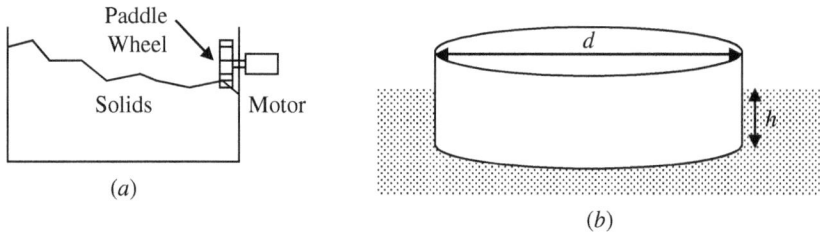

FIGURE 3.12 Illustration of (a) paddle wheel technique to measure the level of free flowing solids and (b) a typical float shape.

Example 3.6 What is the depth of the liquid in a container, if the specific weight of the liquid is 82 lb/ft³; the container weights 45 lb and is 21 in in diameter? A load cell measures a total weight of 385 lb.

Solution Using Eqs. (3.4) and (3.5) we get the following:

$$\text{Weight of liquid} = 385 - 45 = 340 \text{ lb}$$

$$\text{Volume of liquid} = \frac{3.14 \times 21 \times 21 \times d}{12 \times 21 \times 4} \text{ ft} = \frac{340 \text{ lb}}{82/\text{ft}^3}$$

$$\text{Depth } (d) = \frac{4.15 \times 576}{1384.7} \text{ ft} - 1.73 \text{ ft} = 20.7 \text{ in}$$

Paddle wheels driven by electric motors can be used for sensing the level of solids in the form of power, grains, or granules. When the material reaches and covers the paddle wheel, the torque needed to turn the motor greatly increases. The torque can be an indication of the depth of the material. Such a set up is shown in Fig. 3.12a. Some agitation may be required to level the solid particles.

3.4 Application Considerations

A number of factors affect the choice of sensor for level measurement, such as pressure on the liquid, liquid temperature, turbulence, volatility, corrosiveness, accuracy needed, single-point or continuous measurement, direct or indirect, particulates in a liquid, free flowing solids, and so forth.

Floats are often used to sense fluid levels because they are unaffected by particulates and can be used for slurries and with a wide range of liquid specific weights. Flat floats due to their area are less susceptible to turbulence on the surface of the liquid. Figure 3.12*b* shows a commonly used design for a float which can be attached to a level indicator. The float displaces its own weight of liquid as follows:

$$\text{Float weight} = \text{buoyant force} = \frac{\gamma_L \pi d^2 h}{4} \qquad (3.10)$$

where γ_L = specific weight of the liquid
  d = diameter
  h = immersion depth of the float

When the float is used to measure 1 ft or more of liquid depth, any change in h due to large changes in γ_L will have minimal effect on the measured liquid depth.

Displacers must never be completely submerged when measuring liquid depth and must have a specific weight greater than that of the liquid. Care must also be taken to ensure that the displacer is not corroded by the liquid and the specific weight of the liquid is constant over time. The temperature of the liquid may also have to be monitored to make corrections for density changes. Displacers can be used to measure depths up to about 3 m with an accuracy of ±0.5 cm.

Capacitive device accuracy can be affected by the placement of the device, so the manufacturer's installation instructions must be followed. The dielectric constant of the liquid should also be regularly monitored. Capacitive devices can be used in pressurized containers up to 30 MPa and temperatures up to 1000°C, and measure depths up to 6 m with an accuracy of ±1 percent.

Pressure gauge choice for measuring liquid levels can depend on a number of considerations, which are as follows:

1. The presence of particulates that can block the line to the gauge.
2. Damage caused by excessive temperatures in the liquid.
3. Damage due to peak pressure surges.
4. Corrosion of the gauge by the liquid.
5. Differential pressure gauges are needed if the liquid is under pressure.
6. Distance between the tank and the gauge.
7. Use of manual valves for gauge repair.

Differential pressure gauges can be used in pressurized containers up to 30 MPa and temperatures up to 600°C to give accuracies of ±1 percent. The liquid depth depends on its density and the pressure gauge used.

Bubbler devices require certain precautions when being used. To ensure a continuous air or gas supply, the gas used must not react with the liquid. It may be necessary to install a one-way valve to prevent the liquid being sucked back into the gas supply lines if the gas pressure is lost. The bubbler tube must be chosen so that it is not corroded by the liquid. Bubbler devices are typically used at atmospheric pressure, accuracies of about 2 percent can be obtained, depth depends on gas pressure available, and so forth.

Ultrasonic devices can be used with pressurized containers up to 2 MPa and 100°C temperature range for depths of up to 30 m with accuracies of about 2 percent.

Radiation devices are used for point measurement of hazardous materials. Due to the hazardous nature of the material, personnel should be trained in its use, transportation, storage, identification, and disposal.

Parameter Type	Continuous	Slurry	Powder	Range	Accuracy
Sight glass	Y	Y	N	3 ft	± 0.25 in
Float	Y	Y	N	12 ft	± 0.25 in
Bubbler	Y	?	N	50 ft	± 1%
Displacer	Y	?	N	5 ft	± 0.25 in
Capacitance	Y	Y	Y	20 ft	± 1–2%
Resistance	N	?	?	100 ft	± 0.25 in
Ultrasonic	Y	Y	?	12 ft	± 1–2%
Radiation	Y	Y	Y	15 ft	± 1–2%
Resistive tape	Y	N	Y	50 ft	± 3 in
Load cell	Y	Y	Y	60 ft	± 2%
Paddle wheel	N	N	Y	0.5 ft	± 1 in

TABLE 3.2 Level Measurements

Other considerations are that liquid level measurements can be effected by turbulence, readings may have to be averaged, and/or baffles used to reduce the turbulence. Frothing in the liquid can also be a source of error particularly with resistive or capacitive probes. A table of accuracies is given in Table 3.2.

Summary

This chapter introduced the concepts of level measurement. The instruments used for direct and indirect measurement have been described and the application of level measuring instruments considered.

The main points described in this chapter were as follows:

1. The formulas used by instruments for the measurement of liquid levels and free flowing solids with examples.

2. The various types of instruments used to give direct measurement of liquid levels and the methods used to indirectly measure liquid levels.

3. The difference between continuous and single-point level measurements in a liquid.

4. Application considerations when selecting an instrument for measuring liquid and free flowing solid levels.

Problems

3.1 What is the specific weight of a liquid, if the pressure is 4.7 psi at a depth of 17 ft?

3.2 What is the depth of a liquid, if the pressure is 127 kPa and the liquid density is 1.2 g/cm³?

3.3 What is the displaced volume in cubic meters if the buoyancy on an object is 15 lb and the density of the liquid is 785 kg/m³?

3.4   What is the liquid density in grams per cubic centimeter, if the buoyancy is 833 N on a 135 cm³ submerged object?

3.5   The weight of a body in air is 17 lb and submerged in water is 3 lb. What is the volume and specific weight of the body?

3.6   A material has a density of 1263 kg/m³. A block of the material weighs 72 kg when submerged in water. What is its volume and weight in air?

3.7   A container of 4.5-ft diameter is full of liquid. If the liquid has a specific weight of 63 lb/ft³, what is the depth of the liquid if the weight of the container and liquid is 533 lb? Assume the container weighs 52 lb.

3.8   The weight of liquid in a round container is 1578 kg, the depth of the liquid is 3.2 m. If the density of the liquid is 0.83 g/cm³, what is the diameter of the container?

3.9   A capacitive sensor is 3-ft 3-in high and has a capacitance of 25 pF in air and 283 pF when immersed in a liquid to a depth of 2 ft 7 in. What is the dielectric constant of the liquid?

3.10   A capacitive sensor 2.4 m in height has a capacitance of 75 pF in air if the sensor is placed in a liquid with a dielectric constant of 65 to a depth of 1.7 m. What will be the capacitive reading of the sensor?

3.11   A pressure gauge at the bottom of a tank reads 32 kPa. If the tank has 3.2-m diameter, what is the weight of liquid in the container?

3.12   What pounds per square inch is required by a bubbler system to produce bubbles at a depth of 4 ft 7 in water?

3.13   A bubbler system requires a pressure of 28 kPa to produce bubbles in a liquid with a density of 560 kg/m³. What is the depth of the outlet of the bubbler in the liquid?

3.14   A displacer with a diameter of 4.7 cm is used to measure changes in the level of a liquid with a density of 470 kg/m³. What is the change in force on the sensor if the liquid level changes 13.2 cm?

3.15   A displacer is used to measure changes in liquid level. The liquid has a density of 33 lb/ft³. What is the diameter of the dispenser if a change in liquid level of 45 in produces a change in force on the sensor of 3.2 lb?

3.16   A bubbler system requires a pressure of 47 kPa to produce bubbles at a depth of 200 in. What is the density of the liquid in pounds per cubic foot?

3.17   A capacitive sensing probe 2.7-m high has a capacitance of 157 pF in air and 7.4 nF when partially immersed in a liquid with a dielectric constant of 79. How much of the probe is immersed in the liquid?

3.18   A force sensor is immersed in a liquid with a density of 61 lb/ft³ to a depth of 42 in and then placed in a second liquid with a density of 732 N/m³. What is the change in force on the sensor if the diameter of the sensor is 8 cm and the change in depth is 5.9 cm?

3.19   An ultrasonic transmitter and receiver are placed 10.5 ft above the surface of a liquid. How long will the sound waves take to travel from the transmitter to the receiver? Assume the velocity of sound waves is 340 m/s.

3.20   If the liquid in Problem 3.19 is lowered to 6.7 ft, what is the increase in time for the sound waves to go from the transmitter to the receiver?

CHAPTER 4

Flow

Chapter Objectives

This chapter will introduce you to the concepts of fluid velocity, and flow, and its relation to pressure and viscosity. This chapter will help you understand the units used in flow measurement, and become familiar with the most commonly used flow standards.

Topics discussed in this chapter are as follows:

- Reynolds number and its application to flow patterns
- The formulas used in flow measurements
- The Bernoulli equation and its applications
- Difference between flow rate and total flow
- Pressure losses and their effects on flow
- Flow measurements using differential pressure measuring devices and their characteristics
- Open channel flow and its measurement
- Considerations in the use of flow instrumentation

4.1   Introduction

This chapter discusses the basic terms and formulas used in flow measurements and instrumentation. The measurement of fluid and gas flow is a key parameter in process control and industrial applications. Optimum performance of most equipment and operations require specific flow rates that require the use of very accurate sensing techniques. New measuring methods and sensors are continually being introduced to meet the needs of flow measurements. The cost of many liquids and gases are based on the measured flow through a pipeline making it necessary to accurately measure and control the rate of flow for accounting purposes.

4.2   Basic Terms

This chapter will be using terms and definitions from previous chapters as well as introducing a number of new definitions related to flow and flow rate sensing.

Velocity is a measure of speed and direction of an object. When related to fluids it is the rate of flow of fluid particles in a pipe. The speed of particles in a fluid flow varies over the cross section of the flow, i.e., where the fluid is in contact with the

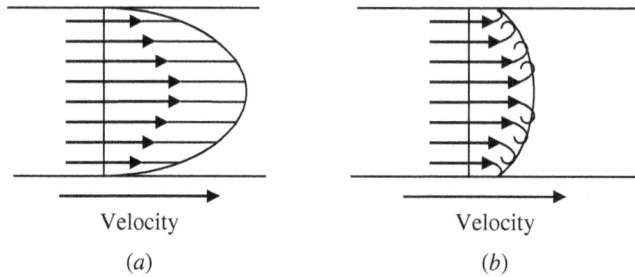

FIGURE 4.1 Flow velocity variations across a pipe with (a) laminar flow and (b) turbulent flow.

constraining walls (the boundary layer) the velocity of the liquid particles is virtually zero; in the center of the flow the liquid particles will have the maximum velocity. Thus, the accurate measurement of flow rates is difficult when average rates of flow are used in flow calculations. The units of flow are normally feet per second (fps), feet per minute (fpm), or meters per second (mps), and so on. In Chap. 2, the pressures associated with fluid flow were defined as static, impact, or dynamic.

Laminar flow of a liquid occurs when its average velocity is comparatively low and the fluid particles tend to move smoothly in layers, as shown in Fig. 4.1a. The velocity of the particles across the liquid takes a parabolic shape.

Turbulent flow occurs when the flow velocity is high and the particles no longer flow smoothly in layers and turbulence or a rolling effect occurs. This is shown in Fig. 4.1b. Note also the flattening of the velocity profile.

Viscosity is a property of a gas or liquid that is a measure of its resistance to motion or flow. A viscous liquid such as syrup has a much higher viscosity than water, and water has a higher viscosity than air. Syrup, because of its high viscosity, flows very slowly and it is very hard to move an object through it. Viscosity (dynamic) can be measured in poise or centipoise, whereas kinematic viscosity (without force) is measured in stokes or centistokes. Dynamic or absolute viscosity is used in the Reynolds and flow equations. Table 4.1 gives a list of conversions. Typically, the viscosity of a liquid decreases as temperature increases.

The *Reynolds number R* is a derived relationship combining the density and viscosity of a liquid with its velocity of flow and the cross-sectional dimensions of the flow and takes the form as follows:

$$R = \frac{VD\rho}{\mu} \tag{4.1}$$

Dynamic Viscosities	Kinematic Viscosities
1 lb s/ft^2 = 47.9 Pa s	1 ft^2/s = 9.29 × 10^{-2} m^2/s
1 centipoise = 10 Pa s	1 stoke = 10.4 m^2/s
1 centipoise = 2.09 × 10^{-5} lb s/ft^2	1 m^2/s = 10.76 ft^2/s
1 poise = 100 centipoise	1 stoke = 1.076 × 10^{-3} ft^2/s

TABLE 4.1 Conversion Factors for Dynamic and Kinematic Viscosities

1 gal/min = 6.309 × 10⁻⁵ m²/s	1 L/min = 16.67 × 10⁻³ m³/s
1 gal/min = 3.78 L/min	1 ft³/s = 449 gal/min
1 gal/min = 0.1337 ft³/min	1 gal/min = 0.00223 ft³/s
1 gal water = 231 in³	1 ft³ water = 7.48 gal

*1 gal water = 0.1337 ft³ = 231 in³; 1 gal water = 8.35 lb; 1 ft³ water = 7.48 gal; 1000 liters water = 1 m³; 1 liter water = 1 kg.

TABLE 4.2 Flow Rate Conversion Factors (US Gallons)*

where V = average fluid velocity
D = diameter of the pipe
ρ = density of the liquid
μ = absolute viscosity

Flow patterns can be considered to be laminar, turbulent, or a combination of both. Osborne Reynolds observed in 1880 that the flow pattern could be predicted from physical properties of the liquid. If the Reynolds number for the flow in a pipe is equal to or less than 2000, the flow will be laminar. From 2000 to about 5000 is the intermediate region where the flow can be laminar, turbulent, or a mixture of both, depending on other factors. Beyond about 5000 the flow is always turbulent.

The Bernoulli equation is an equation for flow based on the conservation of energy, which states that the total energy of a fluid or gas at any one point in a flow is equal to the total energy at all other points in the flow.

Energy factors. Most flow equations are based on energy conservation and relate to the average fluid or gas velocity, pressure, and the height of fluid above a given reference point. This relationship is given by the Bernoulli equation. The equation can be modified to take into account energy losses due to friction and increases in energy as supplied by pumps.

Energy losses in flowing fluids are caused by friction between the fluid and the containment walls and by fluid impacting an object. In most cases these losses should be taken into account. While these equations apply to both liquids and gases they are more complicated in gases by the fact that gases are compressible.

Flow rate is the volume of fluid passing a given point in a given amount of time and is typically measured in gallons per minute (gpm), cubic foot per minute (cfm), liters per min, and so on. Table 4.2 gives the flow rate conversion factors.

Total flow is the volume of liquid flowing over a period of time and is measured in gallons, cubic feet, liters, and so forth.

4.3 Flow Formulas

4.3.1 Continuity Equation

The continuity equation states that if the overall flow rate in a system is not changing with time (see Fig. 4.2a), the flow rate in any part of the system is constant. From which we get the following equation:

$$Q = VA \qquad (4.2)$$

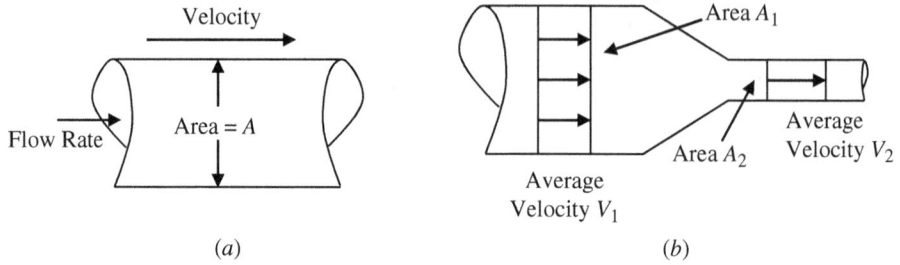

FIGURE 4.2 Flow diagram used for use in the continuity equation: (a) fixed diameter and (b) effects of different diameters on the flow rate.

where Q = flow rate
V = average velocity
A = cross-sectional area of the pipe

The units on both sides of the equation must be compatible, i.e., English units or metric units.

Example 4.1 What is the flow rate through a pipe 9-in diameter, if the average velocity is 5 fps?

Solution

$$Q = \frac{5 \text{ ft/s} \times \pi \times 0.75^2 \text{ ft}^2}{4} = 2.21 \text{ cfs} = \frac{2.21 \text{ gps}}{0.137} = 16.1 \text{ gps} = 16.1 \times 60 \text{ gpm} = 968 \text{ gpm}$$

If liquids are flowing in a tube with different cross-section areas, i.e., A_1 and A_2, as is shown in Fig. 4.2b, the continuity equation gives

$$Q = V_1 A_1 = V_2 A_2 \qquad (4.3)$$

Example 4.2 If a pipe goes from 9-cm diameter to 6-cm diameter and the velocity in the 9-cm section is 2.21 m/s, what is the average velocity in the 6-cm section?

Solution

$$Q = V_1 A_1 = V_2 A_2$$

$$V_2 = \frac{2.21 \text{ m}^3/\text{s} \times \pi \times 4.5^2}{\pi \times 3^2} = 4.97 \text{ m/s}$$

Mass flow rate F is related to volume flow rate Q by

$$F = \rho Q \qquad (4.4)$$

where F is the mass of liquid flowing and ρ is the density of the liquid.
As a gas is compressible, Eq. (4.3) must be modified for gas flow to

$$\gamma_1 V_1 A_1 = \gamma_2 V_2 A_2 \qquad (4.5)$$

where γ_1 and γ_2 are specific weights of the gas in the two sections of pipe.
Equation (4.3) is the rate of weight flow in the case of a gas. However, this could also apply to liquid flow in Eq. (4.3) by multiplying both sides of the equation by the specific weight γ.

4.3.2 Bernoulli Equation

The Bernoulli equation gives the relation between pressure, fluid velocity, and elevation in a flow system. The equation is accredited to Bernoulli (1738). When applied to Fig. 4.3a the following is obtained:

$$\frac{P_A}{\gamma_A} + \frac{V_A^2}{2g} + Z_A = \frac{P_B}{\gamma_B} + \frac{V_B^2}{2g} + Z_B \qquad (4.6)$$

where P_A and P_B = absolute static pressures at points A and B, respectively
γ_A and γ_B = specific weights
V_A and V_B = average fluid velocities
g = gravitation constant
Z_A and Z_B = elevations above a given reference level, i.e., $Z_A - Z_B$ is the head of fluid

The units in Eq. (4.6) are consistent and reduce to units of length (feet in the English system and meters in the SI system of units) as follows:

$$\text{Pressure energy} = \frac{p}{\gamma} = \frac{\text{lb/ft}^2 (\text{N/m}^2)}{\text{lb/ft}^3 (\text{N/m}^3)} = \text{ft (m)}$$

$$\text{Kinetic energy} = \frac{V^2}{g} = \frac{(\text{ft/s})^2 (\text{m/s})^2}{\text{ft/s}^2 (\text{m/s}^2)} = \text{ft (m)}$$

$$\text{Potential energy} = Z = \text{ft (m)}$$

This equation is a conservation of energy equation and assumes no loss of energy between points A and B (see Fig. 4.3a). The first term represents energy stored due to pressure, the second term represents kinetic energy or energy due to motion, and the third term represents potential energy or energy due to height. This energy relationship can be seen if each term is multiplied by mass per unit volume which cancels as the mass per unit volume is the same at points A and B. The equation can be used between any two positions in a flow system. The pressures used in the Bernoulli equation must be absolute pressures.

In the fluid system shown in Fig. 4.3b, the flow velocity V at point 3 can be derived from Eq. (4.6) and is as follows using point 2 as the reference line:

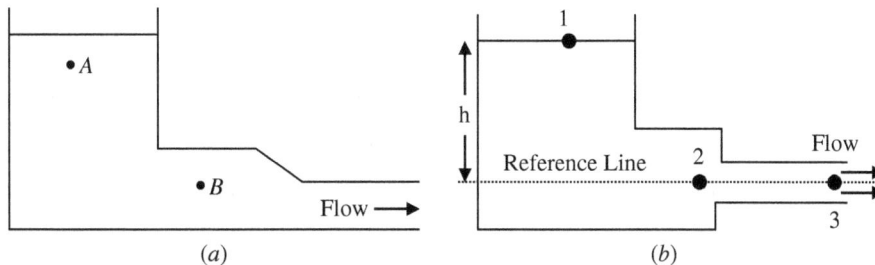

FIGURE 4.3 Container diagrams: (a) the pressures at points A and B are related by the Bernoulli equation and (b) application of the Bernoulli in Example 4.3.

$$\frac{P_1}{\gamma_1} + 0 + h = \frac{P_3}{\gamma_3} + \frac{V_3^2}{2g} + 0$$

$$V_3 = \sqrt{(2gh)} \qquad (4.7)$$

Point 3 at the exit has dynamic pressure but no static pressure above 1 atm, and hence, $P_3 = P_1 = 1$ atm and $\gamma_1 = \gamma_2$. This shows that the velocity of the liquid flowing out of the system is directly proportional to the square root of the height of the liquid above the reference point.

Example 4.3 If h in Fig. 4.3b is 7.5 m, what is the pressure at P_2? Assume the areas at points 2 and 3 are 0.48 m² and 0.3 m², respectively.

Solution

$$V_3 = \sqrt{(2 \times 9.8 \times 7.5)} = 12.12 \text{ m/s}$$

Considering points 2 and 3 with the use of Eq. (4.6)

$$\frac{P_2}{9.8 \text{ kN}} + \frac{V_2^2}{2 \times 9.8} + 0 = \frac{101.3 \text{ kPa}}{9.8 \text{ kN}} + \frac{V_3^2}{2 \times 9.8} + 0 \qquad (4.8)$$

Using the continuity Eq. (4.3) and knowing that the areas at points 2 and 3 are 0.48 m² and 0.3 m², respectively, the velocity at point 2 is given by

$$V_2 = \left(\frac{A_3}{A_2}\right)V_3 = \left(\frac{0.3}{0.48}\right)12.12 \text{ m/s} = 7.58 \text{ m/s}$$

Substituting the values obtained for V_2 and V_3 into Eq. (4.8) gives the following:

$$\frac{P_2}{9.8} + \frac{(7.58)^2}{2 \times 9.8} + 0 = \frac{101.3}{9.8} + \frac{(12.12)^2}{2 \times 9.8} + 0$$

$$P_2 = 146 \text{ kPa } (a) = 44.7 \text{ kPa } (g)$$

4.3.3 Flow Losses

The Bernoulli equation does not take into account flow losses; these losses are accounted for by pressure losses and fall into two categories. First, those associated with viscosity and the friction between the constriction walls and the flowing fluid; and second, those associated with fittings, such as valves, elbows, tees, and so forth.

Outlet losses. The flow rate Q from the continuity equation for point 3 in Fig. 4.3b, e.g., gives

$$Q = V_3 A_3$$

However, to account for losses at the outlet, the equation should be modified to

$$Q = C_D V_3 A_3 \qquad (4.9)$$

where C_D is the discharge coefficient that is dependent on the shape and size of the orifice. The discharge coefficients can be found in flow data handbooks.

Frictional losses. They are losses from liquid flow in a pipe due to friction between the flowing liquid and the restraining walls of the container. These frictional losses are given by

$$h_L = \frac{fLV^2}{2Dg} \qquad (4.10)$$

where h_L = head loss
  f = friction factor
  L = length of pipe
  D = diameter of pipe
  V = average fluid velocity
  g = gravitation constant

The friction factor f depends on the Reynolds number for the flow and the roughness of the pipe walls.

Example 4.4   What is the head loss in a 2-in diameter pipe 120-ft long? The friction factor is 0.03 and the average velocity in the pipe is 11 fps.

Solution

$$h_L = \frac{fLV^2}{2\,Dg} = \frac{0.03 \times 120\ \text{ft} \times (11\ \text{ft/s})^2\ 12}{2\ \text{ft} \times 2 \times 32.2\ \text{ft/s}^2} = 40.6\ \text{ft}$$

This would be equivalent to

$$\frac{40.6\ \text{ft} \times 62.4\ \text{lb/ft}^3}{144} = 17.6\ \text{psi}$$

Fitting losses are losses due to couplings and fittings, which are normally less than those associated with friction and are given by

$$h_L = \frac{KV^2}{2g} \tag{4.11}$$

where h_L = head loss due to fittings
  K = head loss coefficient for various fittings
  V = average fluid velocity
  g = gravitation constant

Values for K can be found in flow handbooks. Table 4.3 gives some typical values for the head loss coefficient factor in some common fittings.

Example 4.5   Fluid is flowing at 4.5 fps through 1-in fittings as follows: 5 x 90° ells, 3 tees, 1 gate valve, and 12 couplings. What is the head loss?

Solution

$$h_L = \frac{(5 \times 1.5 + 3 \times 0.8 + 1 \times 0.22 + 12 \times 0.085)4.5 \times 4.5}{2 \times 32.2}$$

$$h_L = (7.5 + 2.4 + 0.22 + 1.02)0.31 = 3.5\ \text{ft}$$

Threaded ell – 1 in	1.5	Flanged ell – 1 in	0.43
Threaded tee – 1 in in line	0.9	Branch	1.8
Globe valve (threaded)	8.5	Gauge valve (threaded)	0.22
Coupling of union – 1 in	0.085	Bell mouth reducer	0.05

TABLE 4.3 Typical Head Loss Coefficient Factors for Fittings

Circular cylinder with axis perpendicular to flow	0.33 to 1.2
Circular cylinder with axis parallel to flow	0.85 to 1.12
Circular disc facing flow	1.12
Flat plate facing flow	1.9
Sphere	0.1 +

TABLE 4.4   Typical Drag Coefficient Values for Objects Immersed in Flowing Fluid

To take into account losses due to friction and fittings, the Bernoulli Eq. (4.6) is modified as follows:

$$\frac{P_A}{\gamma_A} + \frac{V_A^2}{2g} + Z_A = \frac{P_B}{\gamma_B} + \frac{V_B^2}{2g} + Z_B + h_{L\,friction} + h_{L\,fittings} \tag{4.12}$$

Form drag is the impact force exerted on devices protruding into a pipe due to fluid flow. The force depends on the shape of the insert and can be calculated from

$$F = C_D \gamma \frac{AV^2}{2g} \tag{4.13}$$

where F = force on the object
C_D = the drag coefficient
γ = specific weight
g = acceleration due to gravity
A = cross-sectional area of obstruction
V = average fluid velocity

Flow handbooks contain drag coefficients for various objects. Table 4.4 gives some typical drag coefficients.

Example 4.6   A 5-in diameter ball is traveling through the air with a velocity of 110 fps, if the density of the air is 0.0765 lb/ft³ and $C_D = 0.5$. What is the force acting on the ball?

Solution

$$F = C_D \gamma \frac{AV^2}{2g} = \frac{0.5 \times 0.0765 \text{ lb/ft}^3 \times \pi \times 5^2 \text{ ft}^2 \times (110 \text{ ft/s})^2}{2 \times 32.2 \text{ ft/s}^2 \times 8 \times 144} = 0.98 \text{ lb}$$

4.4   Flow Measurement Instruments

Flow measurements are normally indirect measurements using differential pressures to measure the flow rate. Flow measurements can be divided into the following groups: flow rate, total flow, and mass flow. The choice of measuring device will depend on the required accuracy, and fluid characteristics (gas, liquid, suspended particulates, temperature, viscosity, and so on).

4.4.1   Flow Rate

Differential pressure measurements can be made for flow rate determination when a fluid flows through a restriction. The restriction produces an increase in pressure, which can be directly related to flow rate. Figure 4.4 shows examples of commonly used restrictions: (*a*) orifice plate, (*b*) Venturi tube, (*c*) flow nozzle, and (*d*) Dall tube.

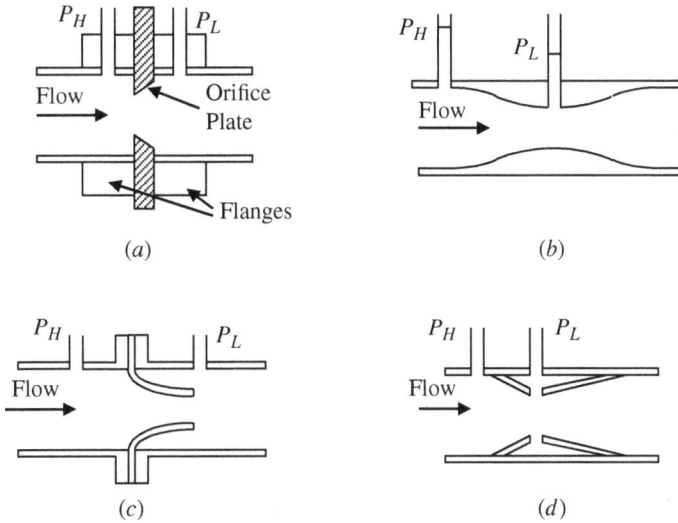

FIGURE 4.4   Types of constrictions used in flow rate measuring devices (a) orifice plate, (b) Venturi tube, (c) flow nozzle, and (d) Dall tube.

The orifice plate is normally a simple metal diaphragm with a constricting hole. The diaphragm is normally clamped between pipe flanges to give easy access. The differential pressure ports can be located in the flange either side of the orifice plate as shown in Fig. 4.4a, or alternatively, at specific locations in the pipe either side of the flange determined by the flow patterns (named vena contracta). A differential pressure gauge is used to measure the difference in pressure between the two ports; the differential pressure gauge can be calibrated in flow rates. The lagging edge of the hole in the diaphragm is beveled to minimize turbulence. In fluids, the hole is normally centered in the diaphragm, see Fig. 4.5a. However, if the fluid contains particulates the hole could be placed at the bottom of the pipe to prevent a buildup of particulates as in Fig. 4.5b. The hole can also be in the form of a semicircle having the same diameter as the pipe, and located at the bottom of the pipe as in Fig. 4.5c.

The Venturi tube shown in Fig. 4.4b uses the same differential pressure principal as the orifice plate. It normally uses a specific reduction in tube size, and is not normally used in larger diameter pipes, where it becomes heavy and excessively long. The advantages of the

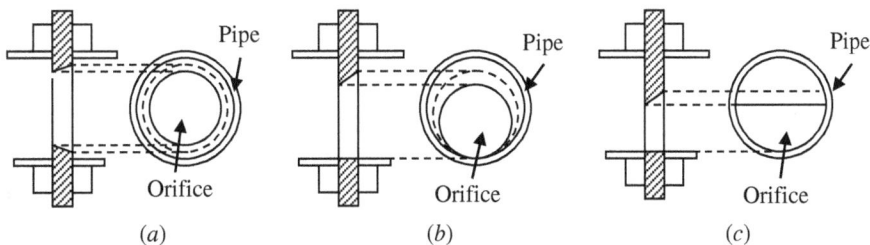

FIGURE 4.5   Orifice shapes and locations used (a) with fluids and (b) and (c) with suspended solids.

Venturi tube are its ability to handle large amounts of suspended solids, it creates less turbulence and hence less insertion loss than the orifice plate. The differential pressure taps in the Venturi tube are located at the minimum and maximum pipe diameters. The Venturi tube has good accuracy but its cost is high.

The flow nozzle is a good compromise on cost and accuracy between the orifice plate and the Venturi tube for clean liquids. It is not normally used with suspended particles. Its main use is the measurement of steam flow. The flow nozzle is shown in Fig. 4.4c.

The Dall tube shown in Fig. 4.4d has the lowest insertion loss but is not suitable for use with slurries.

Typical ratios (beta ratios, which are the diameter of the orifice opening divided by the diameter of the pipe) for the size of the constriction to pipe size in flow measurements is normally between 0.2 and 0.6. The ratios are chosen to give high enough pressure drops for accurate flow measurements but are not high enough to give turbulence. A compromise is made between high beta ratios (d/D) that give low differential pressures and low ratios that give high differential pressures, but can create high losses.

To summarize, the orifice is the simplest, cheapest, easiest to replace, least accurate, more subject to damage and erosion, and has the highest loss. The Venturi tube is more difficult to replace, most expensive, most accurate, high tolerance to damage and erosion, and has the lowest losses of all the three tubes. The flow nozzle is intermediate between the other two, and offers a good compromise. The Dall tube has the advantage of having the lowest insertion loss but cannot be used with slurries.

The elbow can be used as a differential flow meter. Figure 4.6a shows the cross section of an elbow. When a fluid is flowing, there is a differential pressure between the inside and outside of the elbow due to the change in direction of the fluid. The pressure difference is proportional to the flow rate of the fluid. The elbow meter is good for handling particulates in solution, with good wear and erosion resistance characteristics but has low sensitivity.

The pilot static tube shown in Fig. 4.6b is an alternative method of measuring flow rate, but has some disadvantages in measuring flow, in that it really measures fluid velocity at the nozzle. Because the velocity varies over the cross section of the pipe, the pilot static tube should be moved across the pipe to establish an average velocity, or the tube should be calibrated for one area. Other disadvantages are that the tube can become clogged with particulates, and the differential pressure between the impact and static pressures for low flow rates may not be enough to give the required accuracy.

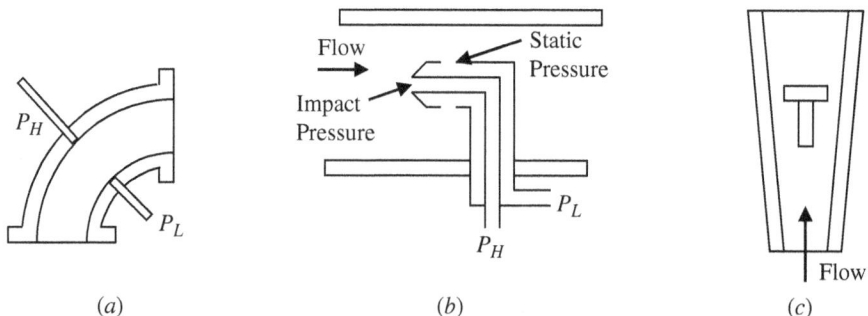

(a) (b) (c)

FIGURE 4.6 Other flow measuring devices are (a) elbow, (b) pilot static tube, and (c) rotameter.

This is the type of device used to measure aircraft air speed. The differential pressures in any of the above devices can be measured using the pressure measuring sensors discussed in Chap. 2 (Pressure).

Variable-area meters, such as the rotameter shown in Fig. 4.6c, are often used as a direct visual indicator for flow rate measurements. The rotameter is a vertical tapered tube with a T (or similar) shaped weight. The tube is graduated in flow rate for the characteristics of the gas or liquid flowing up the tube. The velocity of a fluid or gas flowing decreases as it goes higher up the tube, due to the increase in the bore of the tube. Hence, the buoyancy on the weight reduces the higher up the tube it goes. An equilibrium point is eventually reached where the force on the weight due to the flowing fluid is equal to that of the weight; i.e., the higher the flow rate the higher the weight goes up the tube. The position of the weight is also dependent on its size and density, the viscosity and density of the fluid, and the bore and taper of the tube. The rotameter has a low insertion loss and has a linear relationship to flow rate. In cases where the weight is not visible, i.e., an opaque tube used to reduce corrosion and the like, it can be made of a magnetic material, and tracked by a magnetic sensor on the outside of the tube. The rotameter can be used to measure differential pressures across a constriction or flow in both liquids and gases.

An example of a rotating flow rate device is the turbine flow meter, which is shown in Fig. 4.7a. The turbine rotor is mounted in the center of the pipe and rotates at a speed proportional to the rate of flow of the fluid or gas passing over the blades. The turbine blades are normally made of a magnetic material or ferrite particles in plastic so that they are unaffected by corrosive liquids. As the blades rotate they can be sensed by a Hall device or magneto resistive element (MRE) sensor attached to the pipe. The turbine should be only used with clean fluids such as gasoline. The rotating flow devices are accurate with good flow operating and temperature ranges, but are more expensive than most of the other devices.

The moving vane is shown in Fig. 4.7b. This device can be used in a pipe configuration as shown or used to measure open channel flow. The vane can be spring loaded and able to pivot; by measuring the angle of tilt the flow rate can be determined.

Electromagnetic flow meters can only be used in conductive liquids. The device consists of two electrodes mounted in the liquid on opposite sides of the pipe. A magnetic field is generated across the pipe perpendicular to the electrodes as shown in Fig. 4.8a. The conducting fluid flowing through the magnetic field generates a voltage between the electrodes, which can be measured to give the rate of flow. The meter gives an accurate

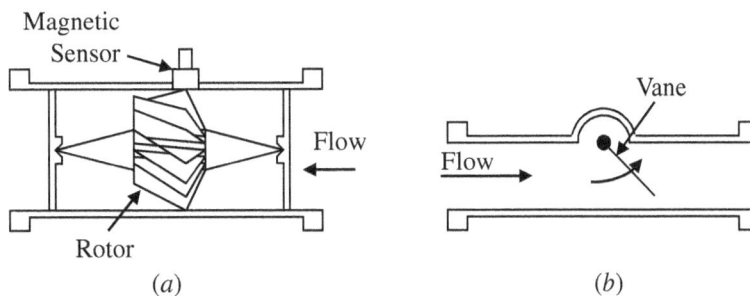

FIGURE 4.7 Flow rate measuring devices: (a) turbine and (b) moving vane.

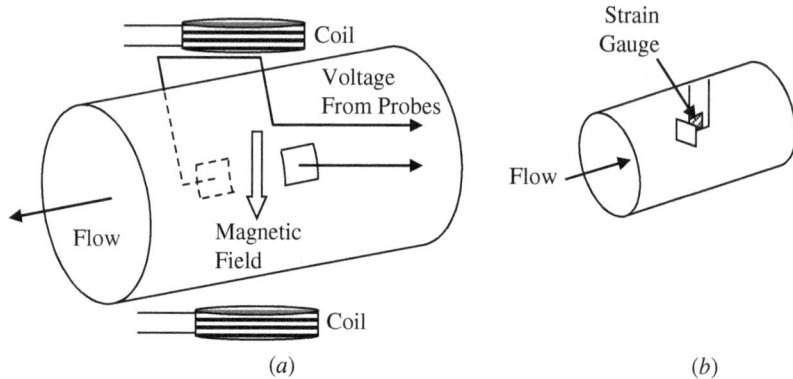

Figure 4.8   Flow measuring devices shown are (a) magnetic flow meter and (b) strain gauge flow meter.

linear output voltage with flow rate. There is no insertion loss and the readings are independent of the fluid characteristics, but it is a relatively expensive instrument.

Pressure flow meters use a strain gauge to measure the force on an object placed in a fluid or gas flow. The meter is shown in Fig. 4.8b. The force on the object is proportional to the rate of flow. The meter is low cost with medium accuracy.

Vortex flow meters are based on the fact that an obstruction in a fluid or gas flow will cause turbulence or vortices, or in the case of the vortex precession meter (for gases and steam flow measurement), the obstruction is shaped to give a rotating or swirling motion forming vortices that can be measured ultrasonically. The frequency of the vortex formation is proportional to the rate of flow and this method is good for high flow rates. At low flow rates the vortex frequency tends to be unstable.

4.4.2   Total Flow

In some applications a meter is required to measure total flow, i.e., to measure the total quantity of fluid flowing or the volume of liquid in a flow.

Positive displacement meters use containers of known size, which are filled and emptied for a known number of times in a given time period to give the total flow volume. Two of the more common instruments for measuring total flow are the piston flow meter and the nutating disc flow meter.

Piston meters consist of a piston in a cylinder. Initially the fluid enters on one side of the piston filling the cylinder, at which point the fluid is diverted to the other side of the piston via valves and the outlet port of the full cylinder is opened. The redirection of fluid reverses the direction of the piston and fills the cylinder on the other side of the piston. The number of times the piston traverses the cylinder in a given time frame gives the total flow. The meter has high accuracy but is expensive.

Nutating disc meters are in the form of a disc that oscillates, allowing a known volume of fluid to pass with each oscillation. The meter is illustrated in Fig. 4.9a. The oscillations can be counted to determine the total volume. This meter can be used to measure slurries but is expensive.

Velocity meters, normally used to measure flow rate, can also be set up to measure total flow by tracking the velocity and knowing the cross-sectional area of the meter to totalize the flow.

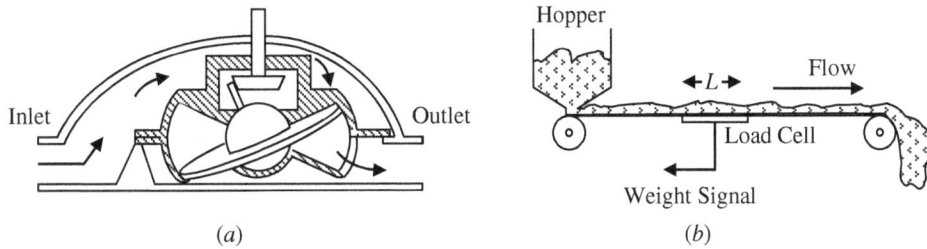

Figure 4.9   Illustrations show (a) the cross section of a nutating disc for the measurement of total flow and (b) conveyer belt system for the measurement of dry particulate flow rate.

4.4.3   Mass Flow

By measuring the flow and knowing the density of a fluid, the mass of the flow can be measured. Mass flow instruments include constant speed impeller turbine wheel-spring combinations that relate the spring force to mass flow and devices that relate heat transfer to mass flow.

Anemometer is an instrument that can be used to measure gas flow rates. One method is to keep the temperature of a heating element in a gas flow constant and measure the power required. The higher the flow rate, the higher the amount of heat required. The alternative method (hot-wire anemometer) is to measure the incident gas temperature, and the temperature of the gas downstream from a heating element; the difference in the two temperatures can be related to the flow rate. Micromachined anemometers are now widely used in automobiles for the measurement of air intake mass. The advantages of this type of sensor is that they are very small, have no moving parts, little obstruction to flow, have a low thermal time constant, and are very cost effective with good longevity.

4.4.4   Dry Particulate Flow Rate

Dry particulate flow rate can be measured as they are being carried on a conveyer belt with the use of a load cell. This method is illustrated in Fig. 4.9b. To measure flow rate it is only necessary to measure the weight of material on a fixed length of the conveyer belt.

The flow rate Q is given by

$$Q = \frac{WR}{L} \tag{4.14}$$

where W = weight of material on length of the weighing platform
  L = length of the weighing platform
  R = speed of the conveyer belt

Example 4.7   A conveyer belt is traveling at 19 cm/s. A load cell with a length of 1.1 m is reading 3.7 kg. What is the flow rate of the material on the belt?

Solution

$$Q = \frac{3.7 \times 19}{100 \times 1.1} \text{ kg/s} = 0.64 \text{ kg/s}$$

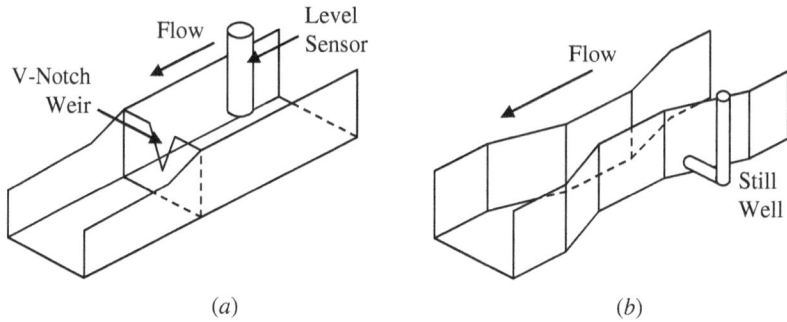

FIGURE **4.10** Open channel flow sensors: (a) weir and (b) Parshall flume.

4.4.5 Open Channel Flow

Open channel flow occurs when the fluid flowing is not contained as in a pipe but is in an open channel. Flow rates can be measured using constrictions as in contained flows. A weir used for open channel flow is shown in Fig. 4.10a. This device is similar in operation to an orifice plate. The flow rate is determined by measuring the differential pressures or liquid levels either side of the constriction.

A Parshall flume is shown in Fig. 4.10b, which is similar in shape to a Venturi tube. A paddle wheel or open flow nozzle are alternative methods of measuring open channel flow rates.

4.5 Application Considerations

Many different types of sensors can be used for flow measurements. The choice of any particular device for a specific application depends on a number of factors, such as reliability, cost, accuracy, pressure range, temperature, wear and erosion, energy loss, ease of replacement, particulates, viscosity, and so forth.

4.5.1 Selection

The selection of a flow meter for a specific application to a large extent will depend on the required accuracy and the presence of particulates, although the required accuracy is sometimes downgraded because of cost. One of the most accurate meters is the magnetic flow meter which can be accurate to 1 percent of full-scale reading or deflection FSD. The meter is good for low flow rates, with high viscosities, and has low energy loss, but is expensive and requires a conductive fluid.

The turbine gives high accuracies and can be used when there is vapor present, but the turbine is better with clean low viscosity fluids. Table 4.5 gives a comparison of flow meter characteristics.

The general purpose and most commonly used devices are the pressure differential sensors used with pipe constrictions. These devices will give accuracies in the 3 percent range when used with solid-state pressure sensors that convert the readings directly into electrical units or the rotameter for direct visual reading. The Venturi tube has the highest accuracy and least energy loss followed by the flow nozzle and the orifice plate. For cost effectiveness the devices are in the reverse order. If large amounts

Meter Type	Range	Accuracy Percent	Comments
Orifice plate	3 to 1	±3 FSD	Low cost, accuracy
Venturi tube	3 to 1	±1 FSD	High cost, good accuracy, low losses
Flow nozzle	3 to 1	±2 FSD	Medium cost, accuracy
Dall tube	3 to 1	±2 FSD	Medium cost, accuracy, low losses
Elbow	3 to 1	±6 – 9 FSD	Low cost, losses, sensitivity
Pilot static tube	3 to 1	±4 FSD	Low sensitivity
Rotameter	10 to 1	±2 of rate	Low losses, line of sight
Turbine meter	10 to 1	±2 FSD	High accuracy, low losses
Moving vane	5 to 1	±10 FSD	Low cost, low accuracy
Electromagnetic	30 to 1	±0.5 of rate	Conductive fluid, low losses, high cost
Vortex meter	20 to 1	±0.5 of rate	Poor at low flow rates
Strain gauge	3 to 1	±2 FSD	Low cost, accuracy
Nutating disc	5 to 1	±3 FSD	High accuracy, cost
Anemometer	100 to 1	±2 of rate	Low losses, fast response

TABLE 4.5 Summary of Flow Meter Characteristics

of particulates are present, the Venturi tube is preferred. The differential pressure devices operate best between 30 and 100 percent of the flow range. The elbow should also be considered in these applications.

Gas flow can best be measured with an anemometer. Solid-state anemometers are now available with good accuracy, are very small in size, and are cost effective.

For open channel applications the flume is the most accurate and best if particulates are present, but is the most expensive.

Particular attention should also be given to manufacturer's specifications and application notes.

4.5.2   Installation

Because of the turbulence generated by any type of obstruction in an otherwise smooth pipe, attention has to be given to the placement of flow sensors. The position of the pressure taps can be critical for accurate measurements. The manufacturer's recommendations should be followed during installation. In differential pressure sensing devices, the upstream tap should be one to three pipe diameters from the plate or constriction and the downstream tap up to eight pipe diameters from the constriction.

To minimize pressure fluctuations at the sensor, it is desirable to have a straight run of 10 to 15 pipe diameters either side of the sensing device. It may also be necessary to incorporate laminar flow planes into the pipe to minimize flow disturbances and dampening devices to reduce flow fluctuations to an absolute minimum.

Flow nozzles may require vertical installation if gases or particulates are present. To allow gases to pass through the nozzle, it should be facing upward and for particulates downward.

4.5.3 Calibration

Flow meters need periodic calibration. This can be done by using another calibrated meter as a reference or by using a known flow rate. Accuracy can vary over the range of the instrument and with temperature and specific weight changes in the fluid, which may all have to be taken into account. Thus, the meter should be calibrated over temperature as well as range, so that the appropriate corrections can be made to the readings. A spot check of the readings should be made periodically to check for instrument drift that may be caused by the instrument going out of calibration, particulate build up, or erosion.

Summary

This chapter discussed the measurement of the flow of fluids in closed and open channels and gases in closed channels. The basic terms, standards, formulas, and laws associated with flow rates are given. Instruments used in the measurement of flow rates are described, as well as considerations in instrument selection for flow measurement are discussed.

The main points described in this chapter were as follows:

1. The relation of the Reynolds number to physical parameters and its use for determining laminar or turbulent flow in fluids.

2. The development of the Bernoulli equation from the concept of the conservation of energy, and modification of the equation to allow for losses in liquid flow.

3. Definitions of the terms and standards used in the measurement of the flow of liquids and slurries.

4. Difference between flow rates, total flow, and mass flow and the instruments used to measure total flow and mass flow in liquids and gases.

5. Various types of flow measuring instruments including the use of restrictions and flow meters for direct and indirect flow measurements.

6. Open channel flow and devices used to measure open channel flow rates.

7. Comparison of sensor characteristics and considerations in the selection of flow instruments for liquids and slurries, and installation precautions.

Problems

4.1 The flow rate in a 7-in diameter pipe is 3.2 ft³/s. What is the average velocity in the pipe?

4.2 A 305 L/min of water flows through a pipe. What is the diameter of the pipe if the velocity of the water in the pipe is 7.3 m/s?

4.3 A pipe delivers 239 gal of water a minute. If the velocity of the water is 27 ft/s, what is the diameter of the pipe?

4.4 What is the average velocity in a pipe if the diameter of the pipe is 0.82 cm and the flow rate is 90 cm³/s?

4.5 Water flows in a pipe of 23-cm diameter with an average velocity of 0.73 m/s. The diameter of the pipe is reduced and the average velocity of the water increases to 1.66 m/s. What is the diameter of the smaller pipe? What is the flow rate?

4.6   The velocity of oil in a supply line changes from 5.1 ft³/s to 6.3 ft³/s when going from a large bore to a smaller bore pipe. If the bore of the smaller pipe has 8.1-in diameter, what is the bore of the larger pipe?

4.7   Water in a 5.5-in diameter pipe has a velocity of 97 gal/s; the pipe splits in two to feed two systems. If after splitting one pipe has 3.2-in diameter and the other 1.8-in diameter, what is the flow rate from each pipe?

4.8   What is the maximum allowable velocity of a liquid in a 3.2-in diameter pipe to ensure laminar flow? Assume the kinematic viscosity of the liquid is 1.7×10^{-5} ft²/s.

4.9   A copper sphere is dropped from a building 273 ft tall. What will be its velocity on impact with the ground? Ignore air resistance.

4.10   Three hundred and fifty gallons of water per min is flowing through a 4.3-in radius horizontal pipe. If the bore of the pipe is reduced to 2.7-in radius and the pressure in the smaller pipe is 93 psig, what is the pressure in the larger section of the pipe?

4.11   Oil with a specific weight of 53 lb/ft³ is exiting from a pipe whose center line is 17 ft below the surface of the oil. What is the velocity of the oil from the pipe if there is 1.5 ft head loss in the exit pipe?

4.12   A pump in a fountain pumps 109 gallons of water a second through a 6.23-in diameter vertical pipe. How high will the water in the fountain go?

4.13   What is the head loss in a 7-in diameter pipe 118-ft long that has a friction factor of 0.027 if the average velocity of the liquid flowing in the pipe is 17 ft/s?

4.14   What is the radius of a pipe, if the head loss is 1.6 ft when a liquid with a friction factor of 0.033 is flowing with an average velocity of 4.3 ft/s through 73 ft of pipe?

4.15   What is the pressure in a 9.7-in bore horizontal pipe, if the bore of the pipe narrows to 4.1 in downstream where the pressure is 65 psig and 28,200 gallons of fluid per hour with a specific gravity (SG) of 0.87 are flowing? Neglect losses.

4.16   Fluid is flowing through the following 1-in fittings; 3 threaded ells, 6 tees, 7 globe valves, and 9 unions. If the head loss is 7.2 ft, what is the velocity of the liquid?

4.17   The drag coefficient on a 6.3-in diameter sphere is 0.35. What is its velocity through a liquid with a SG = 0.79 if the drag force is 4.8 lb?

4.18   A square disc is placed in a moving liquid the drag force on the disc is 6.3 lb when the liquid has a velocity of 3.4 ft/s. If the liquid has a density of 78.3 lb/ft³ and the drag coefficient is 0.41, what is the size of the square?

4.19   The 8×32 in³ cylinders in a positive displacement meter assembly are rotating at a rate of 570 revolutions an hour. What is the average flow rate per min?

4.20   Alcohol flows in a horizontal pipe of 3.2-in diameter; the diameter of the pipe is reduced to 1.8 in. If the differential pressure between the two sections is 1.28 psi, what is the flow rate through the pipe? Neglect losses.

CHAPTER 5

Temperature and Heat

Chapter Objectives

This chapter will help you understand the difference between temperature and heat, the units used for their measurement, thermal time constants, and the most common methods used to measure temperature and heat and their standards.

Topics discussed in this chapter are as follows:

- The difference between temperature and heat
- The various temperature scales
- Temperature and heat formulas
- The various mechanisms of heat transfer
- Specific heat and heat energy
- Coefficients of linear and volumetric expansion
- The wide variety of temperature measuring devices
- Introduction to thermal time constants

5.1   Introduction

Humans make extensive use of temperature measuring devices for maintaining and controlling temperature for comfort, almost all industrial processes require accurately controlled temperatures. Physical parameters and chemical reactions are temperature dependent, and hence temperature control is of major importance. Temperature is without doubt the most measured variable, and for accurate temperature control its precise measurement is required which is not always possible due to the thermal time constant of the measuring device. This chapter discusses the various temperature scales used, their relation to each other, methods of measuring temperature, and the relationship between temperature and heat.

5.2   Basic Terms

5.2.1   Temperature Definitions

Temperature is a measure of the thermal energy in a body, which is the relative hotness or coldness of a medium and is normally measured in degrees using one of the following scales: Fahrenheit (F), Celsius or Centigrade (C), Rankine (R), or Kelvin (K).

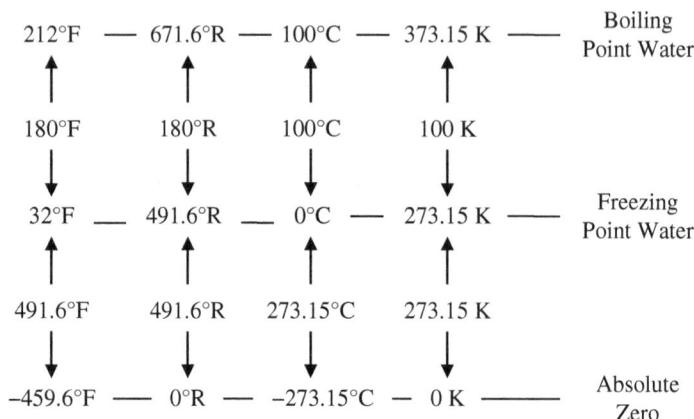

FIGURE **5.1**   Comparison of temperature scales.

Absolute zero is the temperature at which all molecular motion ceases, or the energy of the molecule is zero.

Fahrenheit scale was the first temperature scale to gain acceptance. It was proposed in the early 1700s by Fahrenheit (Dutch). The two points of reference chosen for 0 and 100° were the freezing point of a concentrated salt solution (at sea level) and the internal temperature of oxen (which was found to be very consistent between animals). This eventually led to the acceptance of 32° and 212° (180° range) as the freezing and boiling point, respectively, of pure water at 1 atm (14.7 psi or 101.36 kPa) for the Fahrenheit scale. The temperature of the freezing point and boiling point of water changes with pressure.

Celsius or centigrade scale (C) was proposed in mid 1700s by Celsius (Sweden), who proposed the temperature readings of 0° and 100° (giving a 100° scale) for the freezing and boiling points of pure water at 1 atm.

Rankine scale (R) was proposed in the mid 1800s by Rankine. It is a temperature scale referenced to absolute zero that was based on the Fahrenheit scale, i.e., a change of 1°F = a change of 1°R. The freezing and boiling point of pure water are 491.6°R and 671.6°R, respectively, at 1 atm, see Fig. 5.1.

Kelvin scale (K) named after Lord Kelvin was proposed in the late 1800s. It is referenced to absolute zero but based on the Celsius scale, i.e., a change of 1°C = a change of 1 K. The freezing and boiling point of pure water are 273.15 and 373.15 K, respectively, at 1 atm, see Fig. 5.1. The degree symbol can be dropped when using the Kelvin scale.

5.2.2   Heat Definitions

Heat is a form of energy; as energy is supplied to a system the vibration amplitude of its molecules increases and its temperature increases. The temperature increase is directly proportional to the heat energy in the system.

A *British Thermal Unit* (BTU or Btu) is defined as the amount of energy required to raise the temperature of 1 lb of pure water by 1°F at 68°F and at atmospheric pressure. It is the most widely used unit for the measurement of heat energy.

1 BTU = 252 cal	1 cal = 0.0039 BTU
1 BTU = 1055 J	1 J = 0.000948 BTU
1 BTU = 778 ft-lb	1 ft-lb = 0.001285 BTU
1 cal = 4.19 J	1 J = 0.239 cal
1 ft-lb = 0.324 cal	1 J = 0.738 ft-lb
1 ft-lb = 1.355 J	1 W = 1 J/s

TABLE 5.1 Conversion Related to Heat Energy

A *Calorie unit* (SI) is defined as the amount of energy required to raise the temperature of 1 g of pure water by 1°C at 4°C and at atmospheric pressure. It is also a widely used unit for the measurement of heat energy.

Joules (SI) are also used to define heat energy and is often used in preference to the calorie, where 1 J (Joule) = 1 W (Watt) × s. This is given in Table 5.1, which gives a list of energy equivalents.

Phase change is the transition of matter from the solid to the liquid or the liquid to the gaseous states; matter can exist in any of these three states. However, for matter to make the transition from one state up to the next, i.e., solid to liquid or liquid to gas, it has to be supplied with energy, or energy is removed if the matter is going from gas to liquid to solid.

For example, if heat is supplied at a constant rate to ice at 32°F, the ice will start to melt or turn to liquid, but the temperature of the ice–liquid mixture will not change until all the ice has melted. As more heat is supplied, the temperature will start to rise until the boiling point of the water is reached. The water will turn to steam as more heat is applied but the temperature of the water and steam will remain at the boiling point until all the water has turned to steam, then the temperature of the steam will start to rise above the boiling point. This is illustrated in Fig. 5.2, where the temperature of a substance is plotted against heat input. Material can also change its volume during the

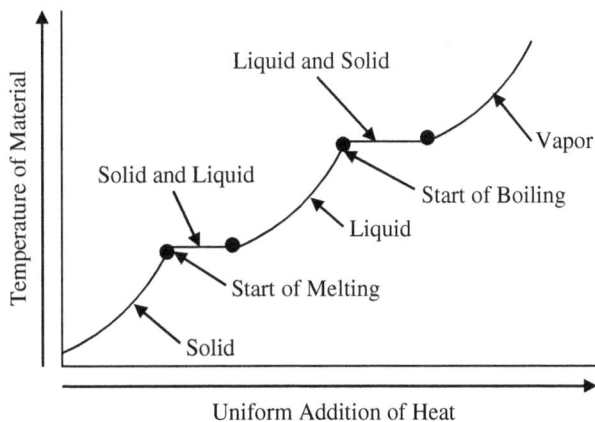

FIGURE 5.2 Showing the relation between temperature and heat energy.

change of phase. Some materials bypass the liquid stage and transform directly from solid to gas or gas to solid; this transition is called *sublimation*.

In a solid, the atoms can vibrate but are strongly bonded to each other so that the atoms or molecules are unable to move from their relative positions. As the temperature is increased, more energy is given to the molecules and their vibration amplitude increases to a point where it can overcome the bonds between the molecules and they can move relative to each other. When this point is reached the material becomes a liquid. The speed at which the molecules move about in the liquid is a measure of their thermal energy. As more energy is imparted to the molecules their velocity in the liquid increases to a point where they can escape the bonding or attraction forces of other molecules in the material and the gaseous state or boiling point is reached.

Specific heat is the quantity of heat energy required to raise the temperature of a given weight of a material by 1°. The most common units are BTUs in the English system, i.e., 1 BTU is the heat required to raise 1 lb (mass) of material by 1°F, and in the SI system, the calorie is the heat required to raise 1 g of material by 1°C. Thus, if a material has a specific heat of 0.7 cal/g °C, it would require 0.7 cal to raise the temperature of a gram of the material by 1°C or 2.93 J to raise the temperature of the material by 1 K. Table 5.2 gives the specific heat of some common materials, the units are the same in either system.

Thermal conductivity is the flow or transfer of heat from a high temperature region to a lower temperature region. There are three basic methods of heat transfer: conduction, convection, and radiation. Although these modes of transfer can be considered separately, in practice two or more of them can be present simultaneously.

Conduction is the flow of heat through a material. The molecular vibration amplitude or energy is transferred from one molecule in a material to the next. Hence, if one end of a material is at an elevated temperature, heat is conducted to the cooler end. The thermal conductivity of a material k is a measure of its efficiency in transferring heat. The units can be in BTUs per hour per ft per °F or watts per meter-Kelvin (W/mK) (1 BTU/ft h °F = 1.73 W/mK). Table 5.3 gives typical thermal conductivities for some common materials.

Convection is the transfer of heat due to motion of elevated temperature particles in a material (liquid and gases). Typical examples are air-conditioning systems, hot water heating systems, and so forth. If the motion is due solely to the lower density of the elevated temperature material, the transfer is called free or natural convection. If the material is moved by blowers or pumps the transfer is called forced convection.

Material	Specific Heat	Material	Specific Heat	Material	Specific Heat
Alcohol	0.58–0.6	Aluminum	0.214	Brass	0.089
Glass	0.12–0.16	Cast iron	0.119	Copper	0.092
Gold	0.0316	Lead	0.031	Mercury	0.033
Platinum	0.032	Quartz	0.188	Silver	0.056
Steel	0.107	Tin	0.054	Water	1.0

*The units are BTU/lb °F or Calories/g °C.

TABLE 5.2 Specific Heats of Some Common Materials*

Material	Conductivity	Material	Conductivity
Air	0.016 (room temp.) (0.028)	Aluminum	119 (206)
Concrete	0.8 (1.4)	Copper	220 (381)
Water	0.36 (room temp.) (0.62)	Mercury	4.8 (8.3)
Brick	0.4 (0.7)	Steel	26 (45)
Brass	52 (90)	Silver	242 (419)

TABLE 5.3   Thermal Conductivity BTU/h ft °F (W/mK)

Material	Linear ($\times 10^{-6}$)	Volume ($\times 10^{-6}$)	Material	Linear ($\times 10^{-6}$)	Volume ($\times 10^{-6}$)
Alcohol	—	61–68	Aluminum	12.8	—
Brass	10	—	Cast iron	5.6	20
Copper	9.4	29	Glass	5	14
Gold	7.8	—	Lead	16	—
Mercury	—	100	Platinum	5	15
Quartz	0.22	—	Silver	11	32
Steel	6.1	—	Tin	15	38

TABLE 5.4   Thermal Coefficients of Expansion per Degree Fahrenheit

Radiation is the emission of energy by electromagnetic waves that travel at the speed of light through most materials that do not conduct electricity. For instance, radiant heat can be felt some distance from a furnace where there is no conduction or convection.

5.2.3   Thermal Expansion Definitions

Linear thermal expansion is the change in dimensions of a material due to temperature changes. The change in dimensions of a material is due to its coefficient of thermal expansion that is expressed as the change in linear dimension (α) per degree temperature change.

Volume thermal expansion is the change in the volume (β) per degree temperature change due to the linear coefficient of expansion. The thermal expansion coefficients for some common materials per degree Fahrenheit are given in Table 5.4. The coefficients can also be expressed as per degree Celsius.

5.3   Temperature and Heat Formulas

5.3.1   Temperature

The need to convert from one temperature scale to another is a common everyday occurrence. The conversion factors are as follows:

To convert °F to °C

$$°C = (°F - 32)5/9 \qquad (5.1)$$

To convert °C to °F

$$°F = (°C \times 9/5) + 32 \tag{5.2}$$

To convert °F to °R

$$°R = °F + 459.6 \tag{5.3}$$

To convert °C to K

$$K = °C + 273.15 \tag{5.4}$$

To convert K to °R

$$°R = 1.8 \times K \tag{5.5}$$

To convert °R to K

$$K = 0.555 \times °R \tag{5.6}$$

Example 5.1   What temperature in K corresponds to 115°F?
From Eq. (5.1)

Solution

$$°C = (115 - 32)5/9 = 46.1°C$$

From Eq. (5.4)

$$K = 46.1 + 273.15 = 319.25 \text{ K}$$

5.3.2   Heat Transfer

The amount of heat needed to raise or lower the temperature of a given weight of a body can be calculated from

$$Q = WC(T_2 - T_1) \tag{5.7}$$

where W = weight of the material
C = specific heat of the material
T_2 = final temperature of the material
T_1 = initial temperature of the material

Example 5.2   What is the heat required to raise the temperature of a 1.5 kg mass by 120°C if the specific heat of the mass is 0.37 cal/g°C?

Solution

$$Q = 1.5 \times 1000 \text{ g} \times 0.37 \text{ cal/g°C} \times 120°C = 66600 \text{ cal}$$

As always, care must be taken in selecting the correct units. Negative answers indicate extraction of heat or heat loss.

Heat conduction through a material is derived from the following relationship:

$$Q = \frac{-kA(T_2 - T_1)}{L} \tag{5.8}$$

where Q = rate of heat transfer
$\quad k$ = thermal conductivity of the material
$\quad A$ = cross-sectional area of the heat flow
$\quad T_2$ = temperature of the material distant from the heat source
$\quad T_1$ = temperature of the material adjacent to heat source
$\quad L$ = length of the path through the material

Note: The negative sign in Eq. (5.8) indicates a positive heat flow.

Example 5.3 A furnace wall 12 ft² in area and 6-in thick has a thermal conductivity of 0.14 BTU/h ft°F. What is the heat loss if the furnace temperature is 1100°F and the outside of the wall is 102°F?

Solution

$$Q = \frac{-kA(T_2 - T_1)}{L}$$

$$Q = \frac{-0.14 \times 12(102 - 1100)}{0.5} = 3353.3 \text{ BTU/h}$$

Example 5.4 The outside wall of a room is 4×3 m and 0.35-m thick. What is the energy loss per hour if the inside and outside temperatures are 35°C and –40°C, respectively? Assume the conductivity of the wall is 0.13 W/mK.

Solution

$$Q = \frac{-kA(T_2 - T_1)}{L}$$

$$Q = \frac{-0.13 \text{ W/mK} \times 4 \text{ m} \times 3 \text{ m} \times (-40 - 35)\text{K}}{0.35 \text{ m}} \times \frac{60 \times 60 \text{ J/s}}{\text{W} \times \text{h}}$$

$$Q = 1203 \text{ kJ/h}$$

Heat convection calculations in practice are not as straight forward as conduction. However, heat convection is given by

$$Q = hA(T_2 - T_1) \qquad (5.9)$$

where Q = convection heat transfer rate
$\quad h$ = coefficient of heat transfer
$\quad A$ = heat transfer area
$\quad T_2 - T_1$ = temperature difference between the source and final temperature of the flowing medium

It should be noted that in practice the proper choice for h is difficult because of its dependence on a large number of variables (such as density, viscosity, and specific heat). Charts are available for h. However, experience is needed in their application.

Example 5.5 How much heat is transferred from a 25×24-ft surface by convection if the temperature difference between the front and back surfaces is 40°F and the surface has a heat transfer rate of 0.22 BTU/h ft²°F?

Solution

$$Q = 0.22 \times 25 \times 24 \times 40 = 39,600 \text{ BTU/h}$$

Heat radiation depends on surface color, texture, shapes involved, and the like. Hence, more information than the basic relationship for the transfer of radiant heat energy given below should be factored in. The radiant heat transfer is given by

$$Q = CA\left(T_2^4 - T_1^4\right) \qquad (5.10)$$

where Q = heat transferred

    C = radiation constant (depends on surface color, texture, and units used, and the like)
    A = area of the radiating surface
    T_2 = absolute temperature of the radiating surface
    T_1 = absolute temperature of the receiving surface

Example 5.6   The radiation constant for a furnace is 0.23×10^{-8} BTU/h ft²°F⁴, the radiating surface area is 25 ft². If the radiating surface temperature is 750°F and the room temperature is 75°F, how much heat is radiated?

Solution

$$Q = 0.23 \times 10^{-8} \times 25 \times [\{750 + 460\}^4 - \{75 + 460\}^4]$$

$$Q = 5.75 \times 10^{-8}[222 \times 10^{10} - 8.4 \times 10^{10}] = 1.2 \times 10^5 \text{ BTU/h}$$

Example 5.7   What is the radiation constant for a wall 5×4 m, if the radiated heat loss is 62.3 MJ/h when the wall and ambient temperatures are 72°C and 5°C?

Solution

$$62.3 \text{ MJ/h} = 17.3 \text{ kW} = C \times 20[\{72 + 273.15\}^4 - \{5 + 273.15\}^4]$$

$$C = 17.3 \times 10^3 / 20(1.419 \times 10^{10} - 0.598 \times 10^{10})$$

$$C = 17.3/16.41 \times 10^7 = 1.05 \times 10^{-7} \text{ W/m}^2\text{K}^4$$

5.3.3   Thermal Expansion

Linear expansion of a material is the change in linear dimension due to temperature changes, and can be calculated from the following formula:

$$L_2 = L_1[1 + \alpha(T_2 - T_1)] \tag{5.11}$$

where L_2 = final length

    L_1 = initial length
    α = coefficient of linear thermal expansion
    T_2 = final temperature
    T_1 = initial temperature

Volume expansion in a material due to changes in temperature is given by

$$V_2 = V_1[1 + \beta(T_2 - T_1)] \tag{5.12}$$

where V_2 = final volume

    V_1 = initial volume
    β = coefficient of volumetric thermal expansion
    T_2 = final temperature
    T_1 = initial temperature

Example 5.8   Calculate the length and volume for a 200 cm on a side copper cube at 20°C, if the temperature is increased to 150°C.

Solution

$$\text{New length} = 200(1 + 9.4 \times 10^{-6} \times [150 - 20] \times 9/5)$$

$$= 200(1 + .0022) = 200.44 \text{ cm}$$

$$\text{New volume} = 200^3(1 + 29 \times 10^{-6} \times [150 - 20] \times 9/5)$$
$$= 200^3(1 + .0068) = 8054400 \text{ cm}^3$$

In a gas, the relation between the pressure, volume, and temperature of the gas is given by

$$\frac{P_1 V_1}{T_1} = \frac{P_2 V_2}{T_2} \tag{5.13}$$

where P_1 = initial pressure
V_1 = initial volume
T_1 = initial absolute temperature
P_2 = final pressure
V_2 = final volume
T_2 = final absolute temperature

5.4   Temperature Measuring Devices

There are several methods of measuring temperature, which can be categorized as follows:

1. Expansion of a material to give visual indication, pressure, or dimensional change

2. Electrical resistance change

3. Semiconductor characteristic change

4. Voltage generated by dissimilar metals

5. Radiated energy

Thermometer is often used as a general term given to devices for measuring temperature. Examples of temperature measuring devices are described in the following discussion.

5.4.1   Thermometers

Mercury in glass was by far the most common direct visual reading thermometer (if not the only one). The device consisted of a small bore graduated glass tube with a small bulb containing a reservoir of mercury. The coefficient of expansion of mercury is several times greater than the coefficient of expansion of glass, so that as the temperature increases the mercury rises up the tube giving a relatively low cost and accurate method of measuring temperature. Mercury also has the advantage of not wetting the glass, and hence, cleanly traverses the glass tube without breaking into globules or coating the tube. The operating range of the mercury thermometer is from −30 to 800°F (−35 to 450°C) (freezing point of mercury −38°F [−38°C]). The toxicity of mercury, ease of breakage, the introduction of cost effective, accurate, and easily read digital thermometers has brought about the demise of the mercury thermometer.

Liquids in glass devices operate on the same principle as the mercury thermometer. The liquids used have similar properties to mercury, i.e., high linear coefficient of expansion, clearly visible, nonwetting, but are nontoxic. The liquid in glass thermometers are used to replace the mercury thermometer and to extend its operating range. These thermometers are accurate and with different liquids (each type of liquid has a limited operating range) can have an operating range from −300 to 600°F (−170 to 330°C).

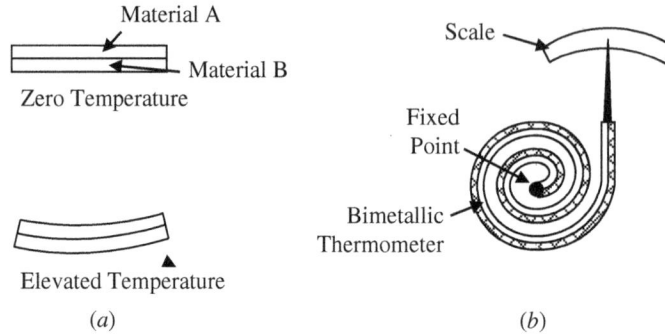

FIGURE 5.3 Illustration of (a) the effect of temperature change on a bimetallic strip and (b) bimetallic strip thermometer.

Bimetallic strip is a type of temperature measuring device that is relatively inaccurate, slow to respond, not normally used in analog applications to give remote indication, and has hysteresis. The bimetallic strip is extensively used in ON/OFF applications not requiring high accuracy, as it is rugged and cost effective. These devices operate on the principle that metals are pliable and different metals have different coefficients of expansion (see Table 5.4). If two strips of dissimilar metals, such as brass and invar (copper-nickel alloy), are joined together along their length they will flex to form an arc as the temperature changes, this is shown in Fig. 5.3a. Bimetallic strips are usually configured as a spiral or helix for compactness, and can then be used with a pointer to make a cheap compact rugged thermometer, as shown in Fig. 5.3b. Their operating range is from −180 to 430°C, and can be used in applications from oven thermometers to home and industrial control thermostats.

5.4.2 Pressure-Spring Thermometers

These thermometers are used where remote indication is required, as opposed to glass and bimetallic devices that give readings at the point of detection. The pressure-spring device has a metal bulb made with a low coefficient of expansion material with a long metal tube, both contain material with a high coefficient of expansion; the bulb is at the monitoring point. The metal tube is terminated with a spiral Bourdon tube pressure gauge (scale in degrees) as shown in Fig. 5.4a. The pressure system can be used to drive a chart recorder, actuator, or a potentiometer wiper to obtain an electrical signal. As the temperature in the bulb increases, the pressure in the system rises, the pressure rise being proportional to the temperature change. The change in pressure is sensed by the Bourdon tube and is converted to a temperature scale. These devices can be accurate to 0.5 percent and can be used for remote indication up to 100 m, but must be calibrated as the stem and Bourdon tube are temperature sensitive.

There are three types or classes of pressure-spring devices. They are as follows:

Class 1	Liquid filled
Class 2	Vapor pressure
Class 3	Gas filled

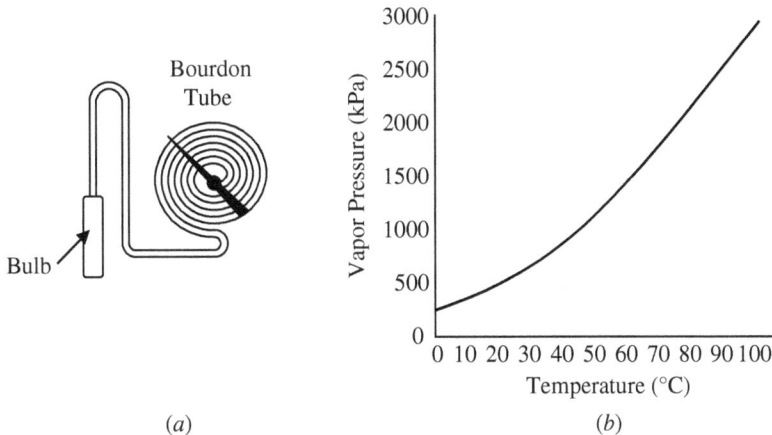

FIGURE 5.4 Illustration of (a) pressure-filled thermometer and (b) vapor pressure curve for methyl chloride.

Liquid-filled thermometer works on the same principle as the liquid in glass thermometer, but is used to drive a Bourdon tube. The device has good linearity and accuracy and can be used up to 550°C.

Vapor-pressure thermometer system is partially filled with liquid and vapor such as methyl chloride, ethyl alcohol, ether, toluene, and so on. In this system, the lowest operating temperature must be above the boiling point of the liquid and the maximum temperature is limited by the critical temperature of the liquid. The response time of the system is slow, being of the order of 20 s. The temperature pressure characteristic of the thermometer is nonlinear as shown in the vapor pressure curve for methyl chloride in Fig. 5.4b.

Gas thermometer is filled with a gas such as nitrogen at a pressure of between 1000 and 3350 kPa at room temperature. The device obeys the basic gas laws for a constant volume system [Eq. (5.13), $V_1 = V_2$] giving a linear relationship between absolute temperature and pressure.

5.4.3 Resistance Temperature Devices

Resistance temperature devices (RTDs) are either a metal film deposited on a former or are wire-wound resistors. The devices are then sealed in a glass-ceramic composite material. The electrical resistance of pure metals is positive, increasing linearly with temperature. Table 5.5 gives the temperature coefficient of resistance of some common metals used in resistance thermometers. These devices are accurate, and can be used to measure temperatures from −300 to 1400°F (−170 to 780°C).

Material	Coeff. Per Degree Celsius	Material	Coeff. Per Degree Celsius
Iron	0.006	Tungsten	0.0045
Nickel	0.005	Platinum	0.00385

TABLE 5.5 Temperature Coefficient of Resistance of Some Common Metals

In a resistance thermometer the variation of resistance with temperature is given by

$$R_{T2} = R_{T1}(1 + \text{Coeff.}\,[T_2 - T_1])\tag{5.14}$$

where R_{T2} is the resistance at temperature T_2 and R_{T1} is the resistance at temperature T_1.

Example 5.9 What is the resistance of a platinum resistor at 250°C, if its resistance at 20°C is 1050 Ω?

Solution

$$\text{Resistance at } 250°C = 1050(1 + 0.00385[250 - 20])$$
$$= 1050(1 + 0.8855)$$
$$= 1979.775\ \Omega$$

Resistance devices are normally measured using a Wheatstone bridge type of system, but are supplied from a constant current source. Care should also be taken to prevent electrical current from heating the device and causing erroneous readings. One method of overcoming this problem is to use a pulse technique. When using this method the current is turned ON for say 10 ms every 10 s, and the sensor resistance is measured during this 10 ms time period. This reduces the internal heating effects by 1000 to 1 or the internal heating error by this factor.

5.4.4 Thermistors

Thermistors are a class of metal oxide (semiconductor material) that typically have a high negative temperature coefficient of resistance, but can also be positive. Thermistors have high sensitivity which can be up to 10 percent change per degree Celsius, making them the most sensitive temperature element available, but with very nonlinear characteristics. The typical response times is 0.5 to 5 s with an operating range from −50 to typically 300°C. Devices are available with the temperature range extended to 500°C. Thermistors are low cost and manufactured in a wide range of shapes, sizes, and values. When in use care has to be taken to minimize the effects of internal heating. Thermistor materials have a temperature coefficient of resistance (α) given by

$$\alpha = \frac{\Delta R}{R_S}\left(\frac{1}{\Delta T}\right)\tag{5.15}$$

where ΔR is the change in resistance due to a temperature change ΔT and R_S the material resistance at the reference temperature.

The nonlinear characteristics are as shown in Fig. 5.5 and makes the device difficult to use as an accurate measuring device without compensation, but its sensitivity and low cost makes it useful in many applications. The device is normally used in a bridge circuit and padded with a resistor to reduce its nonlinearity.

5.4.5 Thermocouples

Thermocouples are formed when two dissimilar metals are joined together to form a junction. An electrical circuit is completed by joining the other ends of the dissimilar metals together to form a second junction. A current will flow in the circuit if the two junctions are at different temperatures as shown in Fig. 5.6a. The current flowing is the result of the difference in electromotive force developed at the two junctions due to their temperature difference. In practice, the voltage difference between the two junctions is measured; the difference in the voltage is proportional to the temperature difference

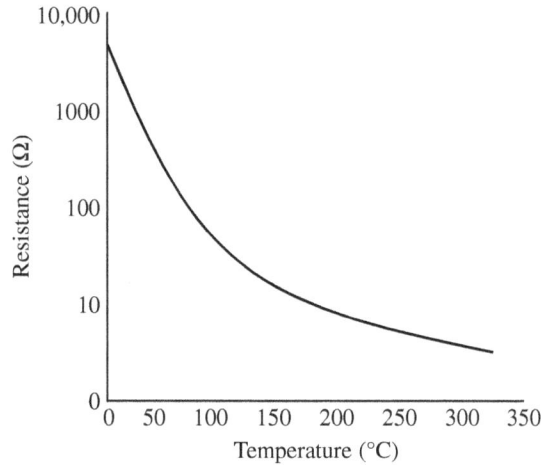

FIGURE 5.5   Thermistor resistance temperature curve.

between the two junctions. Note that the thermocouple can only be used to measure temperature differences. However, if one junction is held at a reference temperature, the voltage between the thermocouples gives a measurement of the temperature of the second junction.

Three effects are associated with thermocouples. They are as follows:

1. *Seebeck effect.* States that the voltage produced in a thermocouple is proportional to the temperature between the two junctions.

2. *Peltier effect.* States that if a current flows through a thermocouple one junction is heated (puts out energy), and the other junction is cooled (absorbs energy).

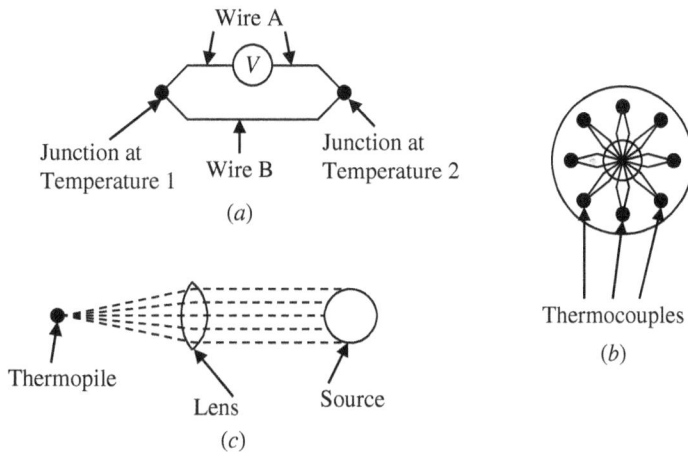

FIGURE 5.6   (a) A thermocouple, (b) a thermopile, and (c) focusing onto a thermopile.

3. *Thompson effect.* States that when a current flows in a conductor along which there is a temperature difference heat is produced or absorbed, depending on the direction of the current and the variation of temperature.

In practice, the Seebeck voltage is the sum of the electromotive forces generated by the Peltier and Thompson effects. There are a number of laws to be observed in thermocouple circuits. First, the law of intermediate temperatures states that the thermoelectric effect depends only on the temperatures of the junctions and is not affected by the temperatures along the leads. Secondly, the law of intermediate metals states that metals other than those making up the thermocouples can be used in the circuit as long as their junctions are at the same temperature; i.e., other types of metals can be used for interconnections and tag strips can be used without adversely affecting the output voltage from the thermocouple. The various types of thermocouples are designated by letters. Tables of the differential output voltages for different types of thermocouples are available from manufacturer's thermocouple data sheets.

Table 5.6 lists some thermocouple materials and their Seebeck coefficient. The operating range of the thermocouple is reduced to the figures in brackets if the given accuracy is required. For operation over the full temperature range, the accuracy would be reduced to about ±10 percent without linearization.

Thermopile is a number of thermocouples connected in series, to increase the sensitivity and accuracy by increasing the output voltage when measuring low temperature differences. Each of the reference junctions in the thermopile is returned to a common reference temperature as shown in Fig. 5.6b. Radiation can be used to sense temperature. The devices used are pyrometers using thermocouples, or color comparison devices.

Pyrometers are devices that measure temperature by sensing the heat radiated from a hot body through a fixed lens, which focuses the heat energy on to a thermopile; this is a noncontact device. Furnace temperatures, e.g., are normally measured through a small hole in the furnace wall. The distance from the source to the pyrometer can be fixed and the radiation should fill the field of view of the sensor. Figure 5.6c shows the focusing lens and thermocouple setup in a thermopile. Pyrometers can also use infrared imaging to measure temperatures.

Figure 5.7 shows plots of the EMF versus temperature of some of the types of thermocouples available.

Type	Approx. Range °C	Seebeck Coefficient (μV/°C)
Copper-Constantan (T)	−140 to 400	40 (−59 to 93) ±1°C
Chromel-Constantan (E)	−180 to 1000	62 (0 to 360) ±2°C
Iron-Constantan (J)	30 to 900	51 (0 to 277) ±2°C
Chromel-Alumel (K)	30 to 1400	40 (0 to 277) ±2°C
Nicrosil-Nisil (N)	30 to 1400	38 (0 to 277) ±2°C
Platinum (rhodium 10%)-Platinum (S)	30 to 1700	7 (0 to 538) ±3°C
Platinum (rhodium 13%)-Platinum (R)	30 to 1700	7 (0 to 538) ±3°C

TABLE 5.6   Operating Ranges for Thermocouples and Seebeck Coefficients

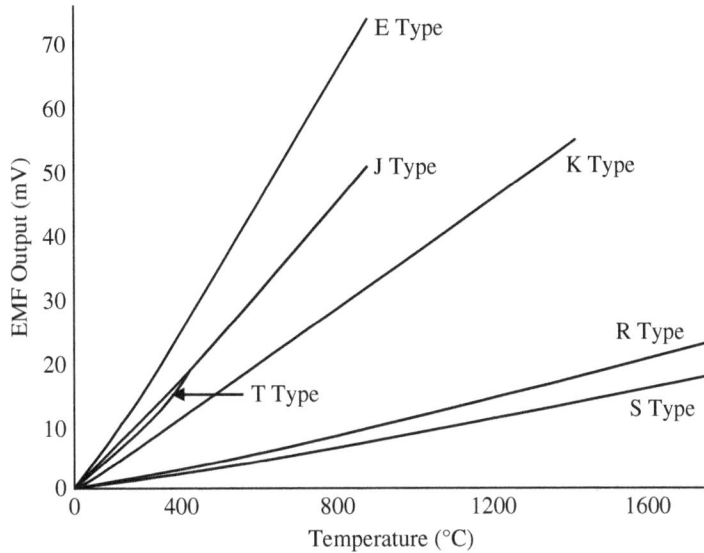

Figure 5.7 Thermocouple EMF versus temperature for various types.

5.4.6 Semiconductors

Semiconductors have a number of parameters that vary linearly with temperature. Normally, the reference voltage of a zener diode or the junction voltage variations are used for temperature sensing. Semiconductor temperature sensors have a limited operating range from −50 to 150°C but are very linear with accuracies of ±1°C or better. Other advantages are that electronics can be integrated onto the same die as the sensor giving high sensitivity, easy interfacing to control systems, and making different digital output configurations possible. The thermal time constant varies from 1 to 5 s, internal dissipation can also cause up to 0.5°C offset. Semiconductor devices are also rugged with good longevity and are inexpensive. For the above reasons, the semiconductor sensor is used extensively in many applications including the replacement of the mercury in glass thermometer.

5.5 Application Considerations

5.5.1 Selection

In process control, a wide selection of temperature sensors are available. However, the required range, linearity, response time, and accuracy can limit the selection. In the final selection of a sensor, other factors may have to be taken into consideration, such as remote indication, error correction, calibration, vibration sensitivity, size, response time, longevity, maintenance requirements, and cost. The choice of sensor devices in instrumentation should not be degraded from a cost stand point. Process control is only as good as the monitoring elements.

Sensor Type		Range (Degree Celsius)	Accuracy (FSD)
Expansion	Mercury in glass	−35 to 430	±1%
	Liquid in glass	−100 to 500	±1%
	Bimetallic	−100 to 500	±5%
Pressure-spring	Liquid filled	−180 to 550	±0.5%
	Vapor pressure	−180 to 550	±2.0%
	Gas filled	−180 to 550	±0.5%
Resistance	Metal resistors	−200 to 800	±5%
	Platinum	−180 to 650	±0.5%
	Nickel	−180 to 320	±1%
	Copper	−180 to 320	±0.2%
Thermistor		0 to 500	±2.5%
Thermocouple		−60 to 540	±0.5%
		−180 to 1300	±1%
Semiconductor IC		−40 to 150	±1%

TABLE 5.7 Temperature Range and Accuracy of Temperature Sensors

5.5.2 Range and Accuracy

Table 5.7 gives the temperature ranges and accuracies of temperature sensors. The accuracies shown are with minimal calibration or error correction. The ranges in some cases can be extended with the use of new materials. In industry, the operating range of the temperature sensing devices is often reduced and devices can be selected to give up to a 10 to 1 improvement in accuracy. Table 5.8 gives a summary of temperature sensor characteristics.

5.5.3 Thermal Time Constant

A temperature detector does not react immediately to a change in temperature. The reaction time of the sensor or thermal time constant is a measure of the time it takes for the sensor to stabilize internally to the external temperature change, and is determined by the thermal mass and thermal conduction resistance of the device. Thermometer bulb size, probe size, or protection well can affect the response time of the reading; i.e., a large bulb contains more liquid for better sensitivity, but this will also increase the time constant taking longer to fully respond to a temperature change.

The thermal time constant is related to the thermal parameters by the following equation:

$$t_c = \frac{mc}{kA} \qquad (5.16)$$

where t_c = thermal time constant
   m = mass
   c = specific heat
   k = heat transfer coefficient
   A = area of thermal contact

Type	Linearity	Advantages	Disadvantages
Bimetallic	Good	Low cost, rugged, and wide range	Local measurement or for ON/OFF switching only
Pressure	Medium		Needs temperature compensation and vapor is nonlinear
Resistance	Very good	Stable, wide range, and accurate	Slow response, low sensitivity, expensive, self-heating, and limited range
Thermistor	Poor	Low cost, small high sensitivity, and fast response	Nonlinear range and self-heating
Thermocouple	Good	Low cost, rugged, and very wide range	Low sensitivity and reference needed
Semiconductor	Excellent	Low cost, sensitive, and easy to interface	Self-heating, range, and power source
Radiation	Good	Noncontact, measure distant sources wide application, and fast	Accuracy

TABLE 5.8 *Summary of Sensor Characteristics*

When the temperature changes rapidly, the temperature output reading of a thermal sensor is given by

$$T - T_2 = (T_1 - T_2)e^{-t/t_c} \tag{5.17}$$

where T = temperature reading
   T_1 = initial temperature
   T_2 = true system temperature
   t = time from when the change occurred

The time constant of a system t_c is considered as the time it takes for the system to reach 63.2 percent of its final temperature value after a temperature change; i.e., a copper block is held in an ice–water bath until its temperature has stabilized at 0°C, it is then removed and placed in a 100°C steam bath; the temperature of the copper block will not immediately go to 100°C, but its temperature will rise on an exponential curve as it absorbs energy from the steam, until after some time period (its time constant) it will reach 63.2°C, aiming to eventually reach 100°C. This is shown in the graph (line A) in Fig. 5.8. During the second time constant, the copper will rise another 63.2 percent of the remaining temperature to get to equilibrium, i.e., (100 – 63.2) 63.2 percent = 23.3°C, or at the end of two time constant periods, the temperature of the copper will be 86.5°C. At the end of three periods the temperature will be 95°C and so on. Also shown in Fig. 5.8 is a second line B for the copper; the time constants are the same but the final aiming temperature is 50°C. The time to stabilize is the same in both cases. Where a fast response time is required, thermal time constants can be a serious problem as in some cases they can be several seconds duration, correction may have to be applied to the

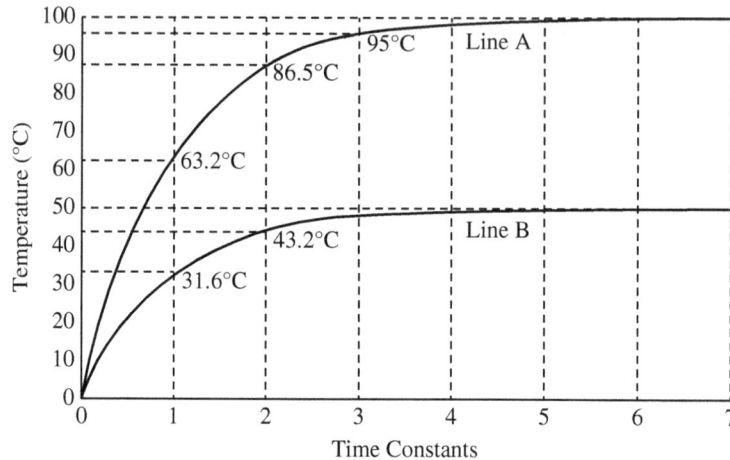

Figure 5.8   A graph showing the response time to changes in temperature.

output reading electronically to correct for the thermal time constant to obtain a faster response. This can be done by measuring the rate of rise of the temperature indicated by the sensor and extrapolating the actual aiming temperature.

The thermal time constant of a body is similar to an electrical time constant which governs the charging and discharging of capacitors and inductors.

5.5.4   Installation

Care must be taken in locating the sensing portion of the temperature sensor. It should be fully encompassed by the medium whose temperature is being measured, and not be in contact with the walls of the container. The sensor should be screened from reflected heat and radiant heat if necessary. The sensor should also be placed downstream from fluids being mixed, to ensure that the temperature has stabilized, but as close as possible to the point of mixing to give as fast as possible temperature measurement for good control. A low thermal time constant in the sensor is necessary for a quick response.

Compensation and calibration may be necessary when using pressure-spring devices with long tubes especially when accurate readings are required.

5.5.5   Calibration

Temperature calibration can be performed on most temperature sensing devices by immersing them in known temperature standards that are the equilibrium points of solid/liquid or liquid/gas mixtures, which is also known as the triple point. Some of these are given in Table 5.9. Most temperature sensing device are rugged and reliable, but can go out of calibration due to leakage during use or contamination during manufacture and should therefore be checked on a regular basis.

5.5.6   Protection

In some applications, temperature sensing devices are placed in wells or enclosures to prevent mechanical damage or for ease of replacement. This kind of protection can greatly increase the system response time, which in some circumstances may be unacceptable.

Calibration Material	Temperature			
	K	**°R**	**°F**	**°C**
Zero thermal energy	0	0	−459.6	−273.15
Oxygen: liquid-gas	90.18	162.3	−297.3	−182.97
Water: solid-liquid	273.15	491.6	32	0
Water: liquid-gas	373.15	671.6	212	100
Gold: solid-liquid	1336.15	2405	1945.5	1063

TABLE 5.9   Temperature Scale Calibration Points

Sensors may need also to be protected from over temperature, so that a second more rugged device may be needed to protect the main sensing device. Semiconductor devices may have built-in over temperature protection. A fail safe mechanism may also be incorporated for system shutdown, when processing volatile or corrosive materials.

Summary

This chapter introduced the concepts of heat and temperature and their relationship to each other. The various temperature scales in use are defined and the conversion equations between the scales are defined. The equations for heat transfer and heat storage are given. Temperature measuring instruments are described and their characteristics compared.

The main points described in this chapter were as follows:

1. Temperature scales and their relation to each other are defined with examples on how to convert from one scale to the other.

2. The transition of material between solid, liquid, and gaseous states or phase changes in materials when heat is supplied.

3. The mechanism and equations of heat energy transfer and the effects of heat on the physical properties of materials.

4. Definitions of the terms and standards used in temperature and heat measurements, covering both heat flow and capacity.

5. The various temperature measuring devices including thermometers, bimetallic elements, pressure-spring devices, RTDs, and thermocouples.

6. Considerations when selecting a temperature sensor for an application, thermal time constants, installation, and calibration.

Problems

5.1   Convert the following temperatures to Fahrenheit: 115°C, 456 K, 423°R.

5.2   Convert the following temperatures to Rankine: −13°C, 645 K, −123°F.

5.3   Convert the following temperatures to Centigrade: 115°F, 356 K, 533°R.

5.4   Convert the following temperatures to Kelvin: –215°C, –56°F, 436°R.

5.5   How many calories of energy are required to raise the temperature of 3 ft³ of water by 15°F?

5.6   A 15-lb block of brass with a specific heat of 0.089 is heated to 189°F and then immersed in 5 gal of water at 66°F. What is the final temperature of the brass and water? Assume there is no heat loss.

5.7   4.3-lb copper block is heated by passing a direct current through it. If the voltage across the copper is 50 V and the current is 13.5 A, what will be the increase in the temperature of the copper after 17 min? Assume there is no heat loss.

5.8   A 129 kg lead block is heated to 176°C from 19°C. How many calories are required?

5.9   One end of a 9-in long × 7-in diameter copper bar is heated to 59.4°F, the far end of the bar is held at 23°C. If the sides of the bar are covered with thermal insulation, what is the rate of heat transfer?

5.10   On a winter's day, the outside temperature of a 17-in thick concrete wall is –29°F; the wall is 15-ft long and 9-ft high. How many BTUs are required to keep the inside of the wall at 69°F? Assume the thermal conductivity of the wall is 0.8 BTU/h ft°F.

5.11   When the far end of the copper bar in Problem 8.9 has a 30-ft² cooling fin attached to the end of the bar, and is cooled by air convection, the temperature of the fin rises to t°F. If the temperature of the air is 23°F and the coefficient of heat transfer of the surface is 0.22 BTU/h, what is the value of t?

5.12   How much heat is lost due to convection in a 25-min period from a 52 × 14-ft wall, if the difference between the wall temperature and the air temperature is 54°F and the surface of the wall has a heat transfer ratio of 0.17 BTU/h ft²°F?

5.13   How much heat is radiated from a surface 1.5 × 1.9 ft, if the surface temperature is 125°F. The air temperature is 74°F, and the radiation constant for the surface is 0.19 × 10⁻⁸ BTU/h ft²°F⁴?

5.14   What is the change in length of a 5-m tin rod, if the temperature changes from 11 to 245°C?

5.15   The length of a 115-ft metal column changes its length to 115 ft 2.5 in when the temperature goes from –40 to 116°F. What is the coefficient of expansion of the metal?

5.16   A glass block measures 1.3 × 2.7 × 5.4 ft at 71°F. How much will the volume increase if the block is heated to 563°F?

5.17   What is the coefficient of resistance per degree Celsius of a material, if the resistance is 2246 Ω at 63°F and 3074 Ω at 405°F?

5.18   A tungsten filament has a resistance of 1998 Ω at 20°C. What will its resistance be at 263°C?

5.19   A chromel–alumel thermocouple is placed in a 1773°F furnace. Its reference is 67°F. What is the output voltage from the thermocouple?

5.20   A pressure-spring thermometer having a time constant of 1.7 s is placed in boiling water (212°F) after being at 69°F. What will be the thermometer reading after 3.4 s?

CHAPTER 6

Humidity, Density, Viscosity, and pH

Chapter Objectives

This chapter will introduce you to humidity, density, viscosity, and pH, and help you understand the units used in their measurement. This chapter will also familiarize you with standard definitions in use, and the instruments used for their measurement.

Topics discussed in this chapter are as follows:

- Humidity ratio, relative humidity, dew point, and its measurement
- Understanding of and use of a psychrometric chart
- Instruments for measuring humidity
- Understand the difference between density, specific weight, and specific gravity
- Instruments for measuring density and specific gravity
- Definition of viscosity and measuring instruments
- Defining and measuring pH values

6.1 Introduction

Many industrial processes such, as textiles, wood, chemical processing, and the like, are very sensitive to humidity; consequently, it is necessary to control the amount of water vapor present in these processes. This chapter discusses four physical parameters.

They are as follows:

1. Humidity
2. Density, specific weight, and specific gravity
3. Viscosity
4. pH values

6.2   Humidity

6.2.1   Humidity Definitions

Humidity is a measure of the relative amount of water vapor present in the air or a gas.

Relative humidity (Φ) is the percentage of water vapor by weight present in a given volume of air or gas compared to the weight of water vapor present in the same volume of air or gas saturated with water vapor, at the same temperature and pressure, i.e.,

$$\text{Relative humidity} = \frac{\begin{array}{c}\text{amount of water vapor present}\\\text{in a given volume of air or gas}\end{array}}{\begin{array}{c}\text{maximum amount of water vapor soluble in}\\\text{the same volume of air or gas } (P \text{ and } T \text{ constant})\end{array}} \times 100 \qquad (6.1)$$

An alternative definition using vapor pressures is as follows:

$$\text{Relative humidity} = \frac{\text{water vapor pressure in air or gas}}{\text{water vapor pressure in saturated air or gas } (T \text{ constant})} \times 100$$

$$(6.2)$$

The term "saturated" means the maximum amount of water vapor that can be dissolved or absorbed by a gas or air at a given pressure and temperature. If there is any reduction of the temperature in the saturated air or gas, water will condense out in the form of droplets, i.e., similar to mirrors steaming up when taking a shower.

Specific humidity, humidity ratio, or absolute humidity can be defined as the mass of water vapor in a mixture in grains (where 7000 grains = 1 lb) divided by the mass of dry air or gas in the mixture in pounds. The measurement units could also be in grams.

$$\text{Humidity ratio} = \frac{\text{mass of water vapor in a mixture}}{\text{mass of dry air or gas in the mixture}} \qquad (6.3)$$

$$= \frac{\text{mass (water vapor)}}{\text{mass (air or gas)}} = \frac{0.622\,P\,(\text{water vapor})}{P\,(\text{mixture}) - P\,(\text{water vapor})} \qquad (6.4)$$

$$= \frac{0.622\,P\,(\text{water vapor})}{P\,(\text{air or gas})} \qquad (6.5)$$

where P (water vapor) is pressure and P (air or gas) is a partial pressure. The conversion factor between mass and pressure is 0.622.

> **Example 6.1**   Examples of water vapor in the atmosphere are as follows:
>
> Dark storm clouds (Cumulonimbus) can contain 10 g/m³ of water vapor.
> Medium density clouds (Cumulus Congestus) can contain 0.8 g/m³ of water vapor.
> Light rain clouds (Cumulus) contain 0.2 g/m³ of water vapor.
> Wispy clouds (Cirrus) contain 0.1 g/m³ of water vapor.
>
> In the case of the dark storm clouds, this equates to 100,000 tons of water vapor per square mile for a 10,000-ft tall cloud.
>
> Dew point is the temperature at which condensation of the water vapor in air or a gas will take place as it is cooled at constant pressure; i.e., it is the temperature at which the mixture becomes saturated

and can no longer dissolve or hold all of the water vapor it contains. The water vapor will now start to condense out of the mixture to form dew or a layer of water on the surface of objects present.

Dry-bulb temperature is the temperature of a room or mixture of water vapor and air (gas) as measured by a thermometer whose sensing element is dry.

Wet-bulb temperature is the temperature of the air (gas) as sensed by a moist element. Air is circulated around the element causing vaporization to take place; the heat required for vaporization (latent heat of vaporization) cools the moisture around the element reducing its temperature.

Psychrometric chart is a somewhat complex combination of several simple graphs showing the relation between dry-bulb temperatures, wet-bulb temperatures, relative humidity, water vapor pressure, weight of water vapor per pound of dry air, BTUs per pound of dry air, and so on. While it may be a good engineering reference tool, it tends to overwhelm the student. For example, Fig. 6.1 shows a psychrometric chart from Heat Pipe Technology, Inc. for standard atmospheric pressure; for other atmospheric pressures the sets of lines will be displaced.

To understand the various relationships in the chart, it is necessary to break the chart down into only the lines required for a specific relationship.

Example 6.2   To obtain the relative humidity from the wet- and dry-bulb temperatures, the three lines shown in Fig. 6.2a should be used. These lines show the wet- and dry-bulb temperatures and the relative humidity lines. For instance, if the dry- and wet-bulb temperatures are measured as 76°F and 57°F, respectively, which when applied to Fig. 6.1 shows the two temperature lines as intersecting on the 30 percent relative humidity line, hence, the relative humidity is 30 percent. When the wet- and dry-bulb temperatures do not fall on a relative humidity line, a judgment call has to be made for the value of the relative humidity.

Example 6.3   If the temperature in a room is 75°F and the relative humidity is 55 percent, how far can the room temperature drop before condensation takes place? Assume no other changes. In this case, it is necessary to get the intersection of the dry-bulb temperature and relative humidity lines, as shown in Fig. 6.2b and then the corresponding horizontal line (weight of water vapor per pound of dry air). Using Fig. 6.1, the intersection falls on 0.01 lb of moisture per pound of dry air, because the weight of water vapor per pound of dry air will not change as the temperature changes. This horizontal line can be followed across to the left until it reaches the 100 percent relative humidity line (dew point). The temperature where these line cross is the temperature where dew will start to form, i.e., 57°F. The wet- and dry-bulb temperatures are the same at this point. Note that in some charts the weight of water vapor in dry air is measured in grains, where 1 lb = 7000 grains, or 1 grain = 0.002285 oz.

Example 6.4   This example compares the weight of water vapor in air at different humidity levels. The question is, how much more moisture does air at 80°F and 50 percent relative humidity contain than air at 60°F and 30 percent relative humidity? Using Fig. 6.3a as a reference, it is necessary to get the intersection of the dry-bulb temperature and the relative humidity lines. The horizontal line where they intersect gives the weight of water vapor per pound of dry air as in the previous example. Using Fig. 6.1, the intersections are 0.0108 and 0.0032 lb, hence, the difference = 0.0076 lb or 53 grains.

Example 6.5   How much heat is required to raise the temperature of air at 50°F and 75 percent relative humidity to 75°F and 45 percent relative humidity? Referring to Fig. 6.3b, the intersection of the dry-bulb temperature and relative humidity lines must be found, and hence, the total dry-heat line passing through the intersection (these lines are an extension of the wet-bulb temperature lines to the total heat per pound dry-air scale). From Fig. 6.1, the intersection of the temperature and relative humidity lines falls on 18.2 and 27.2 BTUs/lb of dry air, giving a difference of 9.0 BTUs/lb of dry air.

Example 6.6   In air at 77°F and 45 percent relative humidity, how much space is occupied by 1 lb of dry air? The lines shown in Fig. 6.4 are used. The intersection of the dry-bulb temperature and relative humidity lines on the cubic feet per pound of dry air line gives the space occupied by 1 lb of dry air. From Fig. 6.1, the lines intersect at 13.65 ft³ giving this as the volume containing 1 lb of dry air.

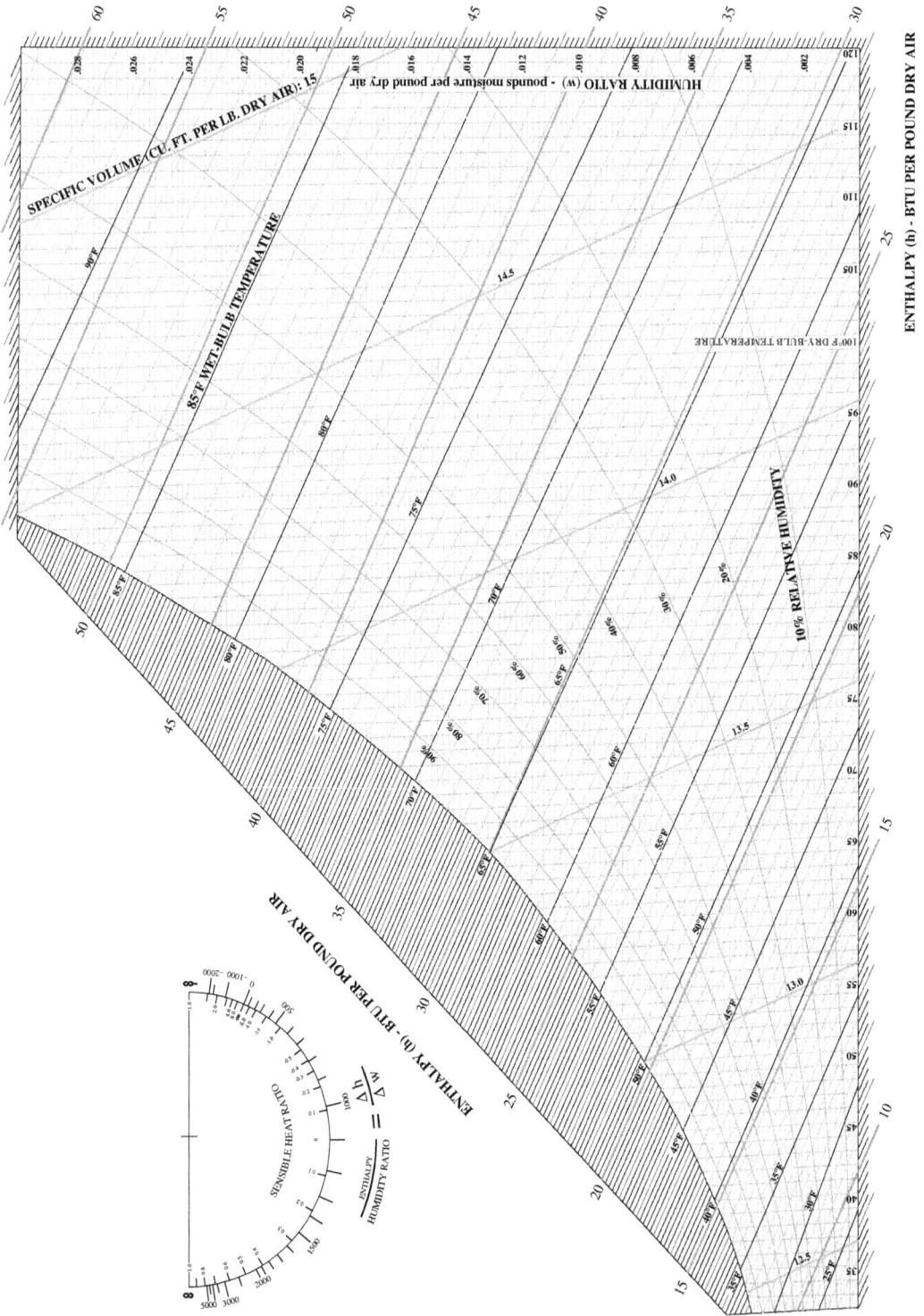

Figure 6.1 Psychrometric chart for air-water vapor mixtures. (Courtesy of Heat Pipe Technology, Inc.)

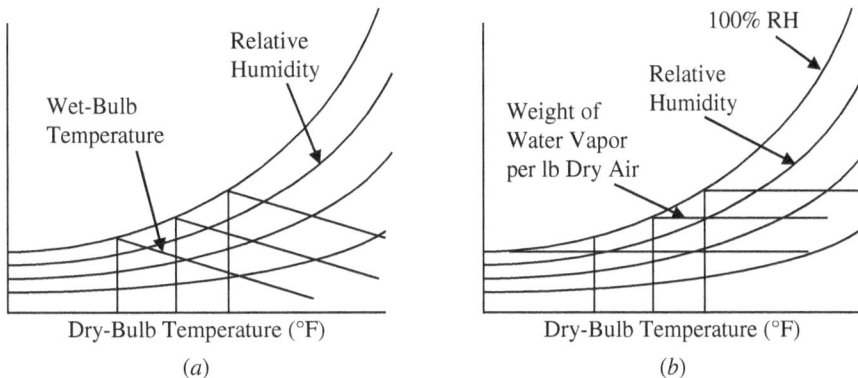

FIGURE 6.2 Lines required for finding (a) the relative humidity and (b) the condensation temperature.

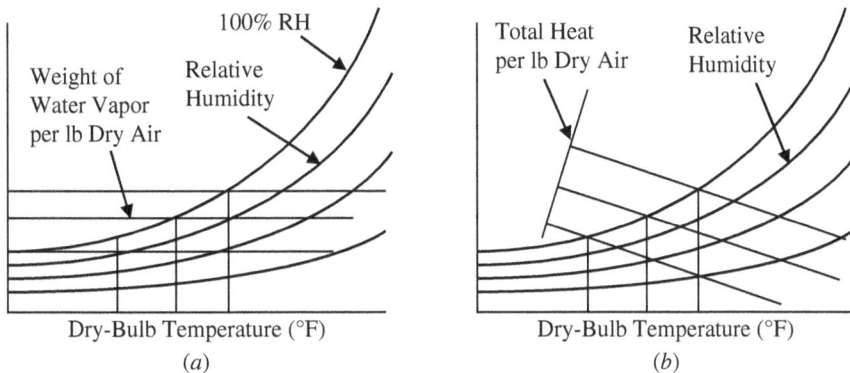

FIGURE 6.3 Lines required finding (a) the weight of water vapor and (b) the heating requirements.

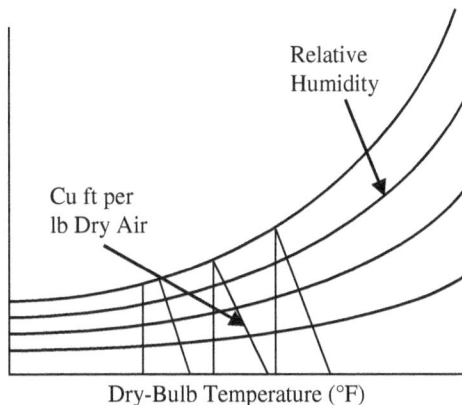

FIGURE 6.4 Lines required finding the volume per pound of dry air.

6.2.2   Humidity Measuring Devices

Hygrometers

Devices that indirectly measure humidity by sensing changes in physical or electrical properties in materials due to their moisture content are called hygrometers. Materials such as hair, skin, membranes, and thin strips of wood change their length as they absorb water. The change in length is directly related to the humidity. Such devices are used to measure relative humidity from 20 to 90 percent, giving accuracies of about ±5 percent. Their operating temperature range is limited to less than 70°C.

Laminate hygrometer is made by attaching thin strips of wood to thin metal strips forming a laminate. The laminate is formed into a helix as shown in Fig. 6.5a, as the humidity changes the helix flexes due to the change in the length of the wood. One end of the helix is anchored, the other is attached to a pointer (similar to a bimetallic strip used in temperature measurements). The scale is graduated in percent humidity.

Hair hygrometer is the simplest and oldest type of hygrometer. It is made using hair as shown in Fig. 6.5b. Human hair lengthens by 3 percent when the humidity changes from 0 to 100 percent. The change in length can be used to control a pointer for visual readings or a transducer such as a linear variable differential transformer (LVDT) for an electrical output. The hair hygrometer has an accuracy of about 5 percent for the humidity range 20 to 90 percent over the temperature range 5 to 40°C.

Resistive hygrometer or resistive humidity sensors consist of two electrodes with interdigitated fingers on an insulating substrate as shown in Fig. 6.6a. The electrodes

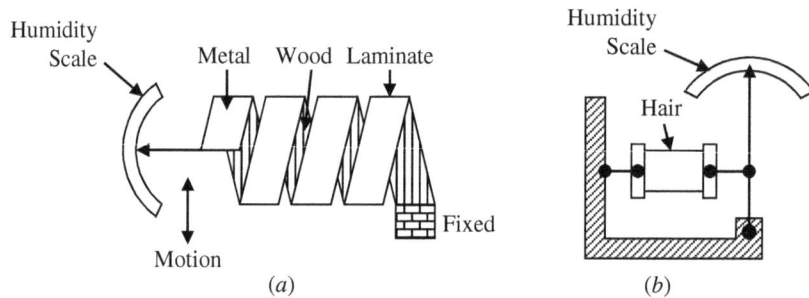

FIGURE 6.5   Two types of hygrometers using (a) metal/wood laminate and (b) hair.

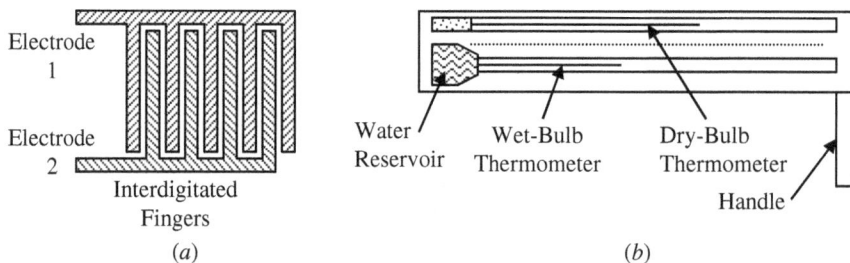

FIGURE 6.6   Methods of measuring humidity: (a) electrical using electrodes with interdigitated fingers covered with a hydroscopic material and (b) sling psychrometer.

are coated with a hygroscopic material (one that absorbs water such as lithium chloride). The hydroscopic material provides a conductive path between the electrodes; the coefficient of resistance of the path is inversely proportional to humidity. Alternatively, the electrodes can be coated with a bulk polymer film that releases ions in proportion to the relative humidity; temperature correction can again be applied for an accuracy of 2 percent over the operating temperature range –40 to 70°C and relative humidity from 2 to 98 percent. An ac voltage is normally used with this type of device; i.e., at 1 kHz a relative humidity change from 2 to 98 percent will typically give a resistance change from 10 MΩ to 1 kΩ. Variations of this device are the electrolytic, and the resistance-capacitance hygrometer.

Capacitive hygrometer. The dielectric constant of certain thin polymer films changes linearly with humidity, so that the capacitance between two plates using the polymer as the dielectric is directly proportional to humidity. The capacitive device has good longevity, a working temperature range of 0 to 100°C, a fast response time, and can be temperature compensated to give an accuracy of ±0.5 percent over the full humidity range.

Piezoelectric or sorption hygrometers use two piezoelectric crystal oscillators; one is used as a reference and is enclosed in a dry atmosphere, and the other is exposed to the humidity to be measured. Moisture increases the mass of the crystal which decreases its resonant frequency. By comparing the frequencies of the two oscillators, the humidity can be calculated. Moisture content of gases from 1 to 25,000 ppm can be measured.

Psychrometers

A psychrometer uses the latent heat of vaporization to determine relative humidity. If the temperature of air is measured with a dry-bulb thermometer and a wet-bulb thermometer, the two temperatures can be used with a psychrometric chart to obtain the relative humidity, water vapor pressure, heat content, and weight of water vapor in the air. Water evaporates from the wet bulb trying to saturate the surrounding air. The energy needed for the water to evaporate cools the thermometer, so that the dryer the day, the more water evaporates and the lower the temperature of the wet bulb.

To prevent the air surrounding the wet bulb from saturating, there should be some air movement around the wet bulb. This can be achieved with a small fan or by using a sling psychrometer, which is a frame holding both the dry and wet thermometers that can rotate about a handle as shown in Fig. 6.6*b*. The thermometers are rotated for 15 to 20 s. The wet-bulb temperature is taken as soon as rotation stops before it can change, and then the dry-bulb temperature (which does not change).

Dew Point Measuring Devices

A simple method of measuring the humidity is to obtain the dew point. This is achieved by cooling the air or gas until water condenses on an object and then measuring the temperature at which condensation takes place. Typically, a mirrored surface, polished stainless steel, or silvered surface is cooled from the back side, by cold water, refrigeration, or Peltier cooling. As the temperature drops, a point is reached where dew from the air or gas starts to form on the mirror surface. The condensation is detected by the reflection of a beam of light by the mirror to a photocell. The intensity of the reflected light reduces as condensation takes place and the temperature of the mirror at that point can be measured.

Moisture Content Measuring Devices

Moisture content of materials is very important in some processes. There are two methods commonly used to measure the moisture content; these are with the use of microwaves or by measuring the reflectance of the material to infrared rays.

Microwave absorption by water vapor is a method used to measure the humidity in a material. Microwaves (1 to 100 GHz) are absorbed by the water vapor in the material. The relative amplitudes of the transmitted and microwaves passing through a material are measured. The ratio of these amplitudes is a measure of the humidity content of the material.

Infrared absorption uses infrared rays instead of microwaves. The two methods are similar. In the case of infrared, the measurements are based on the ability of materials to absorb and scatter infrared radiation (reflectance). Reflectance depends on chemical composition and moisture content. An infrared beam is directed onto the material and the energy of the reflected rays is measured. The measured wavelength and amplitude of the reflected rays are compared to the incident wavelength and amplitude; the difference between the two is related to the moisture content.

Other methods of measuring moisture content are by color changes or by absorption of moisture by certain chemicals and measuring the change in mass, neutron reflection, or nuclear magnetic resonance.

Humidity Application Considerations

Although, wet and dry bulbs were the standard for making relative humidity measurements, more up-to-date and easier to make electrical methods such as capacitance and resistive devices are now available and will be used in practice. These devices are small, rugged, reliable, and accurate with high longevity, and if necessary can be calibrated by the National Institute of Standards and Technology (NIST) against accepted gravimetric hygrometer methods. Using these methods, the water vapor in a gas is absorbed by chemicals that are weighted before and after to determine the amount of water vapor absorbed in a given volume of gas from which the relative humidity can be calculated.

6.3 Density and Specific Gravity

6.3.1 Basic Terms

The density, specific weight, and specific gravity were defined in Chap. 2 as follows:

Density ρ of a material is defined as the mass per unit volume. Units of density are pounds (slug) per cubic foot [lb (slug)/ft³] or kilograms per cubic meter (kg/m³).

Specific weight γ is defined as the weight per unit volume of a material, i.e., pounds per cubic foot (lb/ft³) or newton per cubic meter (N/m³).

Specific gravity (SG) of a liquid or solid is defined as the density of the material divided by the density of water or the specific weight of the material divided by the specific weight of water at 4°C temperature. The specific gravity of a gas is its density/specific weight divided by the density/specific weight of air at 60°F and 1 atm (14.7 psia).

The relation between density and specific weight is given by

$$\gamma = \rho g \tag{6.6}$$

Material	Specific Weight		Density		Specific Gravity
	lb/ft³	kN/m³	slug/ft³	× 10³ kg/m³	
Acetone	49.4	7.74	1.53	0.79	0.79
Ammonia	40.9	6.42	1.27	0.655	0.655
Benzene	56.1	8.82	1.75	0.9	0.9
Gasoline	46.82	7.35	3.4	0.75	0.75
Glycerin	78.6	12.4	2.44	1.26	1.26
Mercury	847	133	26.29	13.55	13.55
Water	62.43	9.8	1.94	1.0	1.0

TABLE 6.1   Density and Specific Weights

where g is the acceleration of gravity 32.2 ft/s² or 9.8 m/s² depending on the units being used.

Example 6.7   What is the density of a material whose specific weight is 27 kN/m³?

Solution

$$\rho = \gamma/g = 27 \text{ kN/m}^3/9.8 \text{ m/s}^2 = 2.75 \times 10^3 \text{ kg/m}^3$$

Table 6.1 gives a list of the density and specific weight of some common materials.

6.3.2   Density Measuring Devices

Hydrometers are the simplest device for measuring the specific weight or density of a liquid. The device consists of a graduated glass tube, with a weight at one end, which causes the device to float in an upright position. The device sinks in a liquid until an equilibrium point between its weight and buoyancy is reached. The specific weight or density can then be read directly from the graduations on the tube. Such a device is shown in Fig. 6.7a.

Thermohydrometer is a combination hydrometer and thermometer, so that both the specific weight/density and temperature can be recorded and the specific weight/density corrected from look up tables for temperature variations to improve the accuracy of the readings.

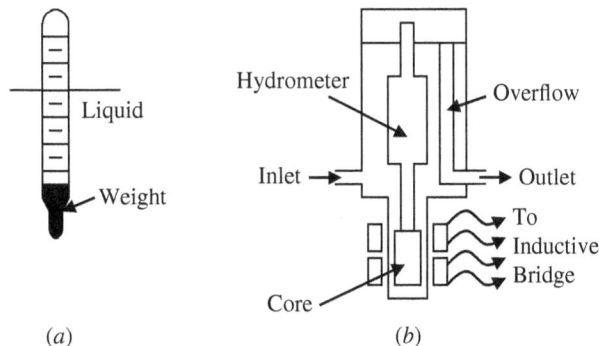

FIGURE 6.7   Illustration of (a) a basic hydrometer and (b) an induction hydrometer.

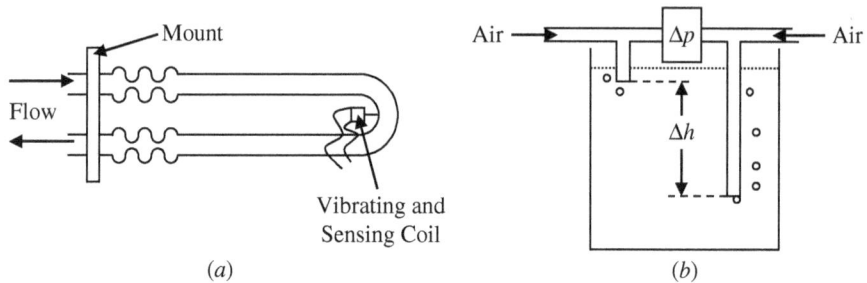

Figure 6.8   Alternative methods for density measurement are (a) vibration sensor and (b) bubbler system.

Induction hydrometers are used to convert the specific weight or density of a liquid into an electrical signal. In this case, a fixed volume of liquid set by the overflow tube is used in the type of setup shown in Fig. 6.7b, the displacement device, or hydrometer has a soft iron or similar metal core attached. The core is positioned in a coil which forms part of a bridge circuit. As the density/specific weight of the liquid changes, the buoyant force on the displacement device changes. This movement can be measured by the coil and converted into a density reading.

Vibration sensors are an alternate method of measuring the density of a fluid (see Fig. 6.8a). Fluid is passed through a U tube which has a flexible mount so that it can vibrate when driven from an outside source. The amplitude of the vibration decreases as the specific weight or density of the fluid increases, so that by measuring the vibration amplitude the specific weight/density can be calculated.

Density and specific gravity of a liquid can also be determined by measuring the pressure at the base of a column of liquid. The density at a depth (h) in the liquid is given by

$$\rho = \frac{p}{gh} \tag{6.7}$$

The specific weight is given by

$$\rho = \frac{p}{h} \tag{6.8}$$

Example 6.8   What is the pressure at the base of a column of liquid if the height of the column is 298 cm and the density of the liquid is 1.26 kg/m³?

Solution

$$p = \rho gh = 1.26 \times 9.8 \times 298/100 = 36.8 \text{ Pa}$$

Figure 6.8b shows the setup of a bubbler system to measure liquid density or specific weight. Two air supplies are used to supply two tubes whose ends are at different depths in a liquid. The difference in air pressures between the two air supplies is directly related to the density of the liquid by the following equation:

$$\rho = \frac{\Delta p}{g \Delta h} \tag{6.9}$$

where Δp is the difference in the pressures and Δh is the difference in the height of the bottoms of the two tubes.

Example 6.9   What is the density of a liquid in a bubbler system if pressures of 500 Pa and 23 kPa are measured at depths of 15 cm and 6.5 m, respectively?

Solution

$$\rho = \gamma/g = 27 \text{ kN/m}^3/9.8 \text{ m/s}^2 = 2.75 \times 10^3 \text{ kg/m}^3$$

Density can be determined from the weight of a known volume of the liquid; i.e., a container of known volume can be filled with a liquid and weighted full and empty. The difference in weight gives the weight of liquid, from which the density can be calculated using

$$\rho = \frac{W_f - W_c}{g \times \text{Vol}} \tag{6.10}$$

where W_f = weight of container + liquid
W_c = weight of container
Vol = volume of the container

Radiation density sensors consist of a radiation source located on one side of a pipe or container and a sensing device on the other. The sensor is calibrated with the pipe or container empty, and then filled. Any difference in the measured radiation is caused by the density of the liquid which can then be calculated.

Gas densities are normally measured by sensing the frequency of vibration of a vane in the gas, or by weighing a volume of the gas and comparing it to the weight of the same volume of air.

6.3.3   Density Application Considerations

Ideally, when measuring the density of a liquid, there should be some agitation to ensure uniform density throughout the liquid. This is to avoid density gradients due to temperature gradients. Excessive agitation should be avoided.

Density measuring equipment is available for extreme temperatures and pressures, i.e., from −150 to 600°F and for pressures in excess of 1000 psi. When measuring corrosive, abrasive, volatile liquids, and the like, radiation devices should be considered.

6.4   Viscosity

Viscosity was introduced in Chap. 4; viscosity and methods of measuring viscosity are discussed in more detail in this chapter.

6.4.1   Basic Terms

Viscosity μ in a fluid is the resistance to its change of shape, which is due to the attraction between molecules in a liquid that resists any change due to flow or motion. When a force is applied to a fluid at rest, the molecular layers in the fluid tend to slid on top of each other as shown in Fig. 6.9a. The force F resisting motion in a fluid is given by

$$F = \frac{\mu A V}{y} \tag{6.11}$$

where A = boundary area being moved
V = velocity of the moving boundaries
y = distance between boundaries
μ = coefficient of viscosity, or dynamic viscosity

The units of measurement must be consistent.

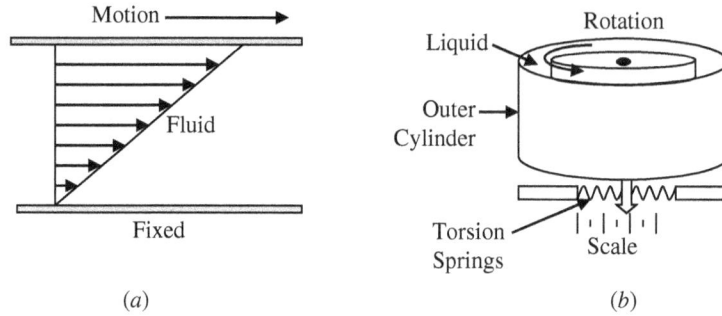

FIGURE 6.9 Illustration of (a) Newtonian laminar flow and (b) a drag-type viscometer.

Sheer stress τ can be calculated from the following formula:

$$\mu = \frac{\tau y}{V} \tag{6.12}$$

where τ is the shear stress or force per unit area.

If F is in pounds, A in square feet, V in feet per second, y in feet, then μ is in pound seconds per square feet. Whereas, if F is in newton, A in square meter, V in meters per second, y in meters, then μ is in newton seconds per square meter. A sample list of fluid viscosities is given in Table 6.2.

The standard unit of viscosity is the poise, whereas a centipoise (poise/100) is the viscosity of water at 68.4°F. Conversions are given in Table 4.1 (1 centipoise = 2.09×10^{-5} lb·s/ft²).

When the temperature of a body increases, more energy is imparted to the atoms making them more active and thus effectively reducing the molecular attraction. This in turn reduces the attraction between the fluid layers lowering the viscosity; i.e., viscosity decreases as temperature increases.

Newtonian fluids are fluids that exhibit only laminar flow as shown in Fig. 6.9a and are consistent with temperature. Only Newtonian fluids will be considered. Non–Newtonian fluid dynamics is complex and considered to be outside the scope of this text.

Fluid	μ (lb·s/ft²)	Fluid	μ (lb·s/ft²)
Air	38×10^{-8}	Carbon dioxide	31×10^{-8}
Hydrogen	19×10^{-8}	Nitrogen	37×10^{-8}
Oxygen	42×10^{-8}	Carbon tetrachloride	20×10^{-8}
Ethyl alcohol	25×10^{-8}	Glycerin	18×10^{-8}
Mercury	32×10^{-8}	Water	21×10^{-8}
Water	1×10^{-2} poise		

TABLE 6.2 Dynamic Viscosities, at 68°F and Standard Atmospheric Pressure

6.4.2   Viscosity Measuring Instruments

Viscometers or viscosimeters are used to measure the resistance to motion of liquids and gases. Several different types of instruments have been designed to measure viscosity, such as the inline falling-cylinder viscometer, the drag-type viscometer, and the Saybolt universal viscometer. The rate of rise of bubbles in a liquid can also be used to give a measure of the viscosity of a liquid.

The falling-cylinder viscometer uses the principle that an object when dropped into a liquid will descend to the bottom of the vessel at a fixed rate; the rate of descent is determined by the size, shape, and density of the object, and the density and viscosity of the liquid. The higher the viscosity, the longer the object will take to reach the bottom of the vessel. The falling-cylinder device measures the rate of descent of a cylinder in a liquid, and correlates the rate of descent to the viscosity of the liquid.

Rotating disc viscometer is a drag-type device. The device consists of two concentric cylinders and the space between the two cylinders is filled with the liquid being measured, as shown in Fig. 6.9b. The inner cylinder is driven by an electric motor and the force on the outer cylinder due to the liquid viscosity is measured by noting its movement against a torsion spring, the viscosity of the liquid can then be determined.

The *Saybolt instrument* measures the time for a given amount of fluid to flow through a standard size orifice or a capillary tube with an accurate bore. The time is measured in Saybolt seconds, which is directly related and can be easily converted to other viscosity units.

Example 6.10   Two parallel plates separated by 0.45 in are filled with a liquid with a viscosity of 7.6×10^{-4} lb·s/ft². What is the force acting on 1 ft² of the plate, if the other plate is given a velocity of 4.4 ft/s?

Solution

$$F = \frac{7.6 \times 10^{-4} \text{ lb·s} \times 1 \text{ ft}^2 \times 4.4 \text{ ft} \times 12}{\text{ft}^2 \times 0.45 \text{ ft·s}} = 0.089 \text{ lb}$$

6.5   pH Measurements

The pH value is used to determine the alkalinity or acidity of a substance. The pH value is discussed and methods of measurement described.

6.5.1   Basic Terms

In many process operations pure and neutral water is required for cleaning or diluting other chemicals; i.e., the water is not acidic or alkaline. Water contains both hydrogen ions and hydroxyl ions. When these ions are in the correct ratio the water is neutral. An excess of hydrogen ions causes the water to be acidic and when there is an excess of hydroxyl ions the water is alkaline. The pH (power of hydrogen) of the water is a measure of its acidity or alkalinity; neutral water has a pH value of 7 at 77°F (25°C). When water becomes acidic the pH value decreases. Conversely, when the water becomes alkaline the pH value increases. The pH values use a log to the base 10 scale; i.e., a change of 1 pH unit means that the concentration of hydrogen ions has increased (or decreased) by a factor of 10 and a change of 2 pH units means the concentration has changed by a factor of 100. The pH value is given by

$$pH = \log_{10}[1/\text{hydrogen ion concentration}] \qquad (6.13)$$

The pH value of a liquid can range from 0 to 14. The hydrogen ion concentration is in grams per liter; i.e., a pH of 4 means that the hydrogen ion concentration is 0.0001 g/L at 25°C. Strong hydrochloric or sulfuric acids will have a pH of 0 to 1.

4% caustic soda	pH =14
Lemon and orange juice	pH = 2 to 3
Ammonia	pH is about 11

Example 6.11   The hydrogen ion content in water goes from 0.15 to 0.0025 g/L. How much does the pH change?

Solution

$$pH_1 = \log\left(\frac{1}{0.15}\right) = 0.824$$

$$pH_2 = \log\left(\frac{1}{0.0025}\right) = 2.6$$

$$\text{Change in pH} = 0.824 - 2.6 = -1.776$$

6.5.2   pH Measuring Devices

The pH is normally measured by chemical indicators or by pH meters. The final color of chemical indicators depends on the hydrogen ion concentration; their accuracy is only 0.1 to 0.2 pH units. For indication of acid, alkali, or neutral water, litmus paper is often used; it turns pink when acidic, and blue when alkaline and stays white if neutral.

A pH sensor normally consists of a sensing electrode and a reference electrode immersed in the test solution which forms an electrolytic cell, as shown in Fig. 6.10a. One electrode contains a saturated potassium chloride (alkaline) solution to act as a reference; the electrode is electrically connected to the test solution via the liquid junction. The other electrode contains a buffer which sets the electrode in contact with the liquid sample. The electrodes are connected to a differential amplifier, which amplifies

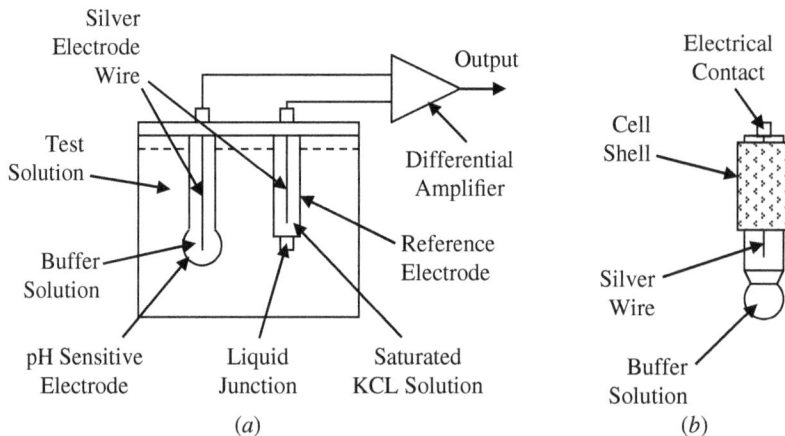

FIGURE 6.10   Representation of the setup of (a) a pH sensor and (b) a pH sensing electrode.

the voltage difference between the electrodes giving an output voltage that is proportional to the pH of the solution. Figure 6.10*b* shows the pH sensing electrode.

6.5.3 pH Application Considerations

The pH of neutral water varies with temperature; i.e., neutral water has a pH of about 7.5 at 32°F and about 6 at 212°F. pH systems are normally automatically temperature compensated. pH test equipment must be kept clean and free from contamination. Calibration of test equipment is done with commercially available buffer solutions with known pH values. Again, cleaning between each reading is essential to prevent contamination.

Summary

This chapter introduced humidity, its relation to dew point and temperature. The psychrometric chart is shown and instructions given on how to read the chart. Humidity terms and instruments are described. Density, specific weight, and specific gravity are defined together with measurement instruments. Viscosity and pH were introduced with definitions and measuring instruments.

The main points described in this chapter were as follows:

1. The definition of and relationship between specific humidity, relative humidity, and dew point.

2. Use of the psychrometric chart for obtaining dew point and the weight of water vapor dissolved in the atmosphere from temperature data.

3. The various types of instruments used for the direct and indirect measurement of humidity.

4. Density, specific weight, and specific gravity are defined with examples in both English and SI units. Instruments for measuring specific weight and specific gravity are given.

5. Instruments used in the measure of density and specific weight.

6. The basic terms and definitions used in viscosity, its relation to flow with examples, and instruments used for measuring viscosity.

7. An introduction to pH terms is given, and its value when determining acidity or alkalinity, and the instruments used to measure pH.

Problems

6.1   The dry bulb of a wet/dry thermometer reads 120°F. What is the relative humidity if the wet bulb reads (*a*) 90°F, (*b*) 82°F, (*c*) 75°F?

6.2   The wet bulb of a wet/dry thermometer reads 85°F. What will the dry bulb read if the relative humidity is (*a*) 80 percent, (*b*) 60 percent, (*c*) 30 percent?

6.3   What is the relative humidity if the wet and dry bulbs in a wet/dry thermometer read 75°F and 85°F, respectively?

6.4   The dry bulb of a wet/dry thermometer reads 60°F and the relative humidity is 47 percent. What will the wet bulb read and what is the absolute humidity?

6.5   The wet and dry bulbs of a wet/dry thermometer read 75°F and 112°F, respectively. What is the relative and absolute humidity?

6.6   If the air temperature is 85°F, what is the water vapor pressure corresponding to a relative humidity of 55 percent and at 55°F with 85 percent relative humidity?

6.7   How much water is required to raise the relative humidity of air from 25 to 95 percent if the temperature is held constant at 95°F?

6.8   How much heat and water are added per pound of dry air to increase the relative humidity from 15 to 80 percent with a corresponding temperature increase from 42 to 95°F?

6.9   How much water does dry air contain at 105°F and 55 percent relative humidity?

6.10   How much heat is required to heat 1 lb of dry air from 35 to 80°F if the relative humidity is constant at 80 percent?

6.11   How much space is occupied by 4.7 lb of dry air if the air temperature is 80°F and the relative humidity is 82 percent?

6.12   The two tubes in a bubbler system are placed in a liquid with a density of 1.395 slugs/ft³. If the bottom ends of the bubbler tubes are 3.5 and 42.7 in below the surface of the liquid, what is the differential pressure supplied to the two bubblers?

6.13   A tank is filled to a depth of 54 ft with liquid having a density of 1.234 slugs/ft³. What is the pressure on the bottom of the tank?

6.14   What would be the specific gravity of a gas with a specific weight of 0.127 lb/ft³, if the density of air under the same conditions is 0.0037 slugs/ft³?

6.15   A square plate 1.2 ft on a side is centrally placed in a channel 0.23-in wide filled with a liquid with a viscosity of 7.3×10^{-5} lb·s/ft². If the plate is 0.01-in thick what force is required to pull the plate along the channel at 14.7 ft/s?

6.16   Two parallel plates 35 ft², 1.7 in apart are placed in a liquid with a viscosity of 2.1×10^{-4} lb·s/ft². If a force of 0.23 lb is applied to one plate in a direction parallel to the plates with the other plate is fixed, what is the velocity of the plate?

6.17   What is the pH of a solution, if there is a concentration of 0.0006 g/L of hydrogen ions?

6.18   What is the change in hydrogen concentration factor if the pH of a solution changes from 3.5 to 0.56?

6.19   What is the concentration of hydrogen ions if the pH is 13.2?

6.20   What is the hydrogen concentration of a neutral solution?

CHAPTER 7

Position, Motion, and Force

Chapter Objectives

This chapter will help you understand and become familiar with position, motion, and force sensors that play a very important part in process control for linear and rotation measurements.

Topics discussed in this chapter are as follows:

- Position, distance, and velocity sensors
- Incremental distance measurement
- Acceleration and vibration sensors
- Magnetic sensors
- Rotation measurement using light sensors
- Force, torque, load cells, balances
- Strain gauges
- Light sensors

7.1 Introduction

There are many sensors other than level, pressure, flow, and temperature that are encountered on a day-to-day basis. These are position, motion, distance, and strain gauge sensors that are being extensively used in process control in today's high-technology industries for precision tool milling, and laser scribing and cutting. These sensors will be discussed in some detail, as the student should be aware of their existence and basic operation.

7.2 Position and Motion Sensing

7.2.1 Basic Position Definitions

Many industrial processes require both linear and angular position and motion measurements. These are required in robotics, rolling mills, machining operations, numerically controlled tool applications, and conveyers. In some applications, it is also necessary to measure speed, acceleration, and vibration. Position sensing devices such as the linear variable differential transformer (LVDT) are used to convert temperature and/or pressure into electrical units. Other types of position sensing devices are used to monitor the position of control valve to give feedback, measure distance and rotation for machine control, and so on.

The definition of terms used in position and motion sensing are as follows:

Absolute position is the distance measured with respect to a fixed reference or degrees of rotation from a fixed point.

Incremental position is a measure of the change in position and is not referenced to a fixed point. If power is interrupted, the incremental position change is lost. An additional position reference such as a limit switch is often used with this type of sensor so that the incremental position can be duplicated. This type of sensing can give very accurate positioning of one component with respect to another and is used when making master plates for tooling and the like.

Rectilinear motion is measured by the distance traversed in a given time, velocity when moving at a constant speed, or acceleration when the speed is changing in a straight line.

Angular position is a measurement of the change in position of a point about a fixed axis measured in degrees or radians, where one complete rotation is 360° or 2π radians. The degrees of rotation of a shaft can be absolute or incremental. These types of sensors are also used in rotating equipment to measure rotation speed as well as shaft position and to measure torque displacement.

Arc-minute is an angular displacement of 1/60 of a degree.

Angular motion is a measure of the rate of rotation. Angular velocity is a measure of the rate of rotation when rotating at a constant speed about a fixed point or angular acceleration when the rotational speed is changing.

Velocity or *speed* is the rate of change of position. This can be a linear measurement, i.e., feet per second (ft/s), meters per second (m/s), and so forth, or angular measurement, i.e., degrees per second, radians per second, rate per minute (r/m), and so forth.

Acceleration is the rate of change of speed, i.e., feet per second squared (ft/s²), meters per second squared (m/s²), and the like for linear motion, or degrees per second squared, radians per second squared, and the like, in the case of rotational motion.

Vibration is a measure of the periodic motion about a fixed reference point or the shaking that can occur in a process due to sudden pressure changes, shock, or unbalanced loading in moving or rotational equipment. Peak accelerations of 100 g can occur during vibrations which can lead to fracture or self-destruction. Vibration sensors are used to monitor the bearings in heavy rollers such as those used in rolling mills; excessive vibration indicates failure in the bearings or damage to rotating parts that can then be replaced before serious damage occurs.

7.2.2 ON/OFF Position Sensing

Position sensing falls into two categories, first as an ON/OFF device sensing the presence or absence of an object, and second to sense the position of a measuring device

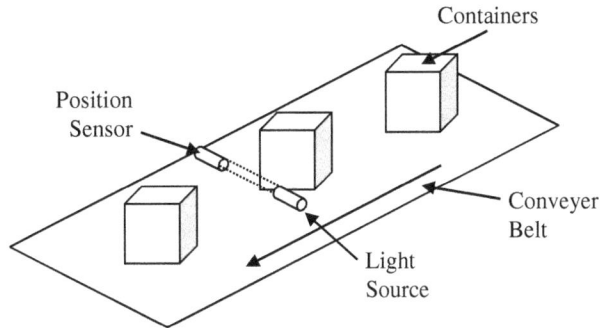

FIGURE 7.1   Conveyer belt with position sensor.

such as a transducer. Sensing used in ON/OFF applications is the breaking of light beams, contact switching, proximity detectors, and magnetic sensing devices. The movement of mechanical sensing devices can be converted into electrical units using potentiometers, linear variable differential transformer devices, or variable capacitance devices.

Light sensors can be used to detect the presence or absence of light. Semiconductor light emitting diodes (LED) are the most common commercially available light sources used in industry for ON/OFF applications; a semiconductor photodiode or similar device would be used as the light detector. Figure 7.1 illustrates the use of beam switching for detecting the position of cartons on a conveyer belt.

Proximity detectors are used to detect objects in close proximity. A pressure contact switch could be used, but mechanical switches have a limited life. The detection of cartons in Fig. 7.1 could also be considered as a proximity detector. An alternative is a reflective strip on the object to reflect the light back to the sensor; this is also the principal used in a bar code reader. The advantage of light proximity detector is that it works with all materials.

Proximity detectors are defined in the International Electrotechnical Commission publication (IEC) 60947-5-2.

Hall effect sensors measure magnetic field strength and detect changes in it, and are also used as an inductive close proximity detector. The Hall effect in a semiconductor occurs when a current flowing in the semiconductor experiences a magnetic field perpendicular to the direction of the current flow. The effect is to cause the current to be deflected and flow perpendicular to both the magnetic field and initial flow direction. Figure 7.2*a* shows the current flow in a Hall device without a magnetic field and in Fig. 7.2*b* in the presence of a magnetic field, the change in current direction produces a voltage (Hall voltage) as shown between the opposite sides of the element.

Magneto resistive element (MRE) is an alternative to the Hall effect device. In the case of the MRE, its resistance changes with magnetic strength and field direction. A current carrying ferromagnetic material will change its resistance in the presence of a magnetic field. The resistivity is at maximum when the current and magnetic field are in the same direction, and minimum when they are perpendicular to each other, see Fig. 7.3*a*. The permalloy element is shown in Fig. 7.3*b*. Because the resistivity changes with magnetic field direction, the MRE can be used in a bridge configuration to sense the direction of a magnetic field.

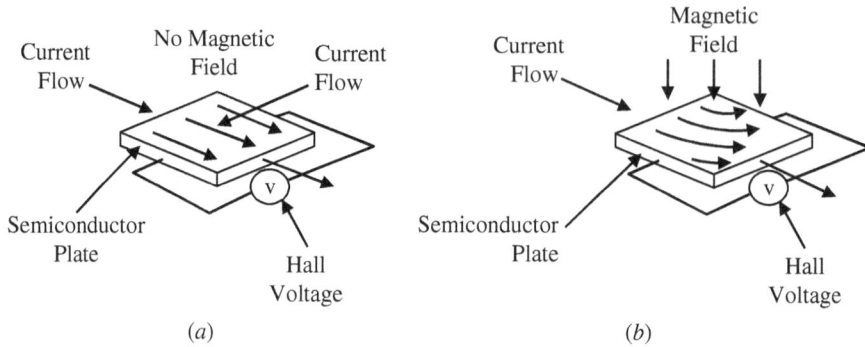

FIGURE 7.2   Hall effect devices (a) without magnetic field and (b) with a magnetic field.

Potentiometers can be used to convert small mechanical distance changes into voltages as in a bellows pressure sensor or are a convenient method of converting the position of a control valve to an electrical variable. The wiper or slider arm of a linear potentiometer can be directly connected to the valve actuator. Where rotation is involved a single or multiturn (up to 10 turns) rotational type of potentiometer can be used. For stability wire wound devices should be used, but in environmentally unfriendly conditions, lifetime of the potentiometer may be limited by dirt, contamination, and wear.

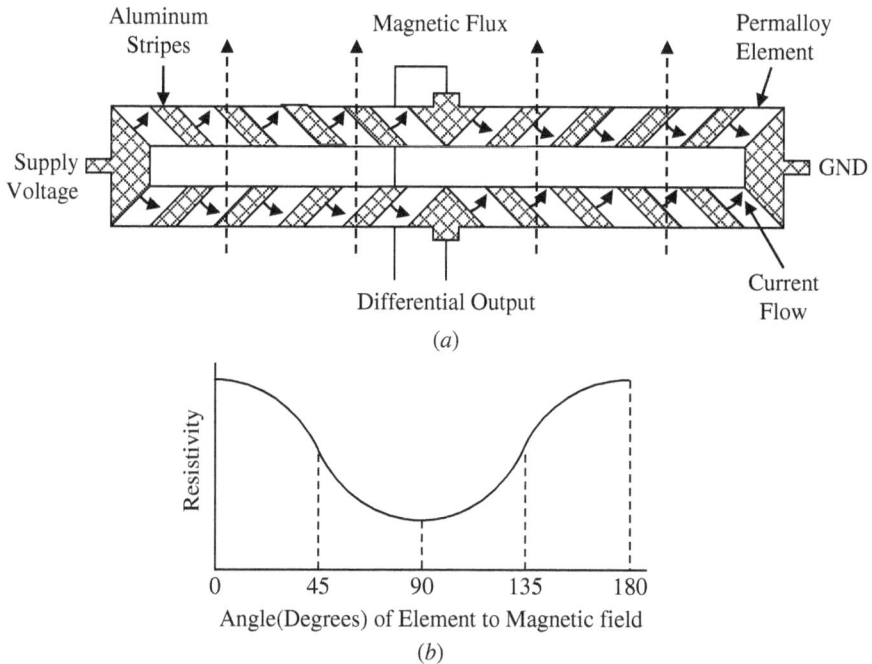

FIGURE 7.3   MRE element: (a) deposited permalloy element and (b) resistance of element versus direction of magnetic field.

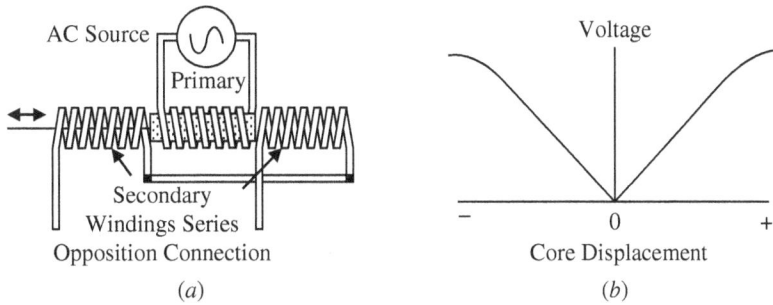

FIGURE 7.4 Demonstration of (a) the LVDT with a movable core and three windings and (b) the secondary voltage versus core displacement for the connections shown.

LVDTs are devices that are used for measuring small distances and are an alternative to the potentiometer. The device consists of a primary coil with two secondary windings one on either side of the primary (see Fig. 7.4a). A movable core when centrally placed in the primary will give equal coupling to each of the secondary coils. When an ac voltage is applied to the primary, equal voltages will be obtained from the secondary windings that are wired in series opposition to give zero output voltage, as shown in Fig. 7.4b. When the core is slightly displaced, an output voltage proportional to the displacement will be obtained. These devices are not as cost-effective as potentiometers, but have the advantage of being noncontact. The outputs are electrically isolated, accurate, and have better longevity than potentiometers.

7.2.3 Motion and Distance Sensing

Light interference lasers are used for very accurate incremental position measurements. Light is ultra-high-frequency electromagnetic wave that travels at 3×10^8 m/s. Light amplitude is measured in lux (lx) or lumens (lm). The wavelength of visible light is from 4 to 7×10^{-7} m. Longer wavelengths of electromagnetic waves are termed "infrared" and shorter wavelengths, "ultraviolet." Light wavelengths are sometimes expressed in terms of angstroms (Å) where $1 \text{ Å} = 1 \times 10^{-10}$ m. The relation between the frequency (f) and the wavelength (λ) of electromagnetic waves is given by

$$c = f\lambda \tag{7.1}$$

where c is the speed of light.

Example 7.1 What is the wavelength of light in Å, if the wavelength in meters is 500 nm?

Solution

$$1 \text{ Å} = 10^{-10} \text{m}$$
$$500 \text{ nm} = 500 \times 10^{-9}/10^{-10} \text{ Å} - 5000 \text{ Å}$$

Intensity is the brightness of light. The unit of measurement of light intensity in the English system is the foot-candle (fc) which is 1 lumen per square foot (lm/ft²), in the SI system the unit is the lux (lx) which is 1 lumen per square meter (lm/m²). The phot (ph) is also used and is defined as 1 lumen per square centimeter (lm/cm^2). The lumen replaces the candela (cd) in the SI system. The dB is also used for the comparison of light intensity as follows:

$$\text{Light level ratio in dB} = 10\log_{10}\left(\frac{\Phi_1}{\Phi_2}\right) \tag{7.2}$$

where Φ_1 and Φ_2 are the light intensity at two different points.

The change in intensity levels for both sound and light from a source is given by

$$\text{Change in levels} = 10\log_{10}\left(\frac{d_1}{d_2}\right) \tag{7.3}$$

where d_1 and d_2 are the distances from the source to the points being considered.

Example 7.2 Two points are 65 and 84 ft from a light bulb. What is the difference in light intensity at the two points?

Solution

$$\text{Difference} = 10\log_{10}\left(\frac{65}{84}\right) = -1.11\,\text{dB}$$

Commercial high-resolution optical sensors are available with the electronics integrated onto a single die to give temperature compensation and a linear voltage output with incident light intensity are commercially available. Such a device is the TSL 250. Also commercially available are infrared (IR) light-to-voltage converters (TSL 260) and light-to-frequency converters (TSL 230). Note, the TSL family is manufactured by Texas Instruments.

X-rays should be mentioned at this point as they are used in the process-control industry and are also electromagnetic waves. X-rays are used primarily as inspection tools; the rays can be sensed by some light-sensing cells and can be very hazardous if proper precautions are not taken.

Monochromatic light (single frequency) can be generated with a laser and collimated into a narrow beam. The beam is reflected by a mirror which generates interference fringes with the incident light as it moves. The fringes can be counted as the mirror moves. The wavelength of the light generated by a laser is about 5×10^{-7} m, so relative positioning to this accuracy over a distance of ½ to 1 m is achievable.

Figure 7.5a shows the set using a laser for incremental distance measurement. Monochromatic light from a laser is collimated into a narrow beam. The light from the beam goes to a beam splitter that splits the beam reflecting light from the laser to the detector. The main beam goes to a mirror attached to the object whose change in position is being measured. The beam reflected by the mirror is then reflected to the detector as shown. The detector compares the phase of the incident light and the light reflected from the object. A change in distance by the object of a quarter wavelength ($\lambda/4$) gives a change of half a wavelength at the detector as shown in Fig. 7.5b so that by comparing the two beams the detector can detect $\lambda/4$ change in position by the object.

Ultrasonic, infrared, laser, and microwave devices can be used for distance and speed measurement. The time for a pulse of energy to travel to an object and be reflected back to a receiver is measured, from which the distance can be calculated; i.e., the speed of ultrasonic waves is 340 m/s, and the speed of light and microwaves is 3×10^8 m/s. Ultrasonic waves can be used to measure distances from 1 to about 50 m, whereas light and microwaves are used to measure longer distances up to 400 km.

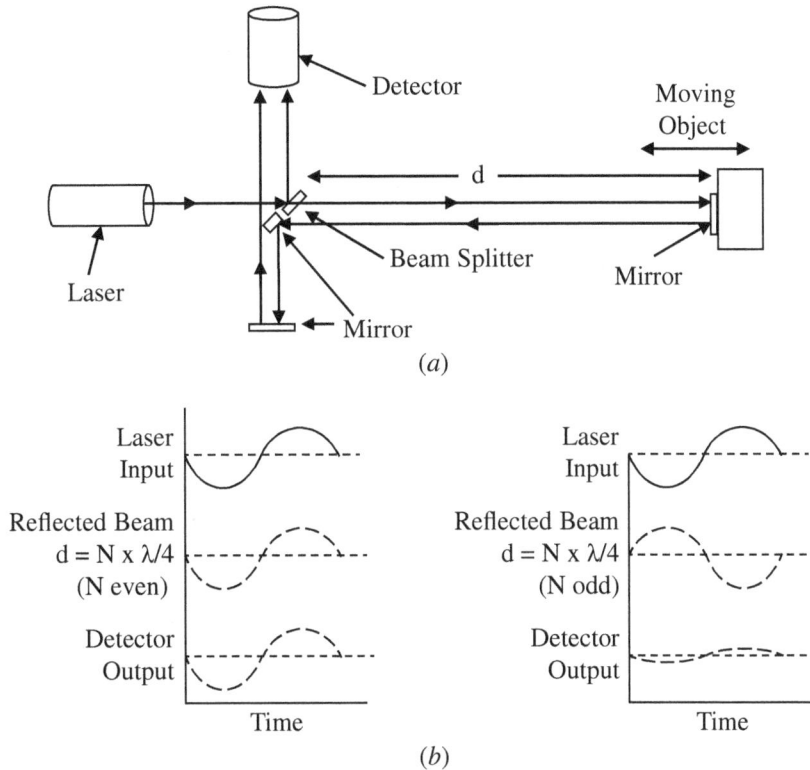

FIGURE 7.5   Laser system (a) to measure distance and (b) waveforms.

Ultrasonic waves can be detected by pressure sensing devices.

If an object is in motion, the Doppler effect can be used to determine its speed. The Doppler effect is the change in frequency of the reflected waves caused by the motion of the object. The difference in frequency between the transmitted and reflected signal can be used to calculate the velocity of the object.

Accelerometers sense speed changes by measuring the force produced by the change in velocity of a known mass (seismic mass), see Eq. (7.4). These devices can be made with a cantilevered mass and a strain gauge for force measurement or can use capacitive measurement techniques. Accelerometers are now commercially available, made using micromachining techniques. The devices can be as small as 500×500 μm, so that the effective loading by the accelerometer on a measurement is very small. The device is a small cantilevered seismic mass, which uses capacitive changes to monitor the position of the mass. Piezoelectric devices similar to the one shown in Fig. 7.6 are also used to measure acceleration. The seismic mass produces a force on the piezoelectric element during acceleration, which causes a voltage to be developed across the element. Accelerometers are used in industry for the measurement of changes in velocity of moving equipment, in the automotive industry as crash sensors for air bag deployment, and in shipping crates where battery operated recorders are used to measure shock during the shipment of expensive and fragile equipment.

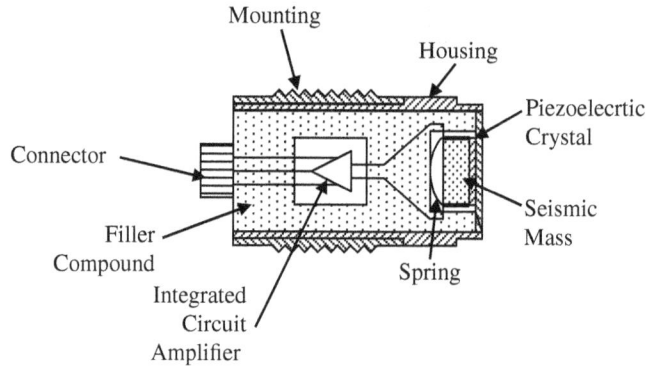

FIGURE 7.6 Piezoelectric accelerometer.

Vibration sensors typically use acceleration devices to measure vibration. Micromachined accelerometers make good vibration sensors for frequencies up to about 1 kHz. Piezoelectric devices make good vibration sensors with an excellent high-frequency response for frequencies up to 100 kHz. These devices have very low mass so that the damping effect is minimal. Vibration sensors are used for the measurement of vibration in bearings of heavy equipment and pressure lines.

7.2.4 Rotation Sensing

In Fig. 7.7, a Hall effect device is used to detect the rotation of a cog wheel. As the cogs move pass the Hall device, the strength of the magnetic field is greatly enhanced causing an increase in the Hall voltage. The device can be used to measure linear as well as rotational position or speed. An MRE device can also be used in this application.

Optical devices detect motion by sensing the presence or absence of light. Figure 7.8 shows two types of optical discs used in rotational sensing. Figure 7.8a shows an incremental optical shaft encoder. Light from the LED shines through windows in the disc on to an array of photodiodes. As the shaft turns, the position of the image moves along the array of diodes. At the end of the array, the image of the next slot is at the start of the array. The relative position of the wheel with respect to its previous location can be obtained by counting the number of photodiodes traversed and multiplying them by the number of slots monitored. The diode array enhances the accuracy of the position of the slots; i.e., the resolution of the sensor is 360° divided by the number of slots in the disc divided by the number of diodes in the array. The slots can also be replaced by

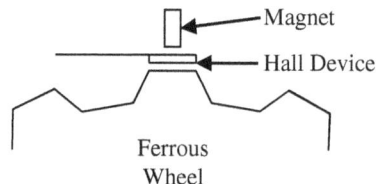

FIGURE 7.7 Application of a Hall effect device for measuring the speed and position of a cog wheel.

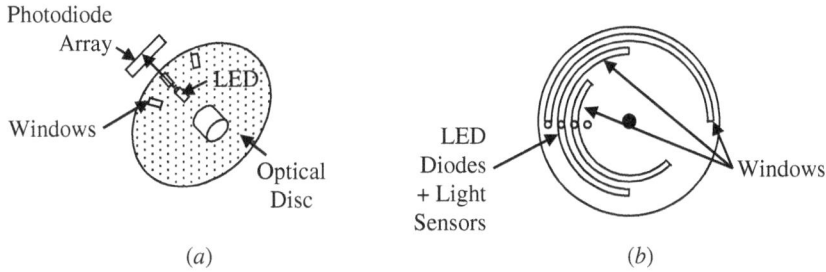

FIGURE 7.8 Representation of (a) an incremental optical disc and (b) an absolute position optical disc.

reflective strips, in which case the light from the LED is reflected back to a photodiode array. Only one slot in the disc is required to measure rate per minute. Figure 7.8b shows an absolute position encoder. An array of LEDs (one for each window) with a corresponding photodetector for each window can give the position of the wheel at any time. Only three windows are shown in Fig. 7.8b, for greater accuracy more slots would be used. The pattern shown on the disc is for the gray code that allows only one bit in a parallel data bus to change at a time. Other patterns may be used on the disc such as the binary code.

Optical devices have many uses in industry other than for the measurement of the position and speed of rotating equipment. They are used for counting objects on conveyer belts on a production line, measurement and control of the speed of a conveyer belt, location and position of objects on a conveyer, location of registration marks for alignment, measurement and thickness control, and detecting for breaks in filaments and so forth.

Power lasers can also be included with optical devices as they are used for scribing, machining, and cutting metal plates, laminates, and the like.

7.2.5   Position Application Consideration

Optical position sensors require clean operating conditions, and in dirty or environmentally unfriendly applications are being replaced by Hall or MRE devices in both rotational and linear applications. These devices are small, sealed, and rugged with very high longevity, and will operate correctly in fluids, in a dirty environment, or in contaminated areas.

Optical devices can be used for reading bar codes on containers, and imaging. Sensors in remote locations can be powered by solar cells that fall into the light sensor category.

7.3   Force, Torque, and Load Cells

7.3.1   Basic Definitions of Force and Torque

Many applications in industry require the measurement of force or load. Force is a vector and acts in a straight line, it can be through the center of a mass, or be offset from the center of the mass to produce a torque, or with two forces a couple. Force can be measured with devices such as strain gages. In other applications where a load or weight is required to be measured the sensor can be a load cell.

Mass is a measure of the quantity of material in a given volume of an object (see Chap. 2).

Force is a term that relates the mass of an object to its acceleration and acts through its center of mass, such as the force required to accelerate a mass at a given rate. Forces are defined by magnitude and direction, and are given by the following:

$$\text{Force}(F) = \text{mass}(m) \times \text{acceleration}(a) \qquad (7.4)$$

Example 7.3 What force is required to accelerate a mass of 27 kg at 18 m/s²?

Solution

$$\text{Force} = 27 \times 18 \text{ N} = 486 \text{ N}$$

Weight of an object is the force on a mass due to the pull of gravity (*g*), which gives

$$\text{Weight}(w) = \text{mass}(m) \times \text{gravity}(g) \qquad (7.5)$$

Example 7.4 What is the mass of a block of metal that weighs 29 lb?

Solution

$$\text{Mass} = 29/32.2 = 0.9 \text{ lb(slug)}$$

Torque occurs when a force acting on a body tends to cause the body to rotate and is defined by the magnitude of the force times the perpendicular distance from the line of action of the force to the center of rotation (see Fig. 7.9*a*). Units of torque are pounds feet (lb·ft), or newton meter (N·m). Torque is sometimes referred to as the moment of the force, and is given by

$$\text{Torque}(t) = F \times d \qquad (7.6)$$

where *F* is the applied force and *d* the distance between fulcrum and force.

A *couple* occurs when two parallel forces of equal amplitude, but in opposite directions are acting on an object to cause rotation, as shown in Fig. 7.9*b* and is given by the following equation:

$$\text{Couple }(c) = 2 \ F \times d/2 = F \times d \qquad (7.7)$$

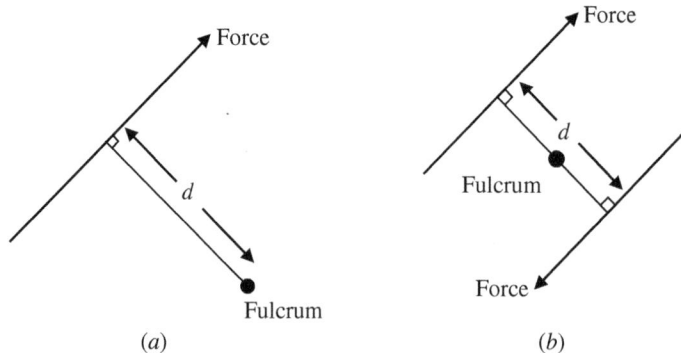

FIGURE 7.9 Types of forces: (a) torque and (b) couple.

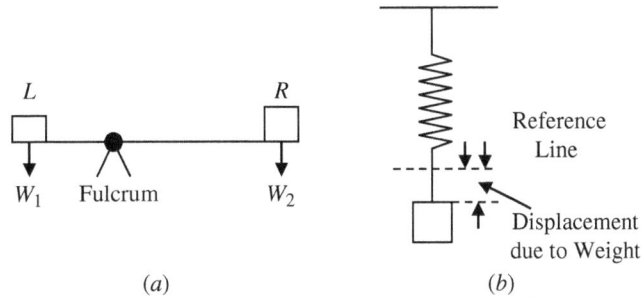

FIGURE 7.10   Force measuring devices: (a) lever balance and (b) spring balance to measure force.

7.3.2   Force and Torque Measuring Devices

Force and weight can be measured by comparison as in a lever type balance which is an ON/OFF system. A spring balance or load cell can be used to generate an electrical signal that is required in most industrial applications.

Analytical or lever balance is a device that is simple and accurate, and operates on the principle of torque comparison. Figure 7.10a shows a diagram of a balance. When in balance the torque on one side of the fulcrum is equal to the torque on the other side of the fulcrum, from which we get the following:

$$W_1 \times L = W_2 \times R \qquad (7.8)$$

where W_1 is a weight at a distance L from the fulcrum and W_2 the counter balancing weight at a distance R from the fulcrum.

Example 7.5   Two pounds of potatoes are being weighted with a balance, the counter weight on the balance is 0.5 lb. If the balance arm from the potatoes to the fulcrum is 6-in long, how far from the fulcrum must the counter balance be placed?

Solution

$$2 \text{ lb} \times 0.5 \text{ ft} = 0.5 \text{ lb} \times d \text{ ft}$$
$$d = 2 \times 0.5/0.5 = 2 \text{ ft}$$

Spring transducer is a device that measures weight by measuring the deflection of a spring when a weight is applied, as shown in Fig. 7.10b. The deflection of the spring is proportional to the weight applied (provided the spring is not stressed), according to the following equation:

$$F = Kd \qquad (7.9)$$

where F = force in pounds or newtons
   K = spring constant in pounds per inch or newtons per meter
   d = spring deflection in inches or meters

Example 7.6   When a container is placed on a spring balance with an elongation constant of 65 lb/in (11.6 kg/cm) the spring stretches 3.2 in (8.1 cm). What is the weight of the container?

Solution

$$\text{Weight} = 65 \text{ lb/in} \times 3.2 \text{ in} = 208 \text{ lb}$$
$$= 11.6 \text{ kg/cm} \times 8.1 \text{ cm} = 93.96 \text{ kg}$$

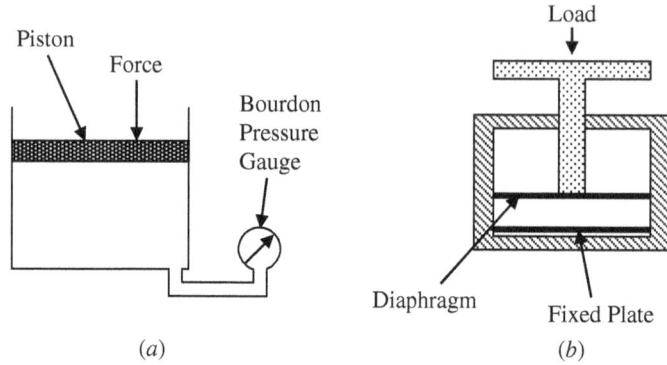

FIGURE 7.11   Force measurements using (a) pressure and (b) capacitive load cell.

Hydraulic and pneumatic load cells can be used to measure force. This can be done by monitoring the pressure in a cylinder when the force (pounds or newtons) is applied to a piston as shown in Fig. 7.11a. The relation between force (F) and pressure (p) is given by

$$F = pA \tag{7.10}$$

where A is the area of the piston.

Example 7.7   What is the force acting on a 14-in (35.6 cm) diameter piston, if the pressure gauge reads 22 psi (152 kPa)?

Solution

$$\text{Force} = \frac{22 \times 3.14 \times 14^2}{4} = 3385 \text{ lb}$$

$$= \frac{152 \times 3.14 \times 35.6 \times 35.6}{4 \times 100 \times 100} = 15.1 \text{ kN}$$

A capacitive load cell is shown in Fig. 7.11b. The capacitance is measured between a fixed plate and a diaphragm. The diaphragm moves toward the fixed plate when force or pressure is applied giving a capacitive change proportional to the force.

It should be noted that a lever type of balance is the only balance that gives consistent weight measurements at any altitude as gravity affects both the mass of the object being weighted and the mass of the reference weight by the same amount.

Strain gauges are deposited resistive or piezoelectric elements. In the resistive gauge, copper or nickel particles are deposited onto a flexible substrate in a serpentine form (see Fig. 7.12c). When the substrate is bent in a concave shape along the axis perpendicular to the direction of the deposited resistor or the substrate is compressed in the direction of the resistor, the particles themselves are forced together and the resistance decreases; if the substrate is bent in a convex shape along this axis or the substrate is under tension in the direction of the resistor, the particles tend to separate and the resistance increases. Bending along an axis parallel to the resistor does not change the elements resistance. Piezoresistive devices using material such as silicon can also be used for strain gauges, but the resistance change is due to the change in electron and hole mobility. These devices are very small and do no not affect the strain measurements. Figure 7.12a shows the use of a strain gauge

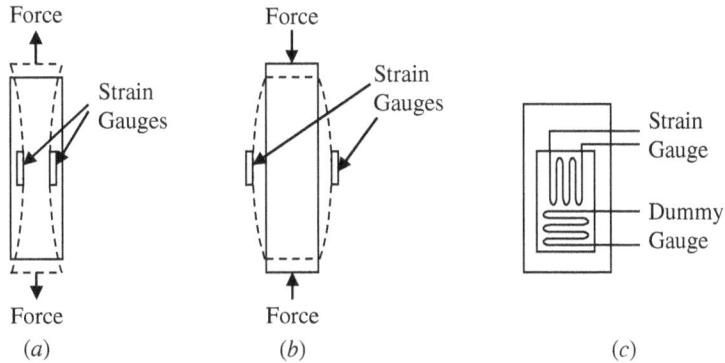

FIGURE 7.12 Examples of (a) a solid object under tensile strain and (b) a solid body under compression, and (c) a strain gauge.

to measure the strain in a solid body under stress from a tensile force. In this case, the material under tension elongates and narrows. Figure 7.12b shows an object under compressive forces, which will shorten and fatten the object. The resistance change in a strain gauge element is proportional to the compression, or tensile force. Because the resistance of the strain gauge element is temperature sensitive, a reference or dummy strain gauge element is also added at right angles to the pressure sensing strain gauge element as is shown Fig. 7.12c and will therefore not sense the deformation as seen by the pressure sensing element.

The resistance change in strain gauges is small and requires the use of a bridge circuit for measurement as shown in Fig. 7.13. The strain gauge elements are mounted in two arms of the bridge and two resistors R_1 and R_2 form the other two arms. The output signal from the bridge is amplified, and impedance matched. The strain gauge elements are in opposing arms of the bridge, so that any change in the resistance of the elements due to temperature changes will not affect the balance of the bridge giving temperature compensation.

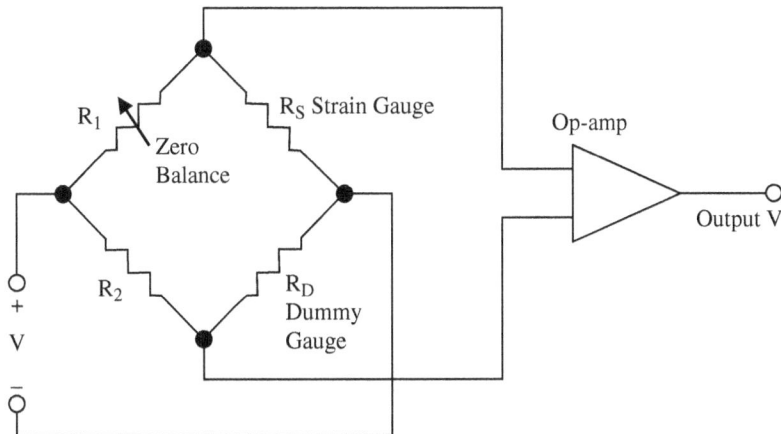

FIGURE 7.13 Resistive bridge for signal conditioning of a strain gauge.

Dynamometer is a device that uses the twist or bending in a shaft due to torque to measure force. One such device is the torque wrench used to tighten bolts to a set level, which can be required in some valve housing. The allowable torque for correct assembly will be given in the manufacture's specification. The twist in a shaft from a motor can also be used to measure the torque output from or load on a motor.

7.3.3   Force and Torque Application Considerations

In most applications, compensation should be made for temperature effects. Electrical transducers can be compensated by using them in a bridge circuit with a compensating device in the adjacent arm of the bridge. Changes in material characteristics due to temperature changes can be compensated using temperature sensors and applying a correction factor to the measurement. Vibration can also be a problem when measuring force, but this can usually be corrected by damping the movement of the measuring system.

Summary

A number of different types of sensors were discussed in this chapter. Sensors for measuring position, speed, and acceleration were introduced. The concepts of force, torque, and load measurements were discussed together with measuring devices.

The main points described in this chapter were as follows:

1. The basic terms and standard used in linear and rotational measurements and the sensors used for the measurement of absolute and incremental position, velocity, and acceleration.

2. Optical and magnetic sensors and their use as position measuring devices in linear as well as rotational applications and sensors used for distance measurement.

3. Definition of force, torque, couples, and load and the use of mechanical forces in weight measurements.

4. Stress in materials, the use of strain gauges for its measurement, and instruments used for measurement.

Problems

7.1   What force is necessary to accelerate a mass of 17 lb at 21 ft/s²?

7.2   What is the acceleration of 81 kg, when acted upon by a force of 55 N?

7.3   What torque does a force of 33 lb produce 13 ft perpendicular from a fulcrum?

7.4   What force is required to produce a torque of 11 N·m, if the force is 13 m from the fulcrum?

7.5   A couple of 53 N·m is produced by two equal force of 15 N, how far are they apart?

7.6   A couple of 38 lb·ft is produced by two equal forces 8 ft apart. What is the magnitude of the forces?

7.7   A balance has a reference weight of 10 kg 0.5 m from the fulcrum. How far from the fulcrum must a weight of 16 kg be placed to be balanced?

7.8 A reference weight of 15 lb is placed 2 ft from the fulcrum to balance a weight 4.7 ft from the fulcrum. What is the weight?

7.9 A spring balance has a spring constant of 3 lb/in. A basket with 6 lb potatoes extends the spring 2.7 in. What is the weight of the basket?

7.10 A spring balance has a spring constant of 2.3 N/m. What is the deflection of the balance when it is loaded with 0.73 N?

7.11 A force of 10 N is applied to a piston 150 cm in diameter in a closed cylinder. What is the pressure in the cylinder?

7.12 What force is applied to a 9.2-in piston in a cylinder, if the pressure in the cylinder is 23 psi?

7.13 What is the wavelength of 13-kHz sound waves?

7.14 What is the frequency of radar waves whose wavelength is 2.5 cm?

7.15 A person is 375 m from a light and a second person is 125 m from the light. What is the difference in the light levels as seen by the two people?

7.16 Where is a dynamometer used?

7.17 The light intensity 20 ft from a light bulb is 3.83 dB higher than at second point. What is the distance of the second point from the bulb?

7.18 What is the change in light intensity when the distance from the light source is increased 35 to 85 ft?

7.19 What is the angular displacement that can be sensed by an angular displacement sensor, if the circular disc on a shaft has 115 slots and the photodiode array has 16 diodes? See Fig. 7.3.

7.20 The rotational speed of a steel cog wheel is being sensed with a magnetic sensor. If the wheel has 63 teeth and is rotating at 1021 rpm, what is the frequency of the output pulses?

CHAPTER **8**

Safety and Alarm

Chapter Objectives

This chapter will help you understand the type of sensors used in safety and alarm systems and the need to protect personnel, the environment, and equipment.

Topics discussed in this chapter are as follows:

- The need for safety and alarm for protection of employees
- Protection of the environment
- The use of sensors for equipment protection
- Sensors for the detection of fire and gas leakage
- Safety precautions (OSHA [Occupational Safety and Health Administration])
- Various types of sensors used for safety and alarm
- Safety and alarm systems

8.1 Introduction

Many hazardous, corrosive, toxic, and environmentally unfriendly chemicals are used in the processing industry. These chemicals require careful monitoring during use, transportation, and handling. In a basic text, it is not possible to cover the hazards and sensors available, their characteristics, and the selection of the most suitable sensor for a specific application, but just to make the student aware of their existence. Analysis labs and control rooms are used for many product evaluations but must meet code; further information can be obtained from the Instrument Society of America (ISA) series RP 60 practices. All processing plants and labs will have an alarm system which can shutdown certain operations if a problem occurs. These systems are regularly tested and are often duplicated to provide built in fail-safe features such as redundancy as protection against equipment and sensor failure.

8.2 Safety Hazards

Safety and alarm sensors can be used for different functions. They are used to protect and warn personnel of a potential hazard or dangerous working conditions, sense potential hazards to the environment that may or may not be related to equipment failure, and in the case of process control they are used to warn of equipment failure, or a process going out of control. There is safety equipment to guard personnel from hazardous chemicals and acids, and restrictions to stop personnel from entering unsafe areas.

113

8.2.1   Personnel Hazards

There are many personnel hazardous conditions that can be caused by fire, smoke, dangerous gases, dangerous chemicals, liquid spills, high noise levels, radiation, and so on. These conditions may not always be process related. Smoke and fire sensors must always be installed. The type of chemical sensors installed will depend on the chemicals and gases being used in the process. Oxygen sensors are also required to ensure that there is an adequate oxygen supply for personnel and that it has not been replaced by cryogenic substances such as nitrogen and helium. Another serious health problem can be from mold growth in hidden or damp places, mold spores can become airborne and contaminate the air and be undetected.

The hazards of electrical shock are always present to personnel and in particular to technicians and engineers serving and maintaining equipment.

A threat in some industrial processing is from radiation. Radiation can be alpha or beta particles, gamma rays, x-rays, or neutrons and are measured in roentgens (R). The amount of radiation absorbed in an object or body is measured in radiation absorbed dose (rad) or gray (Gy). The effective dose takes into account the medical effect of radiation on the body and is measured in roentgen equivalent man (rem) for practical purposes.

$$1 \text{ R} = 1 \text{ rad} = 1 \text{ rem} \tag{8.1}$$

The effective radiation from the earth varies by location from about 0.2 mrem per day to 1 mrem per day. The government has set standards for protection against radiation for radiation workers in the Code of Federal Regulations (10 CFR part 20) to 5 rem. Alerts can be communicated by audible alarms, loud speaker warnings, visual alarms, and personal hazard indicators.

8.2.2   Environmental Hazards

Environmental hazards are air pollution, ground contamination, water pollution, radiation, and so on. The air pollution is caused by fires (chemical fires), emissions, and ordinance, ground contamination by spills, domestic and industrial waste, and water pollution by fall out from air pollution, run off from ground contamination, fertilizer, and so on. These hazards can have very long-term effects and are contaminating our foodstuffs (fruits, vegetables, meat, and fish).

8.2.3   Control Equipment Hazards

Component failure in the PLC (programmable logic controller) or hardware either of which can cause the process to go out of control or in the SIS (safety instrumented system) which can give false alarms and shut the system down. Failure in the PLC or hardware can result in toxic chemical and gas leakage, liquid spills, fires, destruction of processing equipment, and loss of product. The type of hazard varies from industry to industry and in some cases also results in unacceptable radiation levels.

8.2.4   Process Equipment Hazards

Process equipment can fail due to corrosion, fatigue due to constant temperature or pressure changes, vibration, or wear and tear in moving parts and so on. Corrosive substances used in many processes will attack many materials that can result in leakages from storage containers, plumbing, and valves, causing ground contamination, air pollution, and a danger to personnel. Corrosion can also destroy sensors, and instruments resulting in loss of production. Plumbing, gaskets, instruments, and sensing devices must all be inert to the substances being used in the process.

Temperature and pressure changes and vibration can create stress in containers and plumbing causing fatigue and eventual rupture or leakages from joints and valves. Vibration can cause chaffing of electrical wire insulation, breakage in electrical conductors, and be a destroyer of bearings. Good regular maintenance and servicing of moving parts is essential for good longevity.

8.3 Safety Sensors

8.3.1 Smoke and Fire Sensors

The smoke detector is normally used as the indicator of a fire. Smoke particulates can be detected by an ionization chamber or photoelectric device. The ionization chamber typically uses radioisotope such as americium-241 as the radiation source to ionize the gas in a detection chamber; such a device can detect 1 to 2 percent obs/ft (obscuration per foot air flow) in an open chamber when compared to a reference chamber. The photoelectric device uses an LED with a photodetector positioned at right angles to the LED; particulates from the smoke reflect the light from the LED to the detector that can sense 2 to 4 percent obs/ft. A second type of photodetector is the aspirating smoke detector used mainly in commercial buildings which uses a calibrated laser beam and can detect particulates as low as 0.0015 percent obs/ft.

8.3.2 Heat Sensors

There are two basic types of heat sensor, the fixed temperature detector and the rate of temperature rise detector. The fixed detector contains a solid material that melts at 58°C but with the use of newer materials the melting point has now been reduced to 47°C. The rate of temperature rise sensor has a thermocouple that senses convection/radiation temperature and a thermistor that senses ambient temperature. By comparing the two sensors a rapid rate of rise from 6.7 to 8.3°C/min can be detected.

8.3.3 Gas Sensors

Gas sensors can be considered as chemical sensors and can be classified into a number of types as follows:

Semiconductor sensors such as the Taguchi-type sensor are used for the detection of the following gases: oxygen, hydrogen, carbon monoxide, sulphur dioxide, oxides of nitrogen, hydrocarbons, hydrogen sulphide, and ammonia. The Taguchi sensor has an element coated with a metal oxide that combines with the gas to give a change in electrical resistance that can be detected using a bridge technique.

Catalytic combustion devices are mainly used for combustible gases. Platinum coils are set up in a bridge configuration, opposing coils are treated with a catalyst and when heated the combustible gas is oxidized by the catalyst producing a higher temperature and an imbalance in the bridge. This technique is used for the following gases: hydrogen, carbon monoxide, and hydrocarbons.

Photoionization detectors (PIDs) use an ultraviolet (UV) light source to ionize chemicals. It has high sensitivity to volatile organic compounds, ammonia, hydrogen sulphide, phosphine, and so on. Miniature devices are used as clip on devices for hygiene, hazmat, and environmental monitoring.

Other chemical sensors used for gas detection are as follows:

Infrared devices measure the absorption of various wavelengths by different gases. The device can be used for remote sensing and to detect hydrocarbons, carbon dioxide, and water vapor.

Electrochemical sensors use a liquid electrolyte and amperometric cells. They can be used for the detection of oxygen, hydrogen, carbon monoxide, sulphur dioxide, oxides of nitrogen, hydrogen sulphide, ammonia, glucose, and hydrazine.

Ion-selective electrode sensor uses ion-selective membrane to detect acidity, potassium, sodium, chlorine, calcium, magnesium, fluoride, and gold.

Solid electrolyte sensors are similar to the electrochemical sensors but use a solid electrolyte and are often used at high temperatures for the detection of oxygen, hydrogen, carbon monoxide, and hydrocarbons.

Piezoelectric sensors are based on detecting the resonant frequency of the piezoelectric solid and are used in the detection of hydrocarbons and volatile organic compounds.

Optical sensors are primarily used for acids, bases, and hydrocarbons.

Pyroelectric sensors are used for vapor detection.

It should be noted that some of the above sensors are miniature and developed for field use and others can only be used in an analysis laboratory. The area of chemical sensors is an ever changing field as new requirements for sensors and the development of new sensors occurs on a daily basis.

There are many publications available on chemical sensors; two such sources are the Sandia Report SAND2001-0643 Review of Chemical Sensors for In-Situ Monitoring of Volatile Contaminants and the Journal of The Electrochemical Society, 150 (2) S11-S16 (2001) Sensors, Chemical Sensors, Electrochemical Sensors, and ECS.

8.3.4   Artificial Senses Chemical Sensors

Olfaction and electronic nose sensor arrays have been developed for use in several processing industries such as the diagnosis of meat, fruit, and vegetable quality, wine processing and quality, development of new perfumes, and in coffee processing. The use of this type of array has been extended to smart fire detectors, diagnosis of diseases, homeland security, and so on.

8.3.5   Radiation Detectors

A number of devices are commercially available for radiation detection. They are as follows:

Gaseous ionization detector

Geiger-Müller tube

Ionization chamber

Proportional counter

Scintillation counter

Semiconductor detector

8.4 Process Equipment Safety

The International Electrochemical Commission (IEC) has published a number of international standards on safety. Some of these are as follows:

IEC 61511 published in 2003 gives safety standards for the process industry

IEC 62061 machinery standards

IEC 62425 railway signaling systems

IEC 61513 nuclear systems

IEC and ISO 26262 road vehicles

8.4.1 Alarm and Trip Systems

Alarm and trip system regulations, information, and implementation are given in ANSI/ISA-84.01-1996—Application of Safety Instrumented Systems for the Process Control Industry and apply to systems using programmable electronic controllers. The purpose of an alarm system is to bring a malfunction to the attention of operators, maintenance personnel, and to warn employees of possible dangerous working conditions and potential environmental contamination, whereas, a trip system is used to shutdown a system in an orderly fashion when a malfunction occurs, or to switch failed units over to standby units. The elements used in the process-control system are the first line of warning of a failure. In many cases, a trip system remains dormant and on standby until required (or being tested). It may be advantageous to switch between process and standby systems on a regular basis to ensure both are operational, as alarm and trip systems only protect when they work.

The reliability of alarm and trip systems is enhanced by the following:

- The fundamental design
- Quality of components used in the design
- The regular maintenance of the system
- The capabilities of properly trained plant personnel
- The frequency at which they are tested or rotated

8.4.2 Safety Instrumented Systems

Processes are generally provided with two trip systems. The first trip system is an integral part of the process-control system which is used in production, quality control, and so on. The second is a SIS for handling critical trips and is the second line of defense. The sensors and instruments used in the SIS must be totally separate from those used in the process-control system. The SIS has its own sensors, logic, wiring, displays, and control elements so that under failure conditions it will take the process to a safe state to protect the personnel, facility, and environment. To ensure full functionality of the SIS, it must be regularly tested. In the extreme with deadly chemicals, a second SIS can be used in conjunction with the alarm features of the process control-system to try and reach 100 percent protection. The sensors in the SIS will usually be a different type than those used for process control. The control devices are used to accurately sense varying levels in the measured variable, whereas the SIS sensors are used to sense a trip point and need to be much more rugged devices. The use of redundancy in a system must not be used as a justification for low-reliability cheap components. A common SIS is the

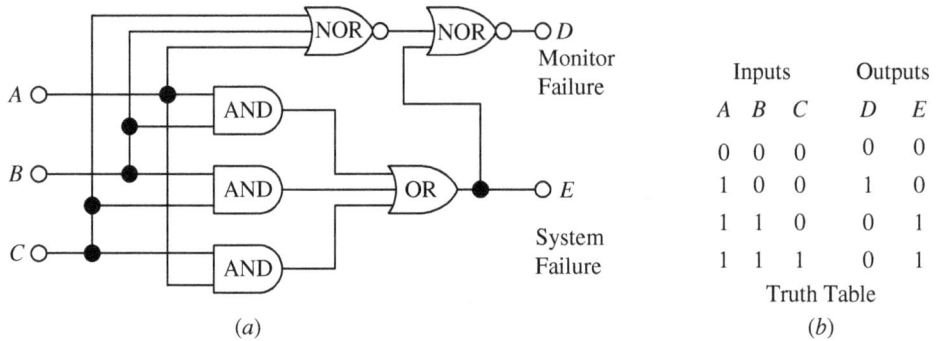

FIGURE 8.1   Illustration of (a) two out of three failure indicator and (b) truth table.

dual redundancy system, which consists of the main process alarm system with two redundant SIS. In this case, a two out of three logic monitoring system determines if a monitor has failed or the system has failed by correlation between the outputs. A two out of three logic circuit is shown in Fig. 8.1a. The truth table is shown in Fig. 8.1b; the inputs are normally low (0). If one input goes high (1) it would indicate a monitor failure but the system output would remain at 0. If 2 or more inputs go high it would indicate a system failure and the system failure output goes from 0 to 1 as shown.

In SIS failure analysis, the rate of component failure is as follows:

Logic   8%
Sensors   42%
Control devices   50%

8.4.3   Power Loss Fail Safe

All systems will fail at some point. If the system is still energized it can be programmed to go into a fail-safe condition, but if the failure is due to a power or hydraulic pressure loss it must still go to a fail-safe state. For a SIS, the fail-safe state on power or pressure loss must be a design criterion; such a design is possible with the use of actuators and valves that will go to a known state with the loss of power (see Chap. 10).

The design considerations for a fail-safe SIS are as follows:

1. Pneumatic valves that will go to a known fail-safe position with the loss of air pressure.

2. Solenoid valves that will deenergize on trip to open all power circuits to prevent power up when electrical energy is restored.

3. When power is restored the system cannot be started without manual supervision.

4. With power loss the system must be able to initiate an alarm system.

5. Apply breaking if necessary on moving equipment.

6. To minimize spills or loss of chemicals, the system should shutdown as quickly as possible to ensure the fast operation of pneumatic valves. It is necessary to vent the air supply as shown in Fig. 8.2.

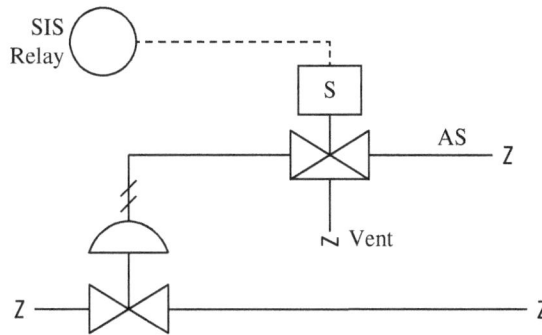

FIGURE 8.2 SIS pneumatic vent.

Another consideration is the failure of the SIS. In processes where very hazardous chemicals are used it may be necessary to have a duplicate SIS for added protection such that either system can detect a failure that will generate a fail-safe condition.

8.4.4 Safety Instrumented System Example

Figure 8.3 shows a simplistic conveyer belt system for automatically filling cartons controlled by a logic controller. Figure 8.4 shows the same system with the addition of SIS sensors. In Fig. 8.3, the PLC senses high and low liquid levels in the operating storage container and turns on the feed when low levels are sensed and turns off the feed when

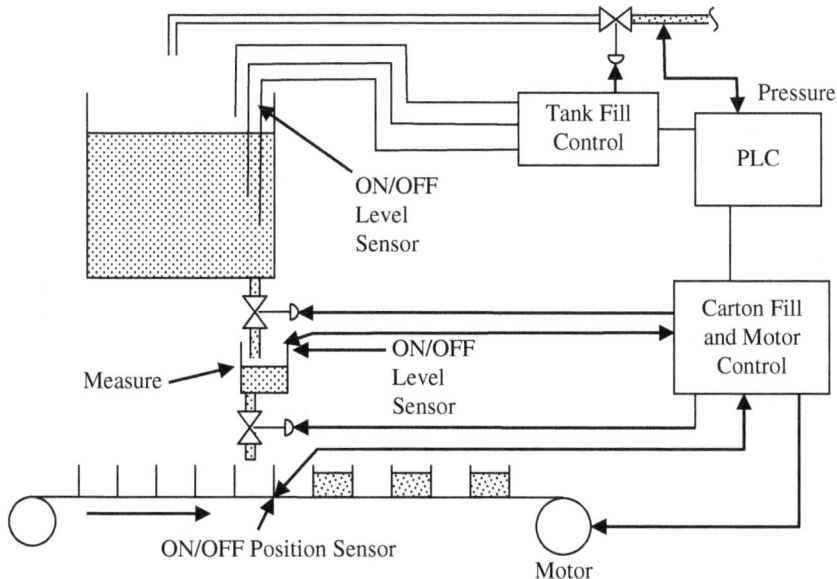

FIGURE 8.3 Automatic carton filler system.

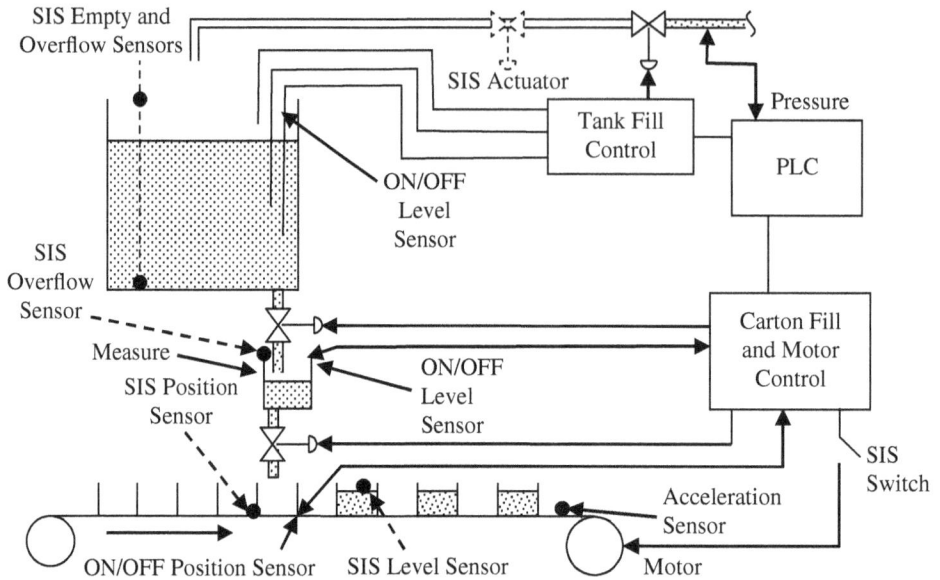

FIGURE **8.4**   Automatic carton filler with SIS.

the set high level is reached. A measuring container is filled to a set level and when the conveyer moves a jar into the filling position, which is sensed by a proximity detector, the measured volume of liquid is emptied into the jar and the measuring container is then refilled while the conveyer moves the next jar into the filling position. The supply line pressure is also monitored to ensure the liquid for the operating storage container is available.

In Fig. 8.4, SIS sensors are used to detect an empty and overfull operating storage container, overflow of the measuring container, correct liquid level in the jars, correct positioning of the jars during filling, and an acceleration sensor to monitor vibration in the conveyer belt or excessive acceleration when moving an empty jar into position for filling.

The sensors used for level measurement in the primary system are shown as conductive probes. The SIS sensors could be float-type sensors in the containers and an ultrasonic sensor for the jars is used to ensure the system is putting the correct volume of liquid in the jars. Bubble sensors could be used for the containers but they require an air supply which would introduce another possible failure element. The acceleration sensor can detect vibration in the conveyer belt, rapid acceleration of the belt, or failure of the conveyer to move. The correct positioning of the jars for filling also needs an SIS sensor.

If the SIS senses a problem it can turn "OFF" the incoming liquid supply using the SIS actuator and valve, and turn "OFF" the power to the conveyer motor.

8.5 Safety and Protection

The safety and alarm, and SIS are used to minimize or prevent system failure from causing hazardous conditions for personnel, the environment, equipment damage and loss of production. However, failure and accidents will occur and hazchem spills and toxic gas leaks are inevitable.

8.5.1 Personnel Protection

Safety procedures and precautions are defined by OSHA (Occupational Safety and Health Administration). Procedures and protective clothing are defined for handling hazardous chemicals and acids, barriers to restrict personnel from rotating and reciprocating machines, furnaces, unsafe areas, and so on, and kill switches on electric motors to prevent someone remotely starting machinery that is being serviced.

Smoke detectors and heat sensors (automatic sprinklers) are now common place in industry for the protection of people. In the case of an electrical fire an inert gas such as nitrogen or foam should be used. Other safety protection features are as follows:

Braking devices on mechanical equipment for emergency stopping

Radiation detectors

Protective clothing

Machine barriers

Fuses and ground fault interrupters

Safety signs

Venting of combustible gases from work areas

Off limit areas to unauthorized personnel

8.5.2 Environmental Protection

For environmental protection scrubbers should be used on all emission stacks. Neutralizing chemicals must be available and used in the case of all hazmat and hazchem spills as well as having a safe method of disposal. Means must be available to contain and neutralize toxic gas escapes, safe disposal of radiation waste, and so on. The types of hazard will vary from industry to industry depending on the processes. For more info, see EPA rules and regulations.

8.5.3 Equipment Protection

Alarm and trip systems are used to shutdown failed equipment in an orderly sequence or switch over to standby units, and give alarms and warnings. SIS is a backup trip system to give added safety features in the case of failure in the operating system and to give additional protection to personnel, environmental, processing equipment, and so on.

Other equipment safety features are as follows:

Easy access to instruments and equipment for ease of servicing and repair

Use of fail-safe valves to automatically go to a safe position if there is loss of power

Ability to isolate valves for repairs

Pressure release valves

Regular maintenance

Good maintenance procedures and knowledgeable mechanics and service engineers

Summary

Introduced in this chapter are the hazards that may be encountered in today's manufacturing facilities from fire, hazmat, and hazchem in use, due to equipment failure and accidents.

The main points described in this chapter were as follows:

1. Personnel safety hazards from fire, dangerous gases, and in some cases radiation.

2. Environmental hazards are due to air pollution from emissions and ground contamination from spills.

3. Equipment hazards that can be due to component failure or corrosion.

4. Sensors for the detection of fire, heat, gas, oxygen, chemicals, and radiation.

5. The functions of electronic controllers for alarm and trip systems to protect personnel, the environment, and the processing equipment as well as equipment safety features and protection for equipment, operators, and mechanics.

6. Safety and protection of personnel, environment, and equipment.

Problems

8.1   Which system is used as the first trip system?

8.2   What is the difference between alarm and trip?

8.3   What are the types of fire sensors?

8.4   What is the sensitivity of smoke detectors?

8.5   Why oxygen sensors are needed?

8.6   Name four types of radiation detectors.

8.7   What is the effective natural radiation from the earth?

8.8   What is the total maximum radiation allowed for nuclear workers?

8.9   Where is a SIS used?

8.10   In hazardous processes how many SIS will probably be used?

8.11   What are the estimated logic failure rates in a SIS?

8.12   Where is a breaking device used?

8.13   What is the use of an accelerometer in a conveyer system?

8.14   Name four types of chemical sensors.

8.15   Name four environmental pollutants.

8.16   Name the types of pollutants that reduce indoor air quality.

8.17   Why is a vent valve used in a SIS?

8.18   What type of actuators should be used in SIS?

8.19   Name the types of heat sensors.

8.20   What types of warning should be used in safety alarms?

CHAPTER 9

Electrical Instruments and Conditioning

Chapter Objectives

This chapter will help you understand nonlinearity and hysteresis in instruments, basic current and voltage amplifiers, why signal conditioning is required in process control, and to familiarize you with signal conditioning methods.

Topics discussed in this chapter are as follows:

- Instrument nonlinearities and hysteresis
- The conversion of sensor signals into pneumatic or electrical signals
- Signal linearization, methods of setting signal zero level, and span
- Nonlinear analog amplifiers
- Current and voltage amplifiers
- The instrument amplifier
- Signal conditioning using bridges
- Considerations using capacitive devices
- Resistance temperature detector (RTD) signal conditioning

9.1 Introduction

Many sensors do not have a linear relationship between physical variable and output signal amplitude. The output signals need to be corrected for the nonlinearity and hysteresis in their characteristic before converting into an electrical signal for transmission and processing. In addition to understanding the operation of measuring and sensing devices, it is necessary to understand methods of linearizing and signal amplification. Measurable quantities are analog in nature; thus sensor signals are usually analog signals but can be sometimes converted directly into digital signals. Analog signals need to be conditioned for transmission to reduce noise and ground loops that will produce inaccurate signals at the receiver.

9.2   Instrument Parameters

9.2.1   Basic Terms

The *accuracy* of an instrument or device is the difference between the indicated value and the actual value. Accuracy is determined by comparing an indicated reading to that of a known standard. Standards can be calibrated devices or obtained from the National Institute of Standards and Technology (NIST). This is the government organization that is responsible for setting and maintaining standards, and developing new standards as new technology requires it. Accuracy depends on linearity, hysteresis, offset, drift, and sensitivity. The resulting discrepancy is stated as a ± deviation from true, and is normally specified as a percentage of full-scale reading or deflection (%FSD). Accuracy can also be expressed as the percentage of span, percentage of reading, or an absolute value.

Example 9.1   A pressure gauge has a range from 0 to 50 psi, the worst case spread in readings is ± 4.35 psi. What is the %FSD accuracy?

Solution

$$\%FSD = \pm(4.35 \text{ psi}/50 \text{ psi}) \times 100 = \pm 8.7\%$$

The *range* of an instrument is the lowest and highest readings it can measure; i.e., a thermometer whose scale goes from −40 to 100°C has a range from −40 to 100°C.

The *span* of an instrument is its range from the minimum to maximum scale value; i.e., a thermometer whose scale goes from −40 to 100°C has a span of 140°C. When the accuracy is expressed as percentage; of span, it is the deviation from true expressed as a percentage of the span.

Reading accuracy is the deviation from true at the point the reading is being taken and is expressed as a percentage, i.e., if a deviation of ±4.35 psi in Example 9.1 was measured at 28.5 psi, the reading accuracy would be $(4.35/28.5) \times 100 = \pm15.26$ percent of reading.

Example 9.2   In the data sheet of a scale capable of weighing up to 200 lb, the accuracy is given as ±2.5 percent of reading. What is the deviation at the 50 and 100 lb readings, and what is the %FSD accuracy?

Solution

$$\text{Deviation at 50 lb} = \pm(50 \times 2.5/100)\text{lb} = \pm1.25 \text{ lb}$$

$$\text{Deviation at 100 lb} = \pm(100 \times 2.5/100) \text{ lb} = \pm2.5 \text{ lb}$$

Maximum deviation occurs at FSD, i.e., ±5 lb or ±2.5 percent of FSD

The *absolute accuracy* of an instrument is the deviation from true as a number not as a percentage; i.e., if a voltmeter has an absolute accuracy of ±3 V in the 100-V range, the deviation is ±3 V at all scale readings, e.g., 10 ± 3 V, 70 ± 3 V, and so on.

Precision refers to the limits within which a signal can be read and may be somewhat subjective. In the analog instrument shown in Fig. 9.1a, the scale is graduated in divisions of 0.2 psi, the position of the needle could be estimated to within 0.02 psi, and hence, the precision of the instrument is 0.02 psi. With a digital scale the last digit may change in steps of 0.01 psi so that the precision is 0.01 psi.

Reproducibility is the ability of an instrument to repeatedly read the same signal over time, and give the same output under the same conditions. An instrument may not be accurate but could have good reproducibility; i.e., an instrument could read 20 psi as having a range from 17.5 to 17.6 psi over 20 readings.

Sensitivity is a measure of the change in the output of an instrument for a change in the measured variable, and is known as the transfer function; i.e., when the output of a pressure transducer changes

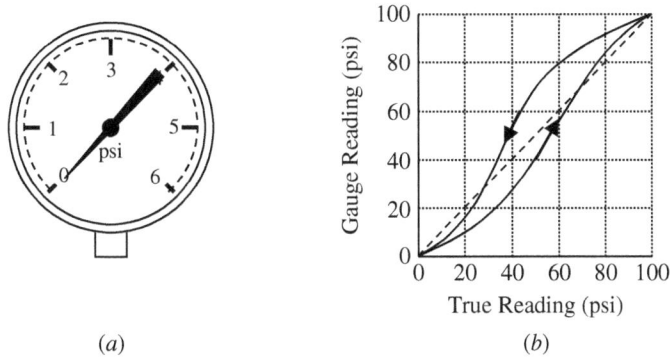

FIGURE 9.1 Gauges: (a) pressure gauge showing graduations and (b) hysteresis curve for an instrument.

by 3.2 mV for a change in pressure of 1 psi, the sensitivity is 3.2 mV/psi. High sensitivity in an instrument is preferred as this gives higher output amplitudes, but this may have to be weighed against linearity, range, and accuracy.

Offset is the reading of the instrument with zero input.

Drift is the change in the reading of an instrument of a fixed variable with time.

Hysteresis is the difference in readings obtained when an instrument approaches a signal from opposite directions; i.e., if an instrument reads a mid-scale value going from zero it can give a different reading than if it read the value after making a full-scale reading. This is due to stresses induced into the material of the instrument by changing its shape in going from zero to full-scale deflection. Hysteresis is illustrated in Fig. 9.1b.

Example 9.3 A pressure gauge is being calibrated. The pressure is taken from 0 to 100 psi and back to 0 psi. The following readings were obtained on the gauge.

Solution

True pressure (psi)	0	20	40	60	80	100	80	60	40	20	0
Gauge reading (psi)	1.2	19.5	37.0	57.3	81.0	104.2	83.0	63.2	43.1	22.5	1.5

Figure 9.2a shows the difference in the readings when they are taken from zero going up to FSD and when they are taken from FSD going back down to zero. There is a difference between the readings of 6 psi or a difference of 6 percent of FSD, i.e., ±3 percent from linear.

Resolution is the smallest amount of a variable that an instrument can resolve, i.e., the smallest change in a variable to which the instrument will respond.

Repeatability is a measure of the closeness of agreement between a number of readings (10 to 12) taken consecutively of a variable, before the variable has time to change. The average reading is calculated and the spread in the value of the readings taken.

Linearity is a measure of the proportionality between the actual value of a variable being measured and the output of the instrument over its operating range. Figure 9.2b shows the pressure input versus voltage output curve for a pressure to voltage transducer with the best-fit linear straight line, as can be seen the actual curve is not a straight line. The maximum deviation of +5 psi from linear occurs at an output of 8 V and −5 psi at 3 V giving a deviation of ±5 psi or an error of ±5 percent of FSD.

The deviation from true for an instrument may be caused by one of the above or a combination of several of the above factors, and can determine the choice of instrument for a particular application.

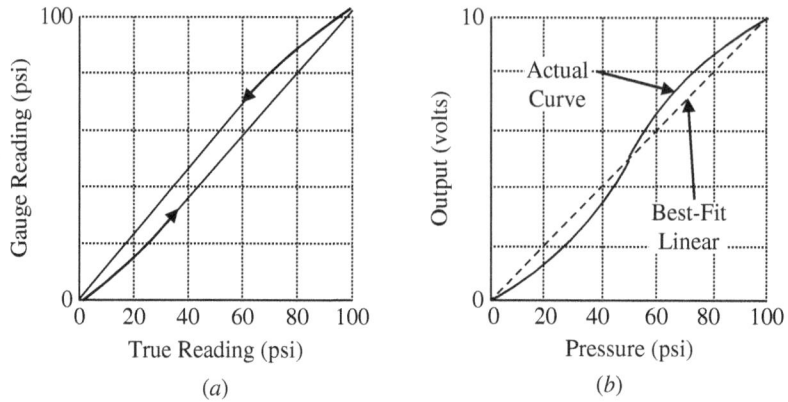

FIGURE 9.2   Instrument inaccuracies: (a) hysteresis error of a pressure gauge and (b) nonlinearity in a pressure to voltage transducer.

9.3   Transducers

9.3.1   Definitions

Sensors are devices that sense a variable, and give an output (mechanical, electrical, etc.) that is related to the amplitude of the variable.

Transducers are systems used to change the output from a sensor into some other energy form so that it can be amplified and transmitted with minimal loss of information.

Converters are used to convert a signal format without changing the type of energy, i.e., an op-amp that converts a voltage signal into a current signal.

9.3.2   Visual Display Considerations

The method of signal conditioning can vary depending on the destination of the signal. For instance, a local signal for a visual display will not require the accuracy of a signal used for process control. Visual displays are not normally temperature compensated or linearized. They often use mechanical linkages which are subject to wear over time giving a final accuracy of between 5 percent and 10 percent of the reading. However, with very nonlinear sensors the scale on the indicator will be nonlinear (see Fig. 9.3) to give a more accurate indication of the reading. These displays are primarily used to give an indication that the system is working within reasonable limits, or is within broadly set limits, i.e., tire pressures, air conditioning systems, etc.

Many processes require variables to be measured to an accuracy of 1 percent or better over their operating range, which can mean accurate sensing, temperature compensation, linearization, zero set, hysteresis, and span adjustment. A central processor that requires electrical signals is normally used for process control. Many sensor outputs are mechanical so that the sensor requires a transducer to convert the mechanical movement into an electrical signal (in some cases pneumatic) for transmission to the central controller. The signal is normally amplified and conditioned by op-amps. Care must also be taken with grounding of the system and screening of the signal cables to minimize noise. Careful selection is needed in the choice of sensors, components, and the use

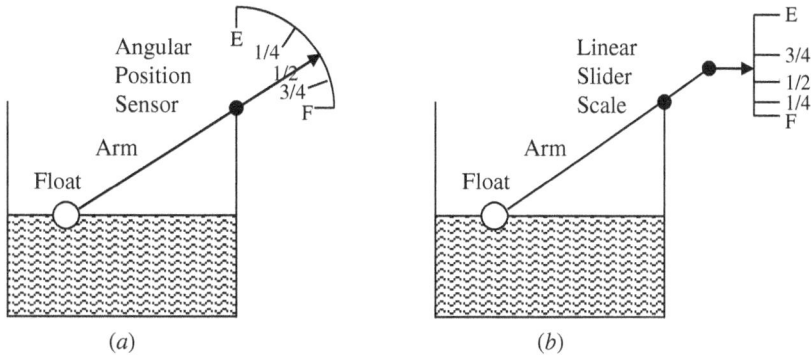

FIGURE 9.3   Scales for a float-type angular arm type of sensor are (*a*) radial and (*b*) straight line.

of impedance matching devices are required to prevent the introduction of errors in conditioning networks.

9.3.3   Mechanical Motion to Electrical Transducers

Pressure sensors can be used to measure pressure, level, flow rates, temperature, force, and density. The bourdon tube, diaphragm, capsule, and bellows pressure sensors, float-type level sensors, and humidity sensors convert the measured variable into mechanical motion.

Figure 9.4*a* shows a pneumatic signal transducer. Air from a 20 psi regulated supply is fed through a constriction to a nozzle and flapper that controls the pressure output. The flapper is mechanically linked to a bellows. When the variable is at its minimum the linkage opens the flapper allowing air to be released, the output pressure to the actuator would then be at its minimum, i.e., 3 psi. As the variable increases the linkage to the flapper causes it to close and the output pressure increases to 15 psi. This gives a linear output pressure range from 3 to 15 psi (20 to 100 kPa) with linear sensor motion which can be used for actuator control. The set zero adjusts the flapper's position and the nozzle can be moved up and down to give a gain or span control. In some cases the mechanical linkage is reversed, so that when the variable is a maximum, the output

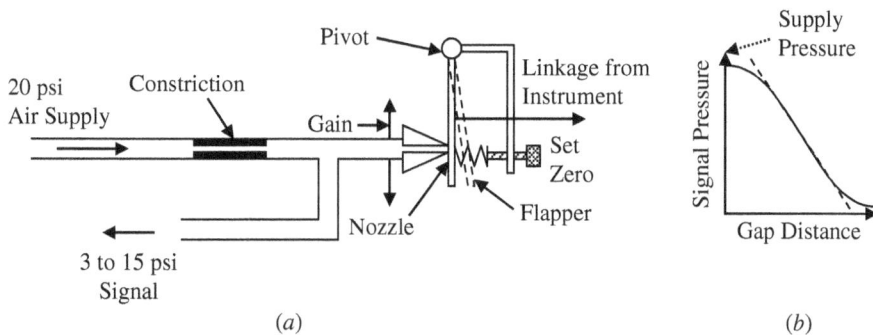

FIGURE 9.4   Illustration of (*a*) mechanical to pneumatic signal transducer and (*b*) the output pressure versus gap distance.

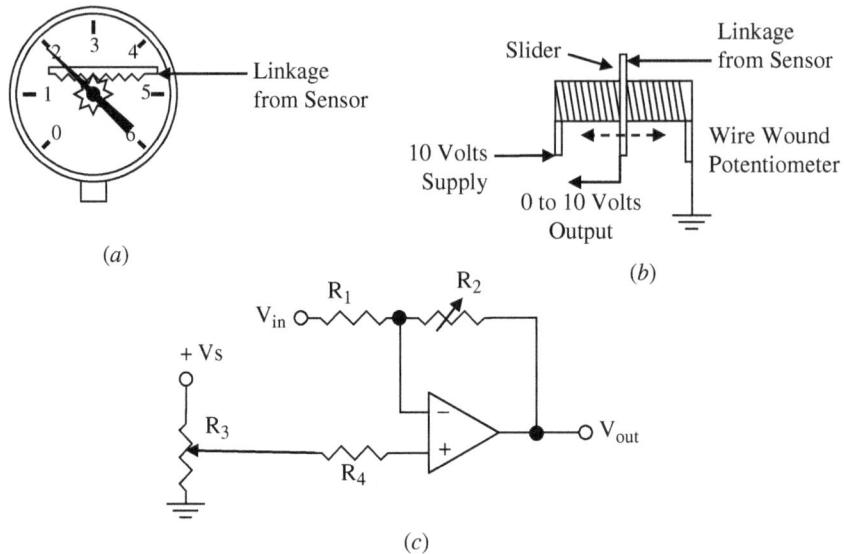

FIGURE 9.5 Mechanical linkage for (a) direct reading indicator, (b) wire wound potentiometer, and (c) a simple circuit for use with the potentiometer.

pressure is 3 psi and 15 psi for the minimum. Figure 9.4b shows the relation between the gap distance and the output pressure. The relationship is linear from 3 to 15 psi. Using 3 psi as a minimum gives an additional advantage in that zero psi indicates a fault condition. Newer systems use electrical signals in preference to pneumatic signals because of higher speeds and microprocessor controllers, and no pressure line required. However, pneumatic actuators are still the standard.

Figure 9.5a shows a mechanical linkage from a sensor to a direct reading indicator as is normally used for pressure sensing. Figure 9.5b shows a mechanical linkage from a sensor to the wiper of a potentiometer. In this case, the variable is converted into an electrical voltage, giving a voltage output from 0 to10 V. The output voltage can be fed to a voltmeter, converted to a current with an amplifier, or digitized to operate a remote sensing indicator, an actuator, or a signal to a controller. Figure 9.5c shows a circuit that could be used for conditioning with set zero and gain control potentiometers.

The set zero can be adjusted by R_3 to give zero output with minimum input, and the span adjusted by R_2 to give the required gain. The supply voltage to the amplifier and + Vs to R_3 will need to be regulated voltages. However, impedance matching devices should be used in instrumentation.

Figure 9.6a gives an alternative method of converting the linear motion output from a bellows into an electrical signal using an LVDT (linear variable differential transformer). The bellows converts the differential pressure between P_1 and P_2 into linear motion, which changes the position of the core in the LVDT. Figure 9.6b shows a circuit that can be used to condition the electrical signal output from an LVDT. As the output from the transformer is ac, diodes are used to rectify the signal. The signal is then smoothed using an RC filter, and the two dc levels are fed to an op-amp for comparison. The set zero and span adjustments are not shown.

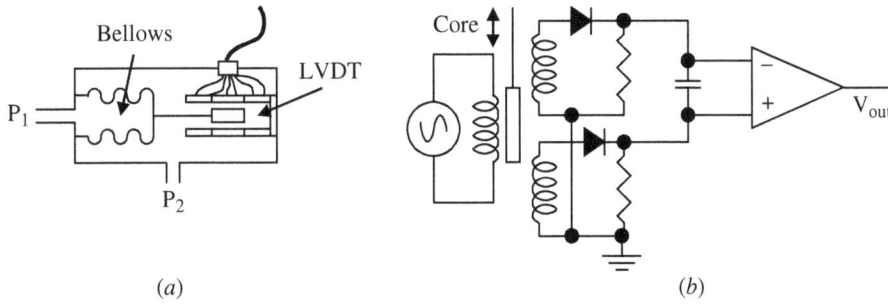

FIGURE 9.6   It shows (a) differential pressure bellows converting pressure into an electrical signal using an LVDT and (b) a signal conditioning circuit for the LVDT.

9.4   Operational Amplifiers

The operational amplifier is the element that amplifies, converts, and conditions the signal from a sensor's transducer into a suitable format for transmission to the controller. The op-amp can be used as a voltage, or current amplifier, a V to I, or I to V converter, a buffer amplifier, or for impedance matching.

9.4.1   Voltage Amplifiers

In Fig. 9.7a, the op-amp is configured as an inverting voltage amplifier. Resistors R_1 and R_2 provide feedback, i.e., some of the output signal is fed back to the input. The large amplification factor in op-amps tends to make some of them unstable and causes dc drift of the operating point with temperature. Feedback stabilizes the amplifier, minimizes dc drift, and sets the gain to a known value.

When a voltage input signal is fed into the negative terminal of the op-amp as in Fig. 9.7a, the output signal will be inverted. In this configuration for a high-gain amplifier, the voltage gain of the stage approximates to

$$\text{Gain} = \frac{-E_{out}}{E_{in}} = \frac{-R_2}{R_1} \tag{9.1}$$

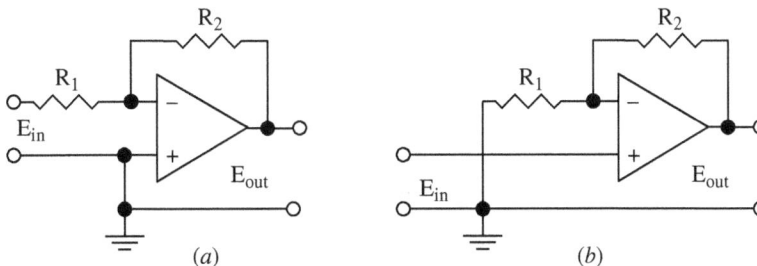

FIGURE 9.7   Circuit diagrams of (a) inverting amplifier and (b) noninverting amplifier.

The voltage gain of the amplifier can be adjusted with different values of R_2 or can be varied by adding a potentiometer in series with R_2. When the input signal is fed into the positive terminal the circuit is noninverting, such a configuration is shown in Fig. 9.7b. The voltage gain in this case approximates to

$$\text{Gain} = \frac{E_{out}}{E_{in}} = 1 + \frac{R_2}{R_1} \tag{9.2}$$

In this configuration the amplifier gain is 1+ the resistor ratio, so that the gain does not directly vary with the resistor ratio. This configuration, does however, gives a high-input impedance (that of the op-amp) and a low-output impedance.

Example 9.4 If in Fig. 9.7a, resistor $R_1 = 1200\ \Omega$ and resistor $R_2 = 150\ k\Omega$, what is the gain and what is the output voltage amplitude if the ac input voltage is 3.5 mV?

Solution

$$\text{Gain} = \frac{R_2}{R_1} = \frac{150}{1.2} = 125$$

$$\text{ac output voltage} = -3.5 \times 125\ mV = -437.5\ mV = -0.44\ V$$

Example 9.5 In Fig. 9.7a and b, $R_1 = 4.7\ k\Omega$ and $R_2 = 120\ k\Omega$. If a dc voltage of 0.15 V is applied to the inputs of each amplifier, what will be the output voltages?

Solution
   In Fig. 9.7a

$$V_{out} = \frac{-120 \times 0.15\ V}{4.7} = -3.83\ V$$

In Fig. 9.7b

$$V_{out} = \left(1 + \frac{120}{4.7}\right) 0.15\ V = +3.98\ V$$

9.4.2   Current Amplifiers

Devices that amplify currents are referred to as current amplifiers. However, in industrial instrumentation a voltage to current converter is sometimes referred to as a current amplifier. Figure 9.8a shows a basic current amplifier. The gain is given by

$$\frac{I_{out}}{I_{in}} = \frac{R_2 R_6}{R_1 R_3} \tag{9.3}$$

where the resistors are related by the equation

$$R_1(R_3 + R_5) = R_2 R_4 \tag{9.4}$$

9.4.3   Differential Amplifiers

A differential amplifier is a dual input amplifier that amplifies the difference between two signals, such that the output is the gain multiplied by the magnitude of the difference

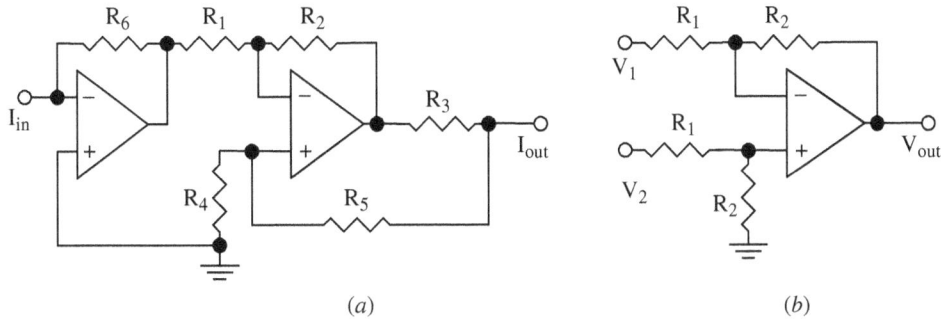

(a)   (b)

FIGURE 9.8 Circuit diagram of the basic configuration of (a) a current amplifier and (b) a differential amplifier.

between the two signals. One signal is fed to the negative input of the op-amp and the other signal is fed to the positive input of the op-amp. Hence the signals are subtracted before being amplified. Figure 9.8b shows a basic differential voltage amplifier. The output voltage is given by

$$V_{out} = \frac{R_2}{R_1}(V_2 - V_1) \tag{9.5}$$

Signals can also be subtracted or added in a resistor network prior to amplification.

9.4.4 Converters

In Fig. 9.8a, the op-amp is used as a current-to-voltage converter. When used as a converter the relation between input and output is called the *transfer function* μ (or ratio). These devices do not have gain as such because of the different input and output units. In Fig. 9.8a, the transfer ratio is given by

$$\mu = \frac{-E_{out}}{I_{in}} = R_1 \tag{9.6}$$

Example 9.6   In Fig. 9.9a, the input current is 165 μA and the output voltage is 2.9 V. What is the transfer ratio and the value of R_1?

Solution

$$\mu = \frac{2.9\ V}{165\ \mu A} = 17.6\ V/mA = 17.6\ kV/A$$

$$R_1 = \frac{2.9\ V}{165\ \mu A} = 17.6\ k\Omega$$

In Fig. 9.9b, the op-amp is used as a voltage-to-current converter. In this case, the transfer ratio is given by

$$\frac{I_{out}}{E_{in}} = \frac{-R_2}{R_1 R_3}\ mhos \tag{9.7}$$

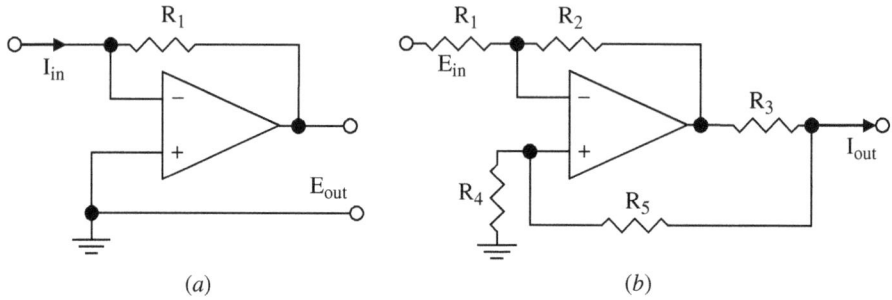

FIGURE 9.9   Examples of (a) current-to-voltage converter and (b) voltage-to-current converter.

Note, in this case the units are in mhos (1/ohms), and the resistors are related by the equation

$$R_1\left(R_3 + R_5\right) = R_2 R_4 \tag{9.8}$$

Example 9.7   In Fig. 9.9b, $R_1 = R_4 = 5$ kΩ, $R_2 = 100$ kΩ. What is the value of R_3 and R_5 if the op-amp is needed to convert an input voltage of 3 V to an output of 20 mA?

Solution

$$\frac{I_{out}}{E_{in}} = \frac{R_2}{R_1 R_3} = \frac{20 \times 10^3}{3} = 6.67 \times 10^{-3}$$

$$\frac{100 \times 10^3}{5 \times 10^3 \times R_3} = 6.67 \times 10^{-3}$$

$$R_3 = \frac{20}{6.67 \times 10^{-3}} = 3 \text{ k}\Omega$$

$$R_5 = \frac{R_2 R_4}{R_1 - R_3} = R_2 - R_3 = (100 - 3) \text{ k}\Omega = 97 \text{ k}\Omega$$

9.4.5   Buffer Amplifiers

An impedance matching op-amp is called a buffer amplifier. Such amplifiers have feedback to give unity voltage gain, high-input impedance (many megaohms) and low-output impedance (<20 Ω). Such an amplifier is shown in Fig. 9.10b. In this context, impedance is used to cover both ac impedance and dc resistance. Circuits have both input and output impedance.

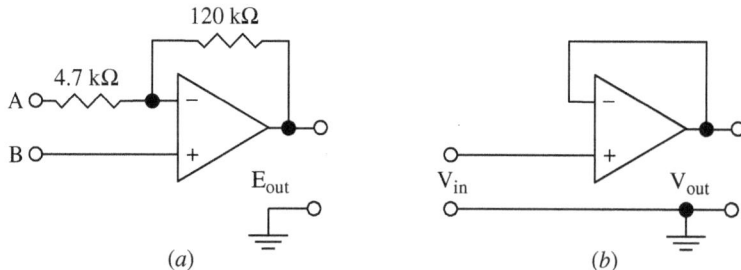

FIGURE 9.10   Schematic diagrams for (a) Example 9.8 and (b) buffer amplifier.

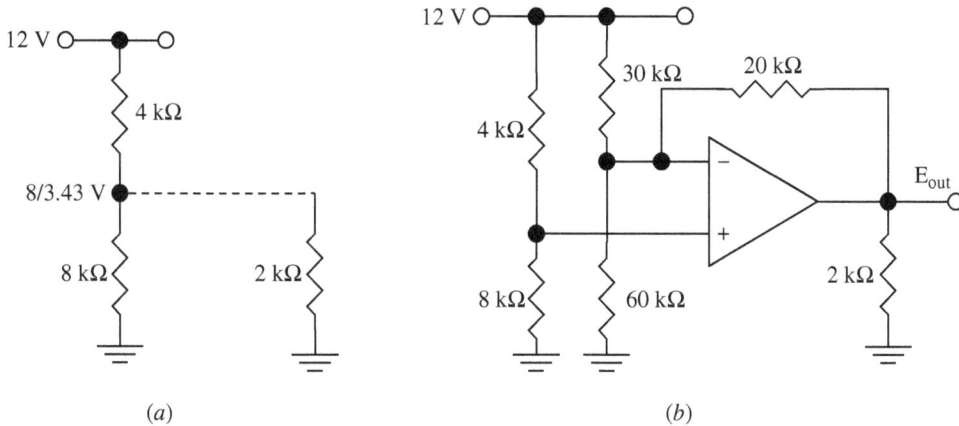

FIGURE 9.11 Circuits show (a) effect of loading on a voltage divider and (b) use of buffer in Example 9.9.

Example 9.8 In the dc amplifier shown in Fig. 9.10a, an input of 130 mV is applied to terminal A and −85 mV is applied to terminal B. What is the output voltage (assume the amplifier was zeroed with 0 V at the inputs)?

Solution

$$E_{out} = \frac{-\Delta V_{in} \times 120}{4.7} = [-130 + (-85)]\text{mV} \times \frac{120}{4.7} = -0.215 \times 25.5 \text{ V} = -5.5 \text{ V}$$

The effect of loading on a circuit can be seen in Fig. 9.11a. The resistor divider gives an output voltage of 8 V and an output impedance of 2.7 kΩ (effectively this impedance is 4 kΩ in parallel with 8 kΩ). If this divider is loaded with a circuit with an input impedance of 2 kΩ, the output voltage will drop from 8 to 3.43 V. A buffer amplifier can be used as shown in Fig. 9.11b to match the input impedance of the second circuit to the resistor divider, thus giving an output voltage of 8 V across the 2 kΩ load.

Example 9.9 In Fig. 9.11b, what is the output voltage of the buffer amplifier? Assume the input impedance of the buffer amplifier is 2 MΩ and its output impedance is 15 Ω.
 2 MΩ in parallel with 8 kΩ has an effective resistance of 7.97 kΩ.

Solution

$$\text{Voltage at input to buffer} = \frac{12 \times 7.97}{7.97 + 4} \text{V} = 7.99 \text{ V}$$

From this, we get that the buffer loading reduces the output voltage from the resistive divider by 0.01 V, about 0.125 percent. The output impedance of the buffer is effectively in series with the 2 kΩ load, so that the output voltage E_{out} is given by

$$E_{out} = \frac{7.99 \times 2000}{2000 + 15} \text{V} = 7.93 \text{ V}$$

Thus, the total loading effect is a reduction of 0.07 V in 8 V, or about 0.9 percent compared to 57.5 percent with direct loading. This error could be totally corrected if the amplifier had a gain of 1.01.

9.4.6 Nonlinear Amplifiers

Many sensors have a logarithmic or nonlinear transfer characteristic and such devices require signal linearization. This can be implemented using amplifiers with nonlinear characteristics. These are achieved by the use of nonlinear elements such as diodes or

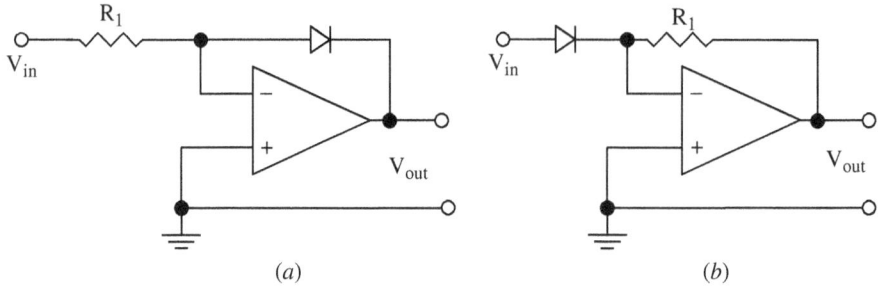

FIGURE **9.12** Circuits of nonlinear amplifiers: (a) log amplifier and (b) antilog amplifier.

transistors in the feedback loop. Figure 9.12 shows two examples of nonlinear amplifiers using a diode in the feedback loop. In Fig. 9.12(a) the amplifier is configured as a logarithmic amplifier, and in Fig. 9.12(b) the amplifier is configured as an antilogarithmic amplifier. Combinations of resistors and nonlinear elements can be chosen to match the characteristics of many sensors for linearization of the output from the sensor.

The op-amp can be used for dc as well as ac amplification or conversion. The only difference being that capacitors are normally used in ac amplifiers between stages, to prevent any dc offset levels present from affecting the biasing or operating levels of the following op-amps. High-gain dc amplifiers are directly coupled and use special op-amps that have a low drift with temperature and feedback for stabilization.

9.4.7 Instrument Amplifier

Because of the very high accuracy requirements in instrumentation, the op-amp circuits shown in Fig. 9.7 are not ideally suited for low-level instrument signal amplification. The op-amp can have different input impedances at the two inputs. The input impedances can be relatively low and tend to load the sensor output, can have different gains at the inverting and noninverting inputs, and common-mode noise can be a problem. Op-amps configured for use as an instrument amplifier are shown in Fig. 9.13. This amplifier has balanced inputs with very high input impedance and good common-mode noise reduction. Gain is set by R_A.

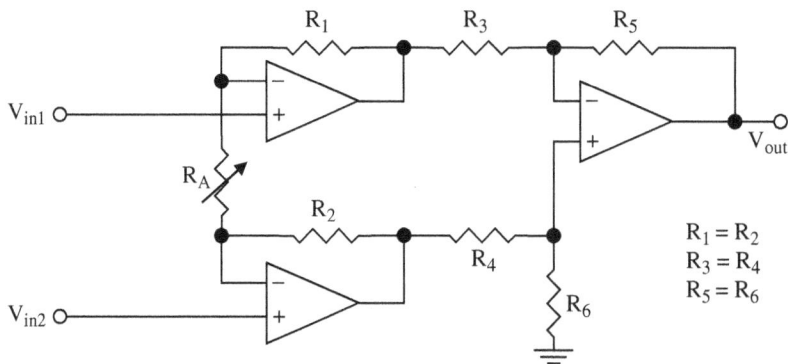

FIGURE **9.13** Circuit schematic of an instrumentation amplifier.

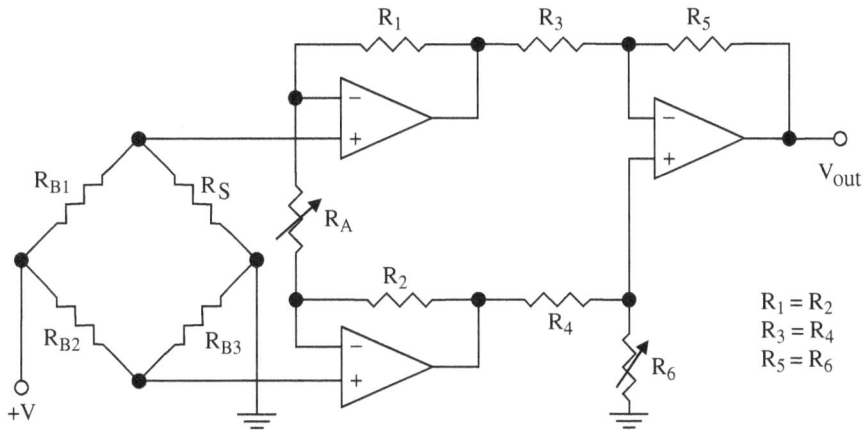

FIGURE 9.14 Instrumentation amplifier used for offset adjustment and to amplify a signal from a bridge.

The output voltage is given by

$$V_{out} = \frac{R_5}{R_3}\left(\frac{2R_1}{R_A}+1\right)(V_{in2}-V_{in1})$$ (9.9)

Figure 9.14 shows a practical circuit using an instrumentation amplifier to amplify the output signal from a resistive bridge. R_6 is used to adjust for any zero signal offset and R_A for gain.

9.5 Signal Conditioning

In most cases, there are several types of sensors that can be chosen to measure a variable. In choosing a sensor, it is necessary to carefully evaluate the sensors characteristics to choose the best sensor to meet the system requirements, rather than use a more cost-effective sensor and use correction to make the sensor suitable. In many cases, the relation between the input and the output of a sensor is nonlinear, temperature sensitive, and offset from zero. Circuit linearization is very hard to achieve and requires the use of specialized networks.

9.5.1 Offset Zero

Figure 9.15a shows the output of a sensor when measuring a variable and the idealized output obtained from a linearization circuit with adjustment of the bias (zero level) as is required on many types of sensor outputs. The output voltage from the linear operating range of the sensor varies from 0.35 to 0.7 V as the process variable varies from low to high over its measurement range. However, the sensor output goes to equipment that requires a voltage from 0 to 3.5 V for the range of the variable. A circuit for changing the output levels is shown in Fig. 9.15b. The reference input to the amplifier is set at 0.35 V to offset the minimum level of the sensor to give zero out at the low end of the range. The gain of the amplifier is set to 10 (with feedback resistors 47k/4.7k) giving

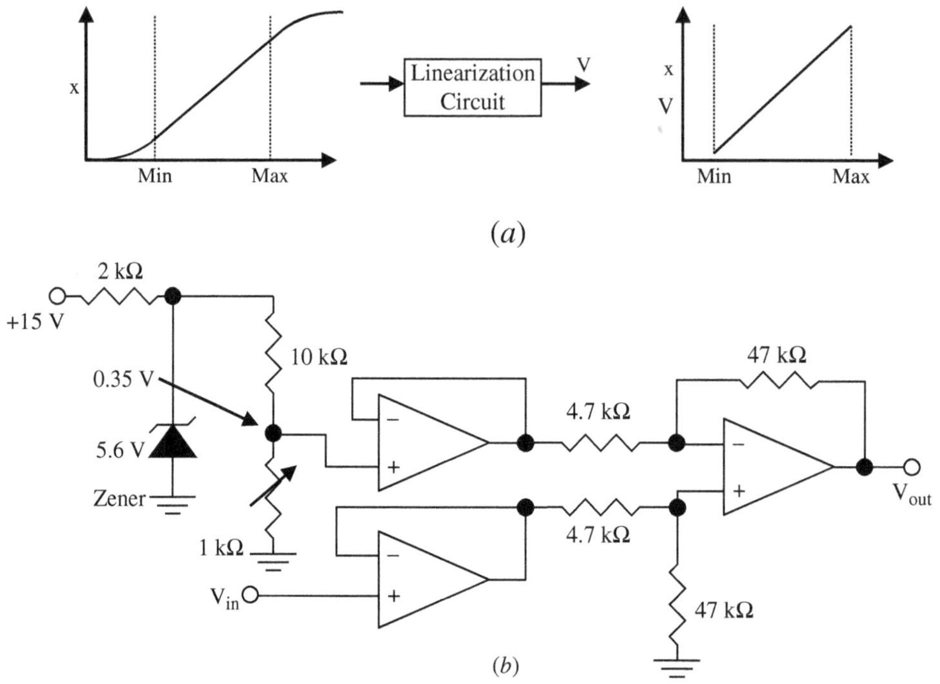

(a)

(b)

Figure 9.15 Representation of (a) the input and ideal output of an ideal linearization circuit and (b) the instrument circuit used for zero adjust.

3.5 V output with 0.75 V input, i.e., $3.5/(0.7 - 0.35) = 10$. Note the use of impedance matching buffers that would be used in instrumentation.

9.5.2 Span Adjustment

The output from the circuit shown in Fig. 9.15b can be used to drive the voltage to current converter shown in Fig. 9.16. Input level and gain of the circuit can be adjusted to give an output of 4 to 20 mA when the output from the sensor goes from minimum to its maximum value. Because the converter inverts the signal phase, the inputs to the final amplifier in Fig. 9.15b should be inverted.

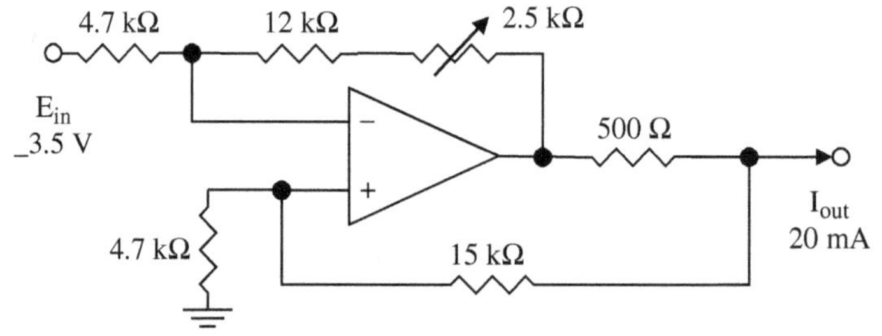

Figure 9.16 Instrument sensor current output circuit.

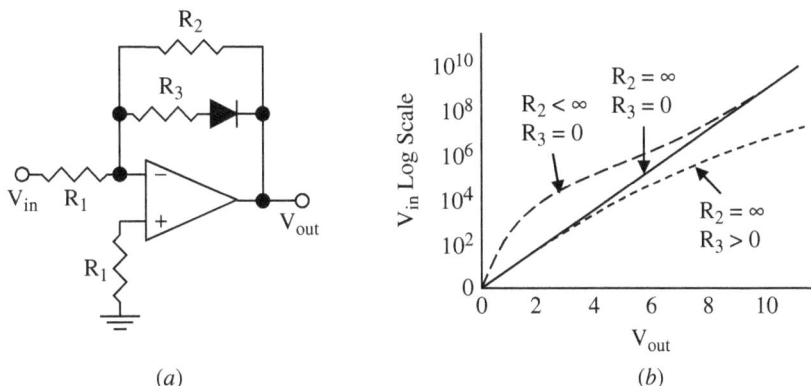

Figure 9.17   Nonlinear amplifier: (a) circuit and (b) characteristics of nonlinear circuit with different feedback values.

9.5.3   Linearization in Analog Circuits

Linearization in analog circuits is difficult unless there is a relatively simple equation to describe the sensors characteristics. In some applications, a much more expensive linear transducer may have to be used due to the inability of analog circuits to linearize the signal conversion.

Figure 9.17a shows the circuit of a logarithmic amplifier. Figure 9.17b shows the variations in characteristics with various resistor values that can be obtained for use in signal linearization. When $R_2 = \infty$ and $R_3 = 0$ the amplifier has a logarithmic relation between input and output. When R_3 is larger than zero the gain is higher at the upper end of the scale as shown. If R_2 is a high value the effect is to reduce the gain at the lower end of the scale. Multiple feedback paths can be used with nonlinear elements and resistors to approximate the amplifier characteristics to those of the sensor. The resistor network can also be used with antilogarithmic amplifiers.

9.5.4   Linearization in Digital Circuits

Linearization in digital circuits can be performed for nonlinear devices by using equations or memory look-up tables. If the relationship between the values of a measured variable and the output of a sensor can be expressed by an equation, the processor can be programmed on the basis of the equation to linearize the data received from the sensor. An example would be a transducer that outputs a current (I) related to flow rate (v) by

$$I = Kv^2 \tag{9.10}$$

where K is a constant.

The current numbers from the sensor are converted into binary, where the relationship still holds. In this case, a linear relationship is required between data and measured variable. This can be obtained by multiplying the I term by itself and then by K. The resulting number is proportional to v^2, or the generated number and measured variable now have a linear relationship. Span and offsets may now require further adjustment.

There are many instances in conversion where there is not an easily definable relationship between variable and transducer output, and it may be difficult or impossible to write a best-fit equation that is adequate for linearization of the variable. In this case, look-up tables are used. The tables correlate transducer output and the true value of the variable. The look-up numbers are stored in memory so that the processor can reproduce the true value of the variable from the transducer reading. This method is extensively used with nonlinear sensors such as thermocouples.

9.5.5 Temperature Correction

Sensors are notoriously temperature sensitive, i.e., their output zero as well as span will change with temperature, and in some cases the change is nonlinear. Variables are also temperature sensitive and require correction. Correction of temperature effects requires a temperature sensitive element to monitor the temperature of the variable and the sensor. The temperature compensation in analog circuits will depend on the characteristics of the sensor used. Because the characteristics of the sensors change from type to type, the correction for each type of sensor will be different. In digital circuits, computers can make the corrections from the sensor and variable characteristics using temperature compensation look-up tables.

Other compensations needed can take the form of filtering to remove unwanted frequencies such as pick up from the 60 Hz line frequency, noise or RF pickup, dampen out undulations or turbulence to give a steady average reading, correction for time constants, and for impedance matching networks.

9.6 Bridge Circuits

In many sensors, such as strain gauges, the variable is measured directly in electrical units. A second element can be used to offset the changes in the sensing element due to temperature. Because of the high sensitivity to small changes in resistance and correction for temperature effects, bridges are extensively used in instrumentation with strain gauges, photoresistive elements, and magnetorestrictive sensors.

9.6.1 DC Bridges

The Wheatstone bridge is the most common resistance network used to measure small changes in resistance, and is often used in instrumentation with resistive types of sensors. The bridge circuit is shown in Fig. 9.18a. Four resistors are connected in the form of a diamond with the supply and measuring instrument forming the diagonals. When all of the resistors are equal the bridge is balanced, i.e., the voltage at A and C are equal ($E/2$), and the voltmeter reads zero.

If R_2 is the resistance of a sensor whose change in value is being measured, the voltage at A will increase with respect to C as the resistance value increases, so that the voltmeter will have a positive reading. The voltage output change does not have a linear relationship to the changes in the value of R_2. However, the bridge is very sensitive to small changes in resistance. A bridge circuit can also be used with temperature compensating resistors, i.e., if R_1 and R_2 are the resistance of the same type of sensing element, such as a strain gauge and reference strain gauge, see Fig. 9.20. The resistance of each gauge will change by an equal percentage with temperature, so that the bridge will remain balanced when the temperature changes.

In many applications, the sensing resistor (R_2) can be remote from a centrally located bridge. In such cases, the resistance of the leads can be zeroed out by adjusting the

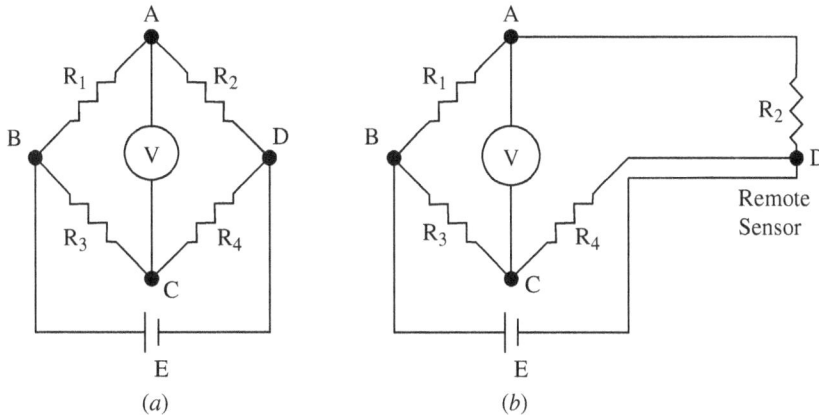

Figure 9.18 Circuit of (a) Wheatstone bridge and (b) compensation for lead resistance used in remote sensing.

bridge resistors. However, any change in lead resistance due to temperature will appear as a sensor value change. To correct for this error lead compensation can be used. This is achieved by using three interconnecting leads as shown in Fig. 9.18b. A separate power lead is used to supply R_2, so that only signal current flows in the signal lead from R_2 to the bridge resistor (R_4), any variations in voltage drop due to the supply current in the lead resistance does not affect the balance of the bridge. The lead resistance between point A and R_2 adds to R_2 and the lead resistance between point D and R_4 adds to R_4 keeping the bridge balanced, and any changes in lead resistance will affect both resistance values equally.

Example 9.10   The resistors in the bridge circuit shown in Fig. 9.18a are all 2.7 kΩ, except R_1 which is 2.2 kΩ. If $E = 15$ V, what will the voltmeter read?

The voltage at point C will be 7.5 V, as $R_3 = R_4$, the voltage at C = ½ the supply voltage. The voltage at A will be given by

Solution

$$E_{AD} = \frac{E \times R_2}{R_1 + R_2} = \frac{15 \text{ V} \times 2.7 \text{ k}\Omega}{2.2 \text{ k}\Omega + 2.7 \text{ k}\Omega} = \frac{40.5 \text{ V}}{4.9} = 8.26 \text{ V}$$

The voltmeter will read 8.26 − 7.5 V = 0.76 V (note meter polarity)

9.6.2   Current-Balanced Bridge

To obtain a linear relationship between changes in the sensing element in a bridge and an output a current-balanced bridge can be used. In this case, a current feedback loop can be introduced to null the bridge as shown in Fig. 9.19. A low value resistor R_5 is connected in series with R_2. A current I is fed through R_5 to develop an offset voltage to keep the bridge balanced. The current I is directly proportional to any changes in the value of R_2.

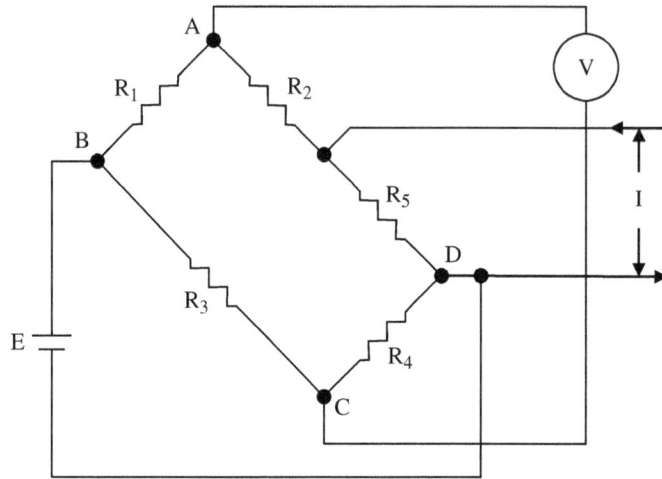

Figure 9.19   Current balanced bridge.

When the bridge is balanced with $I = 0$

$$R_3(R_2 + R_5) = R_1 \times R_4$$

assume R_2 changes to $R_2 + \delta R$
   Then to rebalance the bridge

$$R_3(R_2 + \delta R + R_5) - IR_5 = R_1 \times R_4$$

subtracting the equations we get

$$\delta R = IR_5 / R_3 \qquad\qquad (9.11)$$

showing a linear relationship between changes in the sensing resistor and the feedback current I (R_5 and R_3 are fixed values).

9.6.3   Strain Gauge Sensors

Diaphragms can use strain gauge or capacitive sensing, the movement being too small to control a pneumatic flapper, slider, or potentiometer. The strain gauge elements are resistors made from copper or nickel particles glued onto a nonconducting substrate as shown in Fig. 9.20a. Semiconductor strain gauges are also available that use the piezoresistive effect.

Figure 9.20b shows a circuit using the strain gauge. The strain gauge elements are mounted in two arms of the bridge and two resistors R_1 and R_2 form the other two arms. R_3 and R_5 are the conditioning for the zero offset and span, respectively. The output signal from the bridge is amplified, and impedance matched as shown. The strain gauge elements are in opposing arms of the bridge, so that any change in the resistance of the elements due to temperature changes will not affect the balance of the bridge giving temperature compensation. More gain and impedance matching stages than shown may be required. Current balancing (see Fig. 9.19),

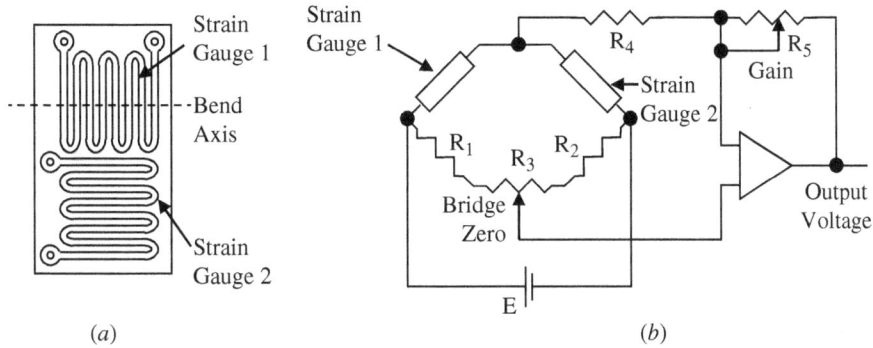

FIGURE 9.20 Representation of (a) configuration for strain gauge elements and (b) resistive bridge for signal conditioning of a strain gauge.

or an A/D converter may be required to make the signal suitable for transmission. Additional linearization may also be needed; this information can be obtained from the manufactures' device specifications.

9.6.4 AC Bridges

The concept of dc bridges can also be applied to ac bridges. The resistive elements are replaced with impedances and the bridge supply is now an ac voltage, as shown in Fig. 9.21a. The differential voltage δV across S is then given by

$$\delta V = E \frac{Z_2 Z_3 - Z_1 Z_4}{(Z_1 + Z_3)(Z_2 + Z_4)} \qquad (9.12)$$

When the bridge is balanced $\delta V = 0$ and Eq. (9.12) reduces to

$$Z_2 Z_3 = Z_1 Z_4 \qquad (9.13)$$

Example 9.11 What are the conditions for the bridge circuit in Fig. 9.21b to be balanced?
To be balanced Eq. (9.13) applies; there are two conditions that must be met for this equation to be balanced because of the phase shift produced by the capacitors. First, the resistive component must balance, and this gives

Solution

$$R_2 R_3 = R_1 R_4 \qquad (9.14)$$

Second, the impedance component must balance, and this gives

$$C_2 R_2 = C_1 R_1 \qquad (9.15)$$

9.6.5 Capacitive Sensors

Capacitive sensing devices can use single-ended sensing or differential sensing. With single-ended sensing, capacitance is measured between the diaphragm and a single capacitor plate in close proximity to the diaphragm. Differential sensing can be used

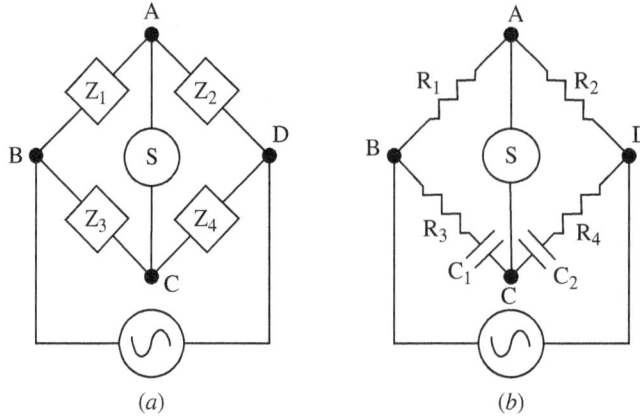

Figure 9.21   AC bridges (a) using block impedances and (b) bridge with R and C components for Example 9.11.

when there is a capacitor plates either side of, and in close proximity to, the diaphragm. See the pressure sensor in Fig. 9.22a. In differential sensing the two capacitors can be used to form two arms of an ac bridge or switch capacitor techniques can be used. For single-ended sensing a fixed reference capacitor can be used. Capacitive sensing can use ac analog or digital techniques. Microminiature pressure sensors can use piezoresistive strain gauge sensing or capacitive sensing techniques. Being a semiconductor-based technology the sensor signal is conditioned on the die, i.e., amplified, impedance matched, linearized, and temperature compensated.

Figure 9.22b shows an ac bridge with offset and span conditioning that can be used with capacitive sensing. Initially, the bridge is balanced for zero offset with potentiometer R_3, the output from the bridge is amplified, and buffered. The output amplitude can be adjusted by the gain control R_5. The signal will need to be converted to a dc signal and further amplified for transmission.

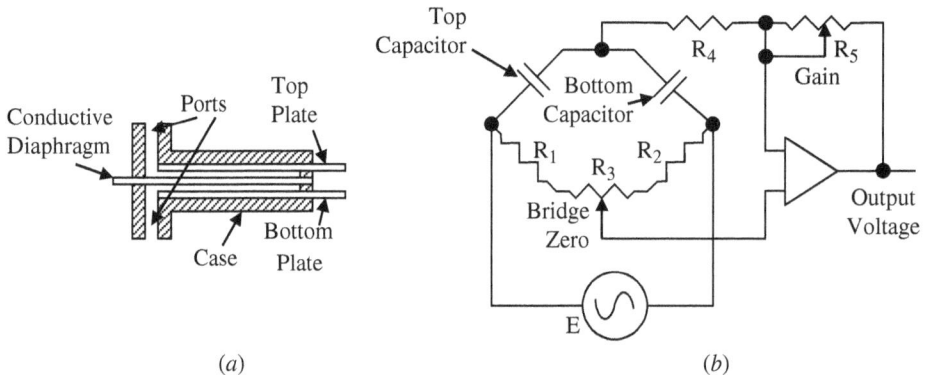

Figure 9.22   Illustration of (a) capacitive diaphragm pressure sensor and (b) an ac bridge for use with a capacitive sensor.

9.6.6 Resistance Sensors

Resistance temperature detectors (RTDs) measure the change in electrical resistance of a wire wound resistor with temperature; typically, a platinum resistance element is used with a resistance of about 100 Ω. Normally, the resistor is driven from a constant current source and the voltage developed across the resistor measured. Care must be taken with these devices to ensure that the current flowing through the devices is low to minimize temperature changes due to internal heating of the resistor. Pulse techniques can be used to prevent internal heating. In this case, the current is turned on for a few milliseconds, the resistance measured and then the current is turned off for, say a second. Figure 9.23a shows the simplest connection to the RTD with just two leads, the meter being connected to the current supply leads. The resistance of long leads between the detector and the resistor contribute to measurement error, as the meter is measuring the voltage drop across the current lead resistance and junctions as well as the RTD.

Figure 9.23b shows a four-wire connection to an RTD. The meter connects direct to the RTD so that only the voltage drop across the RTD is measured; no error is introduced due to the resistance of the current supply contacts or lead resistance to the RTD. Platinum is the material of choice for an RTD. A linearity of 3.6 percent can be obtained from 0 to 850°C without signal conditioning. The RTD has a constant current flowing through it giving an offset zero, so that zero-level correction and span conditioning are required. Direct connection of the resistive element to the controller will be further discussed in Chap. 13.

9.6.7 Magnetic Sensors

Many flow measurements are sensed as differential pressures with the indicator scale graduated in ft^3/min, gal/min, and L/s. However, rotating devices such as the turbine flow meter as described in Chap. 4 can be used for accurate flow measurements. The blade rotation is normally sensed by a magnetic sensor such as a Hall effect or magneto resistive element (MRE) device. The Hall device gives an electrical impulse every time a blade passes under the sensor, whereas the resistance of the MRE device changes. Figure 9.24 shows the circuit used to shape the signal from an MRE into a digital signal or an analog output. The MRE sensor contains four elements (shown in Fig. 7.3) to form a bridge circuit as shown. The Hall or MRE device does not normally require temperature compensation, when they are being used as switches in a digital configuration, but

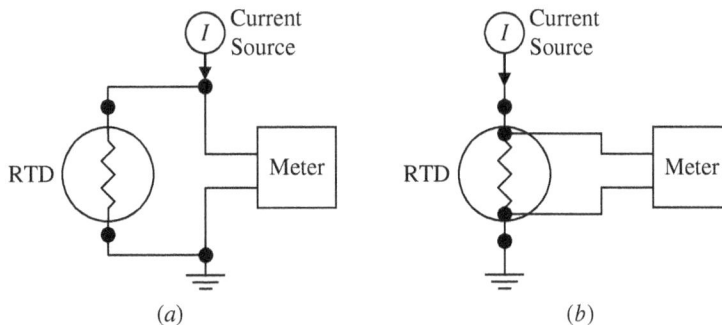

FIGURE 9.23 RTD connections using (a) common supply and meter leads and (b) directly connected meter.

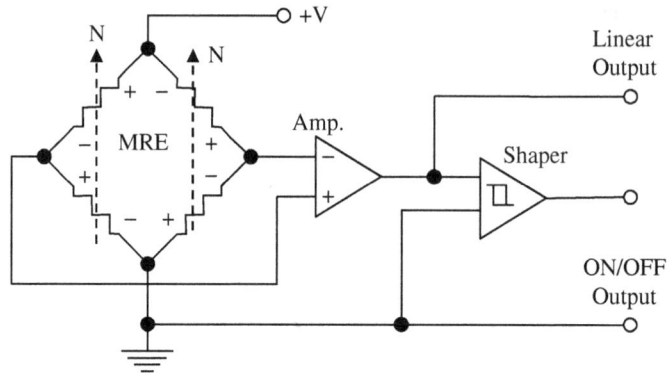

Figure 9.24 Representation of an MRE magnetic field sensing circuit.

the MRE device is temperature sensitive and is used in the bridge configuration as shown when used to give an analog output.

When measuring flow rates some conditioning may be required for the density changes in the liquid with temperature, and for high- and low-flow rates, the conditioning will depend on the requirements of the application and manufactures' specifications.

Summary

In this chapter instrument nonlinearity and hysteresis were introduced. The use of op-amps for amplifying analog signals, linearizing sensor outputs, correcting zero offset, and signal conditioning are discussed.

The main points described in this chapter were as follows:

1. Instrument linearity and hysteresis.

2. Signal inversion and noninversion, methods of applying feedback for gain control and stability.

3. Use of the op-amp as a signal converter, impedance matching, set zero control, and span adjustment.

4. Configuration of op-amps to make an instrument amplifier for accurate signal amplification and noise reduction.

5. The methods used to convert the measured variable sensor signal into a pneumatic or electrical signal. The transducer used can be a potentiometer, LVDT, or capacitive devices.

6. Linearization of signals, changing operating levels and gain control using nonlinear analog amplifiers or equations, and look-up tables when using digital methods.

7. Techniques used in the temperature compensation of strain gauges and other types of devices.

8. Capacitive sensor measurements using ac bridge circuits.

9. The methods used to reduce errors in the measurement of RTD signals.

Problems

9.1   Describe hysteresis in a sensor.

9.2   A spring balance has a span of 10 to 120 kg and the absolute accuracy is ±3 kg. What is its %FSD accuracy and span accuracy?

9.3   A digital thermometer with a temperate range of 129.9°C has an accuracy specification of ± ½ of the least significant bit. What is its absolute accuracy, %FSD accuracy, and resolution?

9.4   A flow instrument has an accuracy of (a) ±0.5 percent of reading, (b) 0.5 percent of FSD. If the range of the instrument is 10 to 100 fps, what is the absolute accuracy at 45 fps?

9.5   A pressure gauge has a span of 50 to 150 psi. Its absolute accuracy is ±5 psi. What is its %FSD and span accuracy?

9.6   What is the output current if the input voltage is 3.8 mV in Fig. 9.25a? Assume R_3 = 1.5 kΩ.

9.7   What is the value of R_3 in Fig. 9.25a for a transfer ratio of 8.5 mA/uV? Assume R_2 = 100 kΩ.

9.8   If in Fig. 9.25b, input A is 17 mV and input B is −21 mV. What is the value of the output voltage if R = 83 kΩ.

9.9   Name two magnetic field sensors.

9.10   What is the difference between sensor, transducer, and converter?

9.11   The output voltage from a sensor varies from a minimum of 0.21 V to a maximum of 0.56 V. Draw a circuit to condition the signal so that the output voltage goes from 0 to 10 V. Assume a reference voltage of 10 V, the resistor from the reference is 10 kΩ, and the input resistor to the amplifier is 5 kΩ.

9.12   What methods are used to convert mechanical sensor movement into electrical or display signals?

9.13   Why are strain gauges normally mounted in pairs at right angles to each other?

9.14   How can a linear relationship between input and output be obtained from a bridge circuit?

9.15   What types of analog circuits are used for linearization?

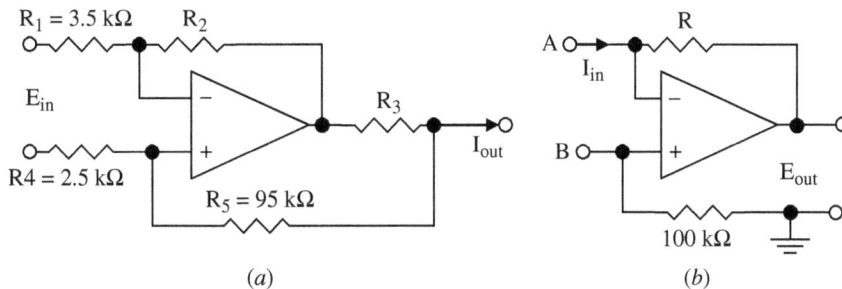

FIGURE 9.25   Circuits for use in (a) Problems 9.6 and 9.7 and (b) Problem 9.8.

9.16   How are temperature corrections made to temperature sensitive sensors?

9.17   What are the pressure ranges used in pneumatic signal transmissions? Why is zero not used?

9.18   How can you use a float level sensor to obtain a linear relation between level and electrical output?

9.19   How is the output from a capacitive type of sensor measured?

9.20   Why should the meter contacts to an RTD be as close to the measuring element as possible?

CHAPTER 10

Regulators, Valves, and Actuators

Chapter Objectives

This chapter is an introduction to pressure regulators, flow control valves, and actuators and will help you understand the operation of control devices and their use in pressure regulation and flow control.

Topics discussed in this chapter are as follows:

- The operation and use of various types of pressure regulators and safety valves
- Flow control valves
- Fail-safe operation in valves
- The various types of valves in use
- Valve characteristics
- Level control
- Feedback loops for flow control
- Positioners
- Smart controllers

This section deals with pressure regulators, level, flow control devices, and actuators used in industrial processing for regulating temperature, pressure, and flow rates. The devices can be self-regulating or under the control of a central processing system that can be monitoring and controlling many variables.

10.1   Introduction

This section will discuss the use of regulators, valves, and actuators to control gas flow, liquid flow, and pressure control. In many processes, this involves the control of many thousands of cubic meters of a liquid or the control of large forces as would be the case in a steel rolling mill from low-level analog, digital, or pneumatic signals. Temperature is also normally controlled by regulating gas and/or liquid flow. Control loops can be local self-regulating loops under pneumatic, hydraulic, or electrical control, or the loops can be processor controlled with position feedback loops. Electrical signals from a controller are low-level signals, which require the use of power switching devices or relays for power control, and possibly optoisolators for isolation. These power control devices are normally at the point of use so that electrically controlled valve actuators and motors can be supplied directly from the power lines.

10.2 Pressure Controllers

10.2.1 Regulators

Gases used in industrial processing such as oxygen, nitrogen, hydrogen, and propane, are stored in high-pressure containers in liquid form outside of the processing facility for safety. The high-pressure gas from above the liquid is reduced in pressure and regulated with gas regulators before it can be safely distributed through the facility in low-pressure lines. The gas lines may have additional regulators at the point of use.

A *spring-controlled regulator* is an internally controlled pressure regulator and is shown in Fig. 10.1a. Initially, the spring holds the inlet valve open and gas under pressure flows into the main cylinder at a rate higher than the gas can exit the cylinder. As the pressure in the cylinder increases, a predetermined pressure is reached where the spring-loaded diaphragm starts to move up, causing the valve to partially close; i.e., the pressure on the diaphragm controls the flow of gas into the cylinder to maintain a constant pressure in the main cylinder and at the output regardless of flow rate (ideally). The output pressure can be adjusted by the spring screw adjustment.

A *weight-controlled regulator* is shown in Fig. 10.1b. The internally controlled regulator has a weight-loaded diaphragm. The operation is the same as the spring-loaded diaphragm except the spring is replaced with a weight. The pressure can be adjusted by the position of a sliding weight on a cantilever arm.

A *pressure-controlled diaphragm regulator* is shown in Fig. 10.2a. The internally controlled regulator has a pressure-loaded diaphragm. Pressure from a regulated external air or gas supply is used to load the diaphragm via a restriction. The pressure to the regulator can then be adjusted by an adjustable bleed valve that in turn is used to set the output pressure of the regulator.

Externally connected spring diaphragm regulator is shown in Fig. 10.2b. The cross section shows an externally connected spring-loaded pressure regulator. The spring holds the valve open until the output pressure, which is fed to the upper surface of the diaphragm, overcomes the force of the spring on the diaphragm, and starts to

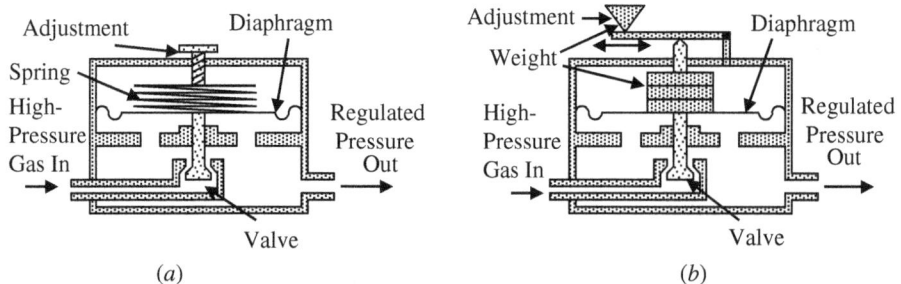

FIGURE 10.1 Self-compensating pressure regulators (a) spring loaded and (b) weight loaded.

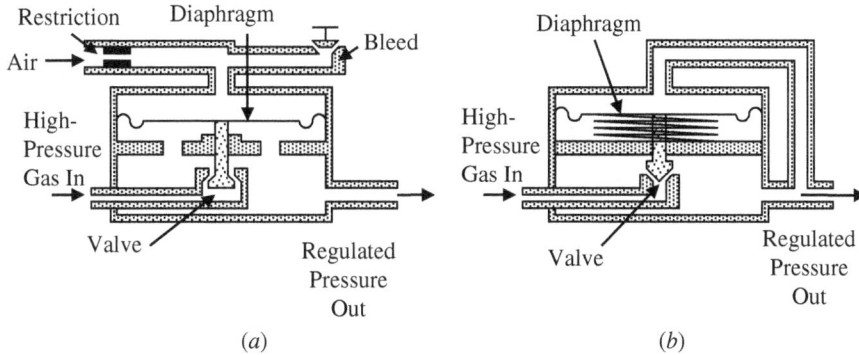

Figure 10.2   Self-compensating pressure regulators: (a) internal pressure-loaded regulator and (b) externally connected spring-loaded regulator.

close the valve, hence regulating the output pressure. Note the valve is inverted from the internal regulator and the internal pressure is isolated from the lower side of the diaphragm. Weight- and air-loaded diaphragms are also available for externally connected regulators.

Pilot-operated pressure regulators can use an internal or external pilot for feedback signal amplification and control. The pilot is a small regulator positioned between the pressure connection to the regulator and the loading pressure on the diaphragm. Figure 10.3a shows such an externally connected pilot regulator. The pressure from the output of the regulator is used to control the pilot, which in turn amplifies the signal and controls the pressure from the air supply to the diaphragm, giving greater control than that available with the internal pressure control diaphragm. A slight change in the output pressure is required to produce a full pressure range change of the regulator giving a high-gain system for good output pressure regulation.

An *instrument pilot-operated pressure regulator* is similar to the pilot-operated pressure regulator but has a proportional band adjustment included giving a gain or sensitivity control feature to provide greater flexibility in control.

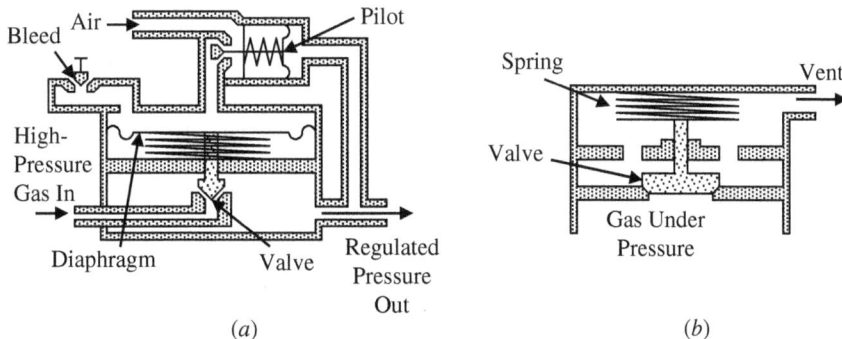

Figure 10.3   Representation of (a) a pilot-operated regulator and (b) an automatic pressure safety valve.

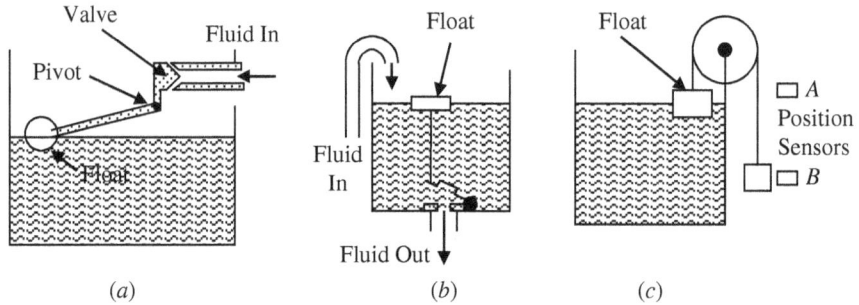

Figure 10.4 Various types of regulators are shown: (a) automatic fluid level controller (b) automatic emptying of a storage tank when full, and (c) means of detecting full level or empty level in a fluid reservoir.

10.2.2   Safety Valves

Safety valves are fitted to all high-pressure containers from steam generators to domestic water heaters (see Fig. 10.3b). The valve is closed until the pressure on the lower face of the valve reaches a predetermined level set by the spring. When this level is reached, the valve moves up allowing the excess pressure to escape through the vent.

10.2.3   Level Regulators

Level regulators are in common use in industry to maintain a constant fluid pressure, or a constant fluid supply to a process, or in waste storage. Level regulators can be a simple float and valve arrangement as shown in Fig. 10.4a to using capacitive sensors as given in Chap. 3 to control a remote pump. The arrangement shown in Fig. 10.4a is used to control water levels in many applications. When the fluid level drops due to use, the float moves downward opening the inlet valve and allowing fluid to flow into the tank. As the tank fills, the float rises causing the inlet valve to close maintaining a constant level and preventing the tank from overflowing.

Figure 10.4b shows an example of a self-emptying reservoir when a predetermined fluid level is reached, as may be used in a waste holding tank. As the tank fills, the float rises to where the connecting link from the float to the valve becomes taut and overcomes the hydrostatic pressure on and lifting the outlet valve. Once lifted, the fluid pressure under the valve balances the pressure above the valve and the buoyancy of the valve will keep it open until the tank is empty, then it will close. Once closed the reservoir will start to fill again and the fluid pressure on the top surface of the valve will hold it closed. The automatic fluid leveler in Fig. 10.4a can be combined with the emptying system in Fig. 10.4b. In this case, the outlet valve is manually or automatically operated to deliver a known volume of liquid to a process or as required in a toilet flush system. The container automatically refills for the next operation cycle.

The position of the weight in Fig. 10.4c is controlled by the float. The position of the weight is monitored by position sensors A and B. When the weight is in position A (container empty) the sensor can be used to turn on a pump to fill the tank, and when sensor B senses the weight (container full) it can be used to turn the pump off. The weight can be made of a magnetic material and the level sensors would be Hall effect or *magneto restrictive element* (MRE) devices.

10.3   Flow Control Actuators

When a change in a measured variable with respect to a reference has been sensed, it is necessary to apply a control signal to a valve actuator to make corrections to an input to bring the measured variable back to its preset value. In most cases, any change in the variables, i.e., temperature, pressure, mixing ingredients, and level, can be corrected by controlling flow rates. Hence, actuators are in general used for flow rate control, and can be electrically, pneumatically, or hydraulically controlled. Actuators can be self-operating in local feedback loops in such applications as temperature sensing with direct hydraulic or pneumatic valve control, pressure regulators, and float level controllers. The most common types of variable aperture valves used for flow control are the globe valve and the butterfly valve, but the weir and ball valve can also be used.

10.3.1   Globe Valve

The globe valve's cross section is shown in Fig. 10.5. The valve actuator is used to control the position of the valve and can be operated electrically using a solenoid, or motor, pneumatically, or hydraulically. The actuator determines the speed of travel and distance the valve shaft travels for flow control applications.

The globe valve can be straight through with single seating as illustrated in Fig. 10.5, or can be configured where the output port is at right angles or 135° to the input port. Globe valves are available with double seating, which is used to reduce the actuator operating force, but is expensive, difficult to adjust and maintain, and does not have a tight seal when shutoff. Many other configurations of the globe valve are available.

Illustrated in Fig. 10.6a is a two-way valve (diverging type), which is used to switch the incoming flow from one exit to another. When the valve stem is up the lower port is closed and the incoming liquid exits to the right, and when the valve is down the upper port is closed and the liquid exits from the bottom. Also available is a converging-type

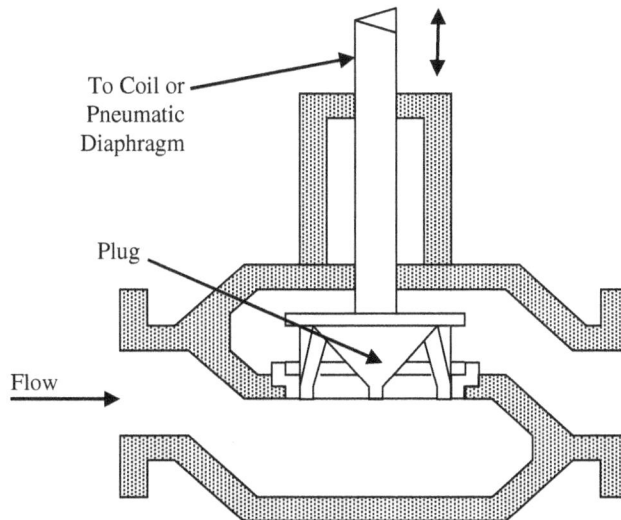

Figure 10.5   Cross section of a globe valve with a linear flow control plug.

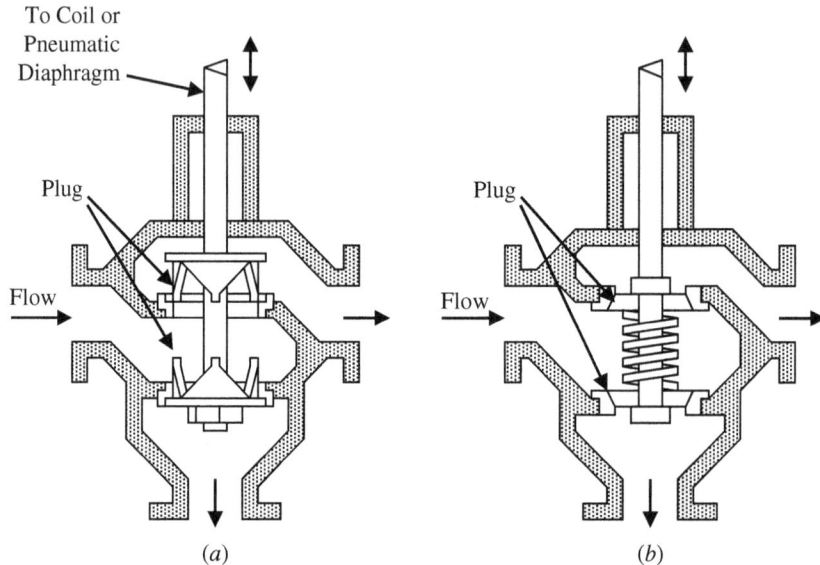

FIGURE **10.6**   Cross sections of globe valve configurations: (*a*) two-way valve and (*b*) three-position valve.

of valve that is used to switch either of two incoming flows to a single output. Figure 10.6*b* illustrates a three-way valve. In the neutral position, both exit ports are held closed by the spring. When the valve stem moves down the top port is opened and when the valve stem moves up from the neutral position the lower port is opened.

Other types of globe valves are the needle valve (less than 1-in diameter), the balanced cage-guided valve, and the split body valve. In the cage-guided valve, the plug is grooved to balance the pressure in the valve body. The valve has good sealing when shutoff. The split body valve is designed for ease of maintenance and can be more cost effective than the standard globe valve, but pipe stresses can be transmitted to the valve and cause it to leak. Globe valves are not well suited for use with slurries.

10.3.2   Flow Control

The globe type valve can be fitted with different shape plugs for flow rate control. The most common plugs are designed for quick opening, linear, or equal percentage flow operation with valve stem travel. In the case of the quick opening valve plug, the flow increases rapidly as soon as the valve starts to open; with the linear plug there is a linear relationship between flow and travel. In equal percent operation, the flow is proportional to the percentage the valve is open, or there is a logarithmic relationship between the flow and valve stem travel. The shape of the plug determines the flow characteristics of the valve and is normally described in terms of percentage of flow versus percentage of lift or travel.

Typical valve plug designs for quick opening, linear, and equal percentage are shown in Fig. 10.7*a* and the characteristics are given in Fig. 10.7*b*. This illustrates some of the characteristics that can be obtained from the large number of plugs that are available. The selection of the type of control plug should be carefully chosen for

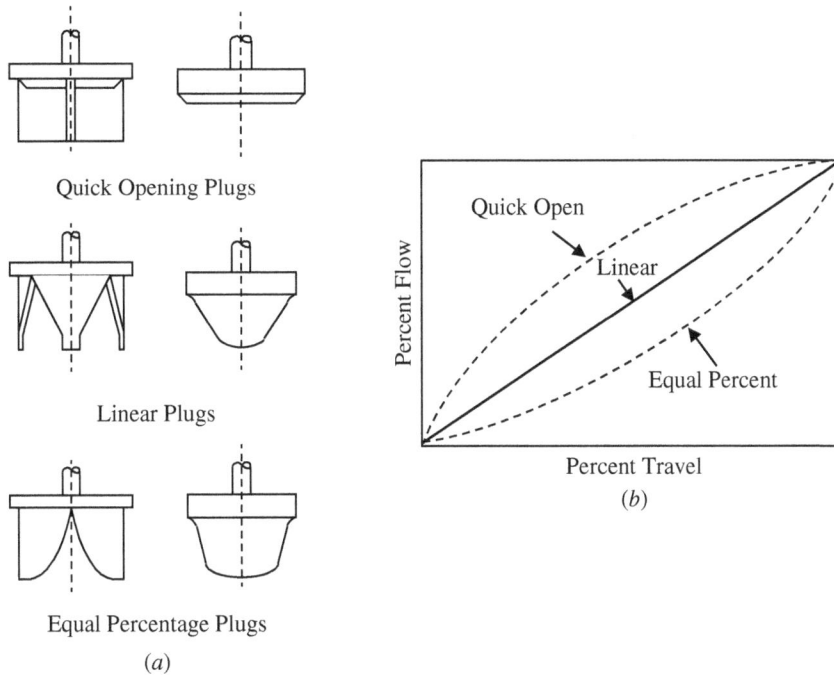

FIGURE **10.7** Various types of plugs used in globe valves for flow control.

any particular application. The type will depend on a careful analysis of the process characteristics, i.e., if the load changes are linear, a linear plug should be used. Conversely, if the load changes are nonlinear, a plug with the appropriate nonlinear characteristic should be used.

10.3.3 Butterfly Valve

The butterfly valve is shown in Fig. 10.8a, and its flow versus travel characteristics are shown in Fig. 10.8b. The relation between flow and lift is approximately equal percentage up to about 50 percent open, after which it is linear. Butterfly valves offer high capacity at low cost, are simple in design, easy to install, and have tight closure. The torsion force on the shaft increases until the valve is open up to 70° and then reduces.

10.3.4 Other Valve Types

A number of other types of valves are in common use. They are the weir diaphragm, ball, and rotary plug valves. The cross sections of these valves are shown in Fig. 10.9.

A *weir-type diaphragm valve* is shown in Fig. 10.9a. The valve is shown open; closure is achieved by forcing a flexible membrane down onto the weir. Diaphragm valves are good for slurries and liquids with suspended solids, are low cost devices, but tend to require high maintenance and have poor flow characteristics.

A one-piece *ball valve* is shown in Fig. 10.9b. The valve is a partial sphere that rotates. The valve tends to be slow to open. Other than the one shown in Fig. 10.9b, the ball valve is available in other configurations with spheres of various shapes for different

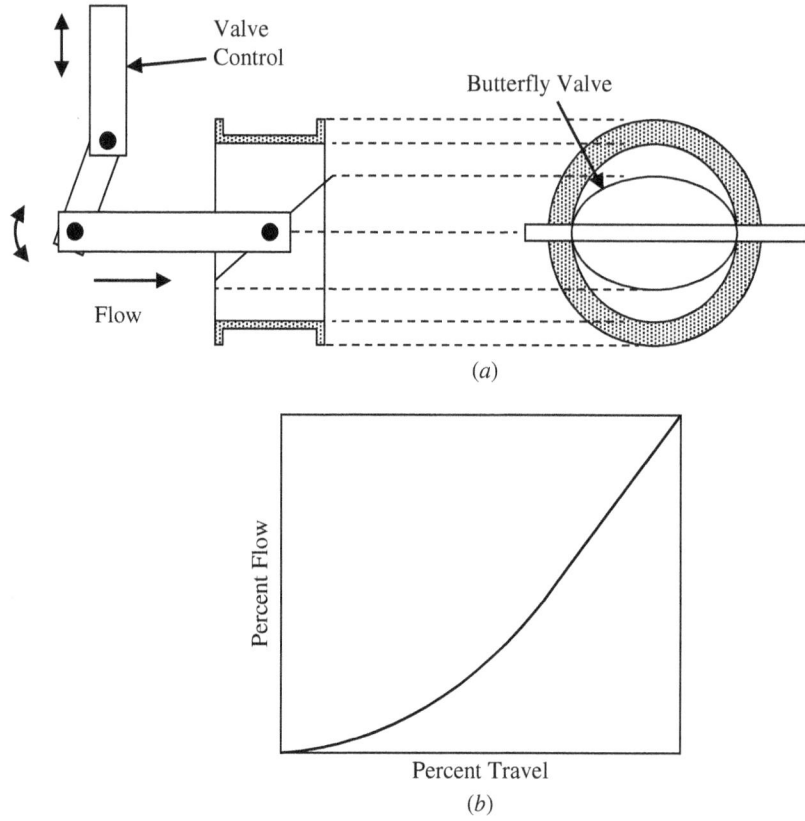

FIGURE 10.8 Cross section of (a) a butterfly valve and (b) its flow versus travel characteristic.

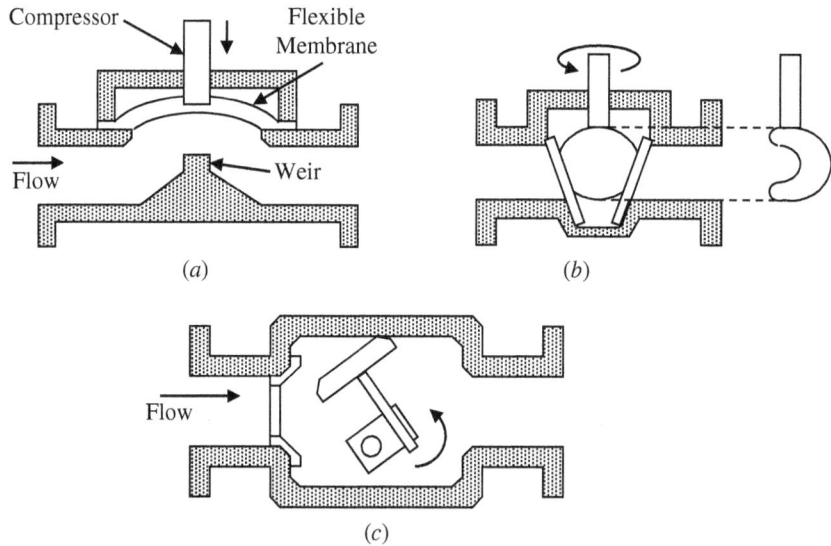

FIGURE 10.9 Different valve types: (a) diaphragm, (b) one-piece ball valve, and (c) rotary plug valve.

flow characteristics. The valve is good for slurries and liquids with solid matter because of its self-cleaning operation. Ball valves have tight turnoff characteristics, are simple in design, and have greater capacity than similar-sized globe valves.

An eccentric *rotary plug valve* is shown in Fig. 10.9c. The valve is medium cost but requires less closing force than many valves and can be used for forward or reverse flow. The valve has tight shutoff with positive seating action, high capacity, and can be used with corrosive liquids.

10.3.5   Valve Characteristics

Other factors that determine the choice of valve type are corrosion resistance, operating temperature ranges, high and low pressures, fluid velocities, and fluids containing solids. Correct valve installation is essential; vendor recommendations must be carefully followed. In situations where sludge or solid particulates can be trapped upstream of a valve, a means of purging the pipe must be available. To minimize disturbances and obtain good flow characteristics, a clear run of 1 to 5 pipe diameters up and down stream of the valve should be allowed.

Valve sizing is based on pressure loss. Valves are given a C_V number that is based on test results. The C_V number is the number of gallons of water flowing per minute through a fully open valve at 60°F (15.5°C) that will cause a pressure drop of 1 psi (6.9 kPa). It implies that when flowing through the fully opened valve, it will have a pressure drop of 1 psi (6.9 kPa), i.e., a valve with a C_V of 25 will have a pressure drop of 1 psi when 25 gallons of water per minute is flowing through it. For liquids, the relation between pressure drop P_d (pounds per square inch), flow rate Q (US gallon per minute), and C_V is given by

$$C_v = Q \times \sqrt{(SG/P_d)} \qquad (10.1)$$

where SG is the specific gravity of the liquid.

Example 10.1   What is the C_V of a valve, if there is a pressure drop of 3.5 psi when 2.3 gal per second of a liquid with a SW of 60 lb/ft³ are flowing?

Solution

$$C_v = 2.3 \times 60 \sqrt{\frac{60}{62.4 \times 3.5}} = 138 \times 0.52 = 72.3$$

Table 10.1 gives a comparison of some of the valve characteristics; the values shown are typical of the devices available and may be exceeded by some manufacturers with new designs and materials.

10.3.6   Valve Fail Safe

An important consideration in many systems is the position of the actuators when there is a loss of power, i.e., will chemicals or the fuel to the heaters continue to flow or will total system shutdown occur? Figure 10.10 shows examples of pneumatically or hydraulically operated globe valve designs that can be configured to go to the open or closed position during a system failure. The modes of failure are determined by simply changing the spring position and the pressure port.

In Fig. 10.10a, the globe valve is closed by applying pressure to the pressure port to oppose the spring action. If the system fails, i.e., if there is a loss of pneumatic pressure,

Parameter	Globe	Diaphragm	Ball	Butterfly	Rotary Plug
Size	1 to 36 in	1 to 20 in	1 to 24 in	2 to 36 in	1 to 12 in
Slurries	No	Yes	Yes	No	Yes
Temperature range °C	−200 to 540	−40 to 150	−200 to 400	−50 to 250	−200 to 400
Quick-opening	Yes	Yes	No	No	No
Linear	Yes	No	Yes	No	Yes
Equal %	Yes	No	Yes	Yes	Yes
Control range	20:1 to 100:1	3:1 to 15:1	50:1 to 350:1	15:1 to 50:1	30:1 to 100:1
Capacity C_v (d = Dia)	10 to 12 × d^2	14 to 22 × d^2	14 to 24 × d^2	12 to 35 × d^2	12 to 14 × d^2

TABLE 10.1   Valve Characteristics

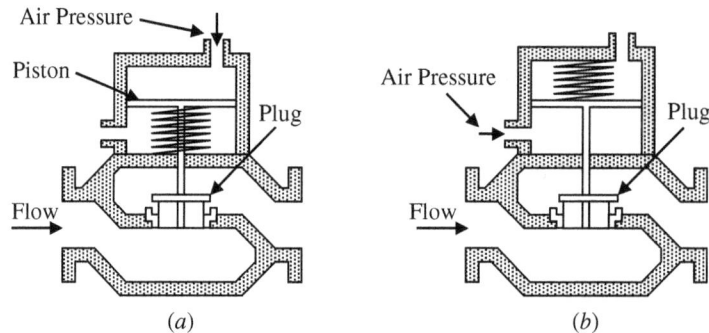

FIGURE 10.10   Fail-safe pneumatic- or hydraulic-operated valves, with the loss of operating pressure the valve in (a) opens and (b) closes.

the spring acting on the piston will force the valve to its open position. In Fig. 10.10b, the spring is removed from below the piston to a position above the piston and the inlet and exhaust ports are reversed. In this case, the valve is opened by the applied pressure working against the spring action. If the system fails and there is a loss of control pressure, the spring action will force the piston down and close the valve. Similar fail-safe electrically and hydraulically operated valves are available. Two-way and three-way fail-safe valves are also available which can be configured to be in a specific position when the operating system fails.

10.4   Actuators

Actuators are used to open and close valves or hold a valve in an intermediate position for flow or level control. Air and electrically operated valves are the most common, with air being the simplest only requiring an air supply and is still the industry standard in the operation of most valves. The hydraulically controlled valve requires the addition of high-pressure supply and return lines. The actuator-operated valve position is set by a central process controller. The actuator also has the ability to inform the controller of the valves position and correct operation of the valve. The response time of an actuator can vary between several seconds and 60+ seconds depending on its size, which must be taken into account by the proportional, integral, and derivative (PID) action controller.

10.4.1 Operation

Flow rates are controlled by a variable position valve that can be set in any position between fully open and fully closed. The position of the valve can be controlled automatically, electrically, hydraulically, or pneumatically.

The industry standard for pneumatic pressure control signals are in the range of 3 to 15 psi and for electrical signals the range is 4 to 20 mA (in the case of HVAC 0 to 10 V). Process-control systems are now computer based and the control signals are normally electrical, so that for the control of pneumatic actuators converters are used at the actuator to convert the 4 to 20 mA control signal to 3 to 15 psi that is required by the actuator. Figure 10.11 shows two control loops; the loop in Fig. 10.11a is the original pneumatic loop and Fig. 10.11b is an electrical loop with an I to P converter to operate the pneumatic actuator.

Control signals can be used to directly control the actuator, but with wear over time, friction, and hysteresis errors can occur in the position of the valve for given amplitude control signal. To eliminate these errors and to confirm the valve is operating correctly direct feedback from the actuator to the controller is used.

This feedback loop is known as the "positioner." In the case of a pneumatic actuator, the feedback is provided by mechanical linkage from the actuator drive shaft to the

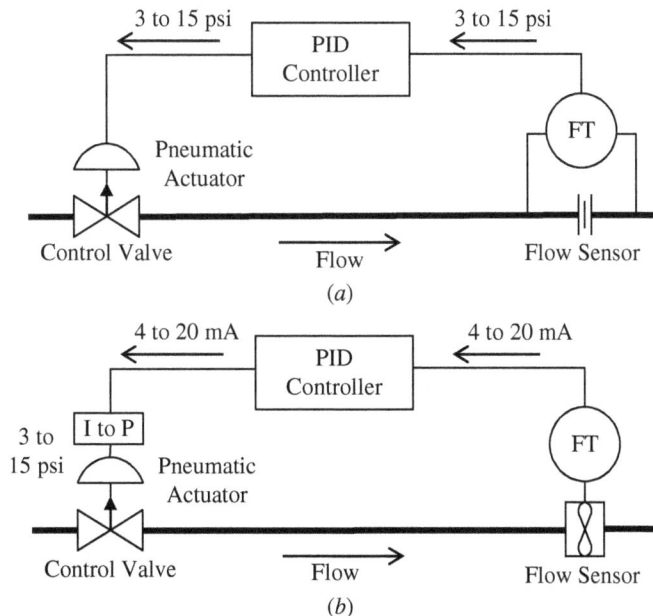

Figure 10.11 It shows (a) pneumatic control loop and (b) electrical control loop with I to P converter.

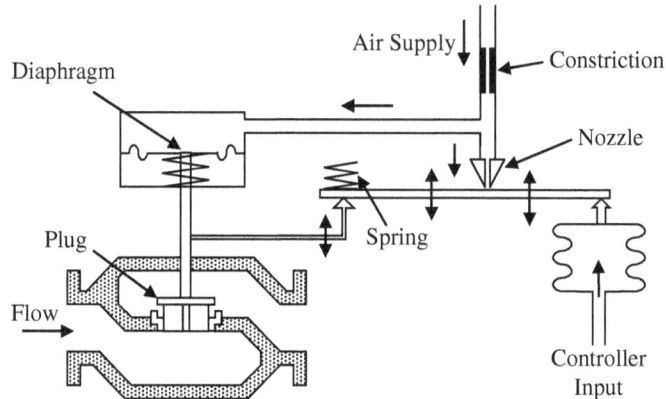

FIGURE **10.12** Pneumatic valve positioner.

bleed on the air supply controlling the actuator, this is shown in Fig. 10.12. The smart positioner compares the position of the actuator drive shaft to the position indicated by the controller, and adjusts the air pressure to the actuator until the shaft position is the same position as requested by the controller.

The positioner is used in the smart controller as shown in Fig 10.13. The smart controller communicates with the PLC using one of the protocols such as the HART, Fieldbus Foundation, or Profibus. The PLC sends set point information to the flow controller that monitors the flow rate and sends the signal to the positioner to ensure accurate flow rate, without having to readjust for wear and hysteresis.

In Fig. 10.14, a globe valve operated by an electric motor is shown. The screw driven by the motor can move the plug in the valve up or down. A potentiometer wiper is attached to the valve stem and gives a voltage directly proportional to the amount the valve is open. This voltage can be compared to the set point from the controller, and can operate as a smart positioner. The output from the comparator is amplified and can be used to drive the motor using high-current solid-state switches. Such a setup would be used in a smart controller. Advantages of the motor drive are that the motor

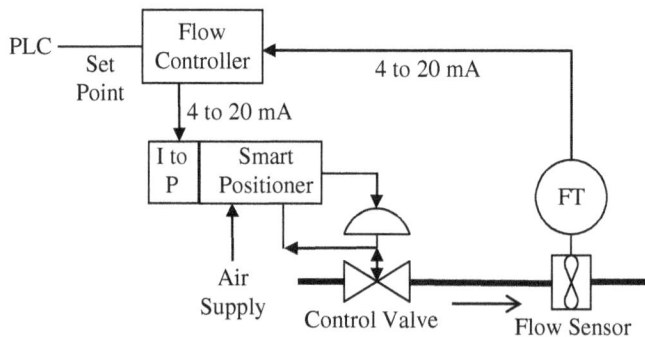

FIGURE **10.13** Smart controller.

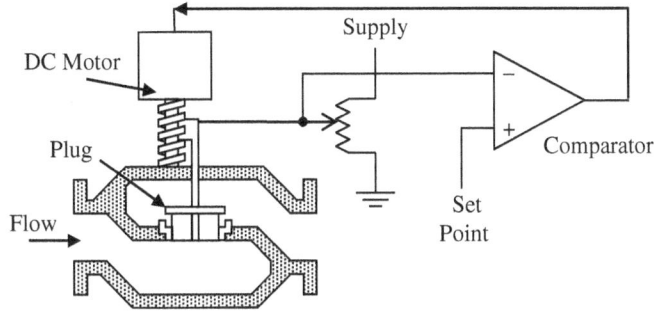

Figure 10.14 Electrical actuator with positioner.

can drive the valve so as to increase or reduce the flow and has a faster response than a pneumatic actuator. A disadvantage is that limit or torque switches are needed to turn the power to the motor OFF when the valve is fully open or closed.

10.4.2 Control Valves

A very large number of types and sizes of control valves are in use ranging in size from 1/10 of an inch to several feet. The valve design depends on many factors such as the volume and flow rate of the liquid to be controlled, the pressure of the liquid, and the required temperature operating range. Other questions are what is the valve controlling, liquids or gases, what is the density and viscosity of the liquid, are the liquids or gases corrosive, and are there particulates in the liquids. Valves can be configured with single or double seating to reduce actuator drive requirements. However, the main controlling actions are sliding stem or rotary.

The most common types of sliding stem control valves in use are the globe and angled globe. Common types of rotary valves are the ball and butterfly.

10.5 Electronic Devices

A number of electronic devices are available for switching high currents for motor control.

They are as follows:

1. Silicon-controlled rectifier (SCR)
2. Bidirectional ac switch (TRIAC)
3. Darlington bipolar junction transistors (BJT)
4. Power metal-oxide semiconductor field-effect transistor (MOSFET)
5. Insulated gate bipolar transistor (IGBT)
6. MOS-controlled thyristor (MCT)

A comparison of the power devices characteristics is given in Table 10.2.

Device	Power Handling	Saturation (Volt)	Turn-on Time	Turn-off Time
SCR	2 kV 1.5 kA	1.6	20 μs	N/A
TRIAC	2 kV 1 kA	2.1	20 μs	N/A
BJT	1.2 kV 800 A	1.9	2 μs	5 μs
MOSFET	500 V 50 A	3.2	90 ns	140 ns
IGBT	1.2 kV 800 A	1.9	0.9 μs	200 ns
MCT	600 V 60 A	1.1	1.0 μs	2.1 μs

Table 10.2 Comparison of Power Device Characteristics

10.6 Application Considerations

10.6.1 Valves

The selection of control valves for a particular application depends on many variables such as the corrosive nature of the gas or fluid, temperature of operation, pressures involved, high or low flow rates, volume of flow, liquid viscosity, and the amount of suspended solids.

Valves are the final element in a control loop and are critical in controlling the "manipulated variable" to provide the correct flow for good process control. The valve is subject to operation in very harsh conditions and is one of the most costly elements in the process-control system. The choice and correct installation requires both knowledge and experience. Careful attention must be made to the system requirements and manufacturers' specifications, only then can a careful valve selection be made (additional information can be obtained from the ISA 75 series of standards).

Some of the factors affecting the choice of valves are as follows:

1. Type of valve for two-way or three-way use, and so on

2. Fail-safe considerations

3. Valve size from flow requirements, care must be taken to avoid both oversizing and undersizing

4. Materials used in the valve construction, from considerations of pressure, size, and corrosion. Materials used in valves range from PVC to brass to steel

5. *Tightness of shutoff*: Valves are classified by quality of shutoff by leakage at maximum pressure. Valves are classified into six classes depending on leakage from 0.5 percent of rated capacity to 0.15 mL/min for a 1-in diameter valve

6. Acceptable pressure drop (C_v) across the valve

7. Valve body for linear or rotary motion, i.e., globe, diaphragm versus ball, butterfly, and so forth

8. Valve accessibility and ease of maintenance

The type of valve or plug depends on the nature of the process. In the case of a fast reaction with small load changes, control is only slightly affected by valve characteristics.

When the process is slow with large load changes, valve characteristics are important; i.e., if the load change is linear, a valve with a linear characteristic should be used, in the case of a nonlinear load change, a valve with an equal percentage change may be required. In some applications, valves are required to be completely closed when OFF. Other considerations are maintenance, serviceability, fail-safe features, pneumatic, hydraulic, solenoid or motor control, and the need for feedback. The above is a limited review of actuator valves, as previously noted, the manufacturer's data sheets should be consulted when choosing a valve for a particular application.

Summary

This chapter discussed the type of valves used to control flow or the manipulated variable, the types of actuators used for valve control, and the feedback loops used under the control of the system controller to control flow rates.

The main points described in this chapter were as follows:

1. The type of self-regulating gas pressure regulators used in process control, the internal and external loading of the regulators using springs, weights, pressure, and pressure amplifiers.

2. Various methods of automatically controlling liquid levels.

3. A wide variety of control valves are available for flow control. A comparison of their characteristics is given and some options available when choosing a control valve for a specific application are also discussed.

4. Flow control valves are designed with a variety of plugs for different control characteristics for different applications such as linear, quick opening valves, and equal percentage valves. The plug with the proper characteristics should be chosen for the specific application.

5. Fail-safe valve configurations are needed to prevent the flow of material during a system failure or loss of power. Valve configurations are shown for valves to fail in the open position or in the closed position.

6. Pneumatic actuators and the use of "positioners" for accurate valve positioning.

7. Potentiometers for electrical position feedback are shown with a "positioner" for feedback control.

8. The electronic power control devices that are now available for efficient power control.

Problems

10.1 What is the prime use for a regulator?

10.2 What is an actuator?

10.3 What is an instrument pilot-operated pressure regulator?

10.4 What do you understand by fail-safe "open"?

10.5 What are the methods used to load valve regulators?

10.6 What are the power sources for actuators?

10.7 What is a "positioner"?

10.8 How is the position of a valve communicated back to the controller?

10.9 What is a smart controller?

10.10 Where would you use a safety valve?

10.11 What is the range of currents used in control signals?

10.12 What is the pressure range used in pneumatic signals?

10.13 What are the common types of plugs?

10.14 How are liquid levels controlled?

10.15 Name the various types of valve families.

10.16 Name the various valve configurations that can be found within the globe valve family.

10.17 A valve has a C_v of 88. What is pressure drop in the valve when 1.8 gal/sec of a liquid with a specific weight (SW) of 78 lb/ft³ is flowing?

10.18 Describe a three-position globe valve.

10.19 Describe a butterfly valve.

10.20 Name three types of high-voltage power devices used to switch high currents.

CHAPTER 11

Process Control

Chapter Objectives

This chapter is an introduction to different process-control concepts and will help you understand and become familiar with the different process-control actions.

Topics discussed in this chapter are as follows:

- Concepts of signal control and controller modes
- Concept of lag time, error signals, and correction signals
- ON/OFF types of process-controller action
- Proportional, derivative, and integral action in process controllers
- ON/OFF pneumatic control systems
- ON/OFF electric controllers
- Pneumatic proportional, integral, and derivative (PID) controllers
- Analog electronic implementation of proportional, derivative, and integral action
- An electronic PID loop
- Digital controller system

11.1 Introduction

Control systems vary extensively in complexity and industrial application. Industrial controllers, e.g., in the petrochemical industry, automotive industry, soda processing industry, and the like, have completely different types of control features and functions. The control loops can be very complex requiring microprocessor supervision, down to very simple loops such as those used for controlling water temperature or heating, ventilation, and air-conditioning (HVAC) for comfort. Some of the functions need to be very tightly controlled, with tight tolerance on the variables and a quick response time, while in other areas the tolerances and response times are not so critical. These systems are closed-loop systems. The output level is monitored against a set reference level and any difference detected between the two is amplified and used to control an input variable that will maintain the output at the set reference level.

11.2 Basic Terms

Some of these terms have already been defined, but apply to this chapter. Hence, the terms are redefined here for completeness.

Measured variable is an output process variable that must be held within given limits.

Controlled variable is an input variable to a process that is varied by a valve to keep the output variable (measured variable) within its set limits.

Lag time is the time required for a control system to return a measured variable to its set point after there is a change in the measured variable, which could be the result of a loading change or set point change, and so on.

Dead time is the elapse time between the instant an error occurs and when the corrective action starts.

Dead-band is a set hysteresis between detection points of the measured variable when it is going in a positive or a negative direction. This band is the separation between the turn ON set point and the turn OFF set point of the controller and is sometimes used to prevent rapid switching between the turn ON and turn OFF points.

Set point is the desired amplitude of an outpoint variable from a process.

Error signal is the difference between a set reference point and the amplitude of the measured variable.

Transient is a temporary variation of a load parameter after which the parameter returns to its nominal level.

Variable range is the acceptable limits within which the measured variable must be held and can be expressed as a minimum and a maximum value, or a nominal value (set point) with ± spread (percent).

Control parameter range is the range of the controller output required to control the input variable to keep the measured variable within its acceptable range.

Offset is the difference between the measured variable and the set point after a new controlled variable level has been reached. It is that portion of the error signal which is amplified to produce the new correction signal and produces an "Offset" in the measured variable.

11.3 Control Modes

The two basic modes of process control are ON/OFF action and "continuous control" action. In either case, the purpose of the control is to hold the measured variable output from a process within set limits by varying the controlled input variable to the process.

In the case of ON/OFF control (discrete control or two-position control), the output of the controller changes from one fixed condition (ON) to another fixed position (OFF). Control adjustments are the set point and in some applications a dead-band is used. In continuous control (modulating control) action the feedback controller determines the error between a set point and a measured variable. The error signal is then used to produce an actuator control signal to operate a valve and reduce the error signal. This type of control continuously monitors the measured variable and has three modes of operation which are proportional, integral, and derivative. Controllers can use one of the functions, two, or all three of the functions as required.

11.3.1 ON/OFF Action

The simplest form of control in a closed-loop system is ON/OFF action. The measured variable is compared to a set reference. When the variable is above the reference the system is turned ON and when below the reference the system is turned OFF or vice versa, depending on the system design. This could make for rapid changes in switching between states. However, such systems normally have a great deal of inertia or momentum that produces over swings and introduces long delays or lag times before the variable again reaches the reference level. Figure 11.1a shows an example of a simple room heating system. The top graph shows the room temperature or measured variable and the lower graph shows the actuator signal. The room temperature reference is set at 75°F. When the air is being heated, the temperature in the center of the room has already reached 77°F before the temperature at the sensor reaches the reference temperature of 75°F and similarly as the room cools, the temperature in the room will drop to 73°F before the temperature at the sensor reaches 75°F. Hence, the room temperature will go from about 72°F to 78°F due to the inertia in the system.

ON/OFF action has many applications in industry such as filling containers on a conveyer belt, packaging goods, and so on.

11.3.2 Differential Action

Differential or delayed ON/OFF action is a mode of operation where the simple ON/OFF action has hysteresis or a dead-band built in. Figure 11.1b shows an example of a room heating system similar to that shown in Fig. 11.1a except that instead of the thermostat turning ON and OFF at the set reference of 75°F, the switching points are delayed by ±3°F. As can be seen in the top graph, the room temperature reaches 78°F before the thermostat turns OFF the heating system and the room temperature falls to 72°F before the system is turned ON giving a built-in hysteresis of 6°F. There is, of course, still some inertia. Hence, the room temperature will go from about 70°F to about 80°F.

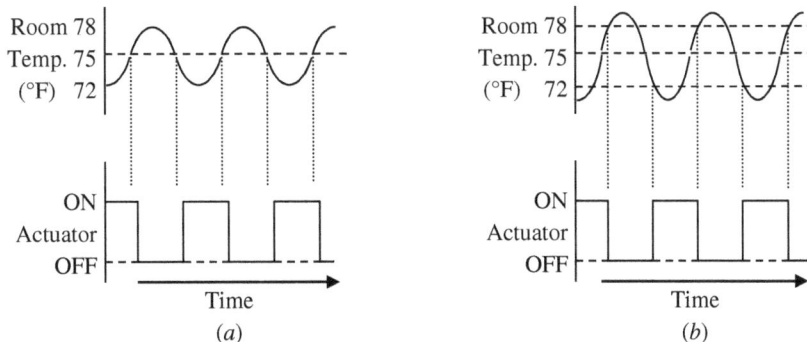

FIGURE 11.1 A room heating system with (a) simple ON/OFF action of a room heating system and (b) differential ON/OFF action.

11.3.3   Proportional Action

The most common of all continuous industrial process-control action is proportional control action. The amplitude of the output variable from a process is measured and converted to an electrical signal. This signal is compared to a set reference point. Any difference in amplitude between the two (error signal) is amplified and fed to a control valve (actuator) as a correction signal. The control valve controls one of the inputs to the process. Changing this input will result in the output amplitude changing until it is equal to the set reference or the error signal is zero. The amplitude of the correction signal is transmitted to the actuator controlling the input variable and is proportional to the percentage change in the output variable amplitude measured with respect to the set reference. The industrial process-control system has low inertia; overshoot, and response times must be minimized for fast recovery and to keep processing tolerances within tight limits. In order to achieve these goals fast reaction and settling times are needed. There may also be more than one variable to be controlled and more than one output being measured in a process.

The change in output level may be a gradual change, a large on-demand change, or caused by a change in the reference level setting. An example of an on-demand change would be cleaning stations using hot water at a required fixed temperature, as shown in Fig. 11.2a. At one point in time the demand could be very low with a low flow rate as would be the case if only one cleaning station were in use. If cleaning commenced at several of the other stations, the demand could increase in steps or there could be a sudden rise to a very high flow rate. The increased flow rate would cause the water temperature to drop. The drop in water temperature would cause the temperature sensor to send a correction signal to the actuator controlling the steam flow so as to increase the steam flow to raise the temperature of the water to bring it back to the set reference level (see Fig. 11.2b). The rate of correction will depend on the inertia in the system, gain in the feedback loop, allowable amount of overshoot, and so forth.

In a closed-loop feedback system settings are critical. If the system has too much gain, i.e., the amplitude of the correction signal is too great, it will cause the controlled variable to over correct for the error, which in turn will give a false error signal in the reverse direction. The actuator will then try to correct for the false error signal.

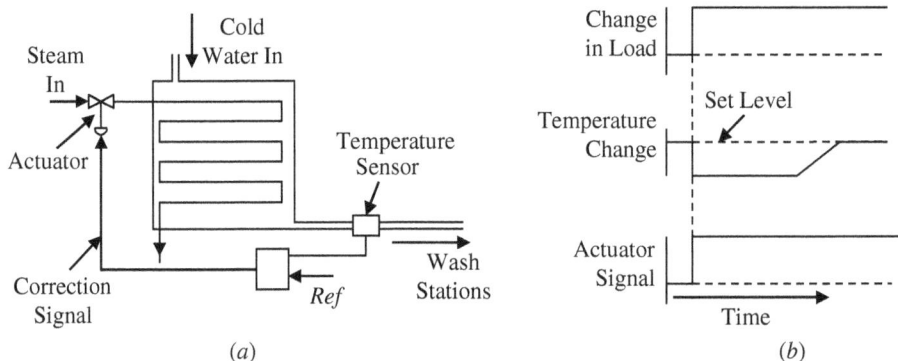

FIGURE 11.2   Water heater (a) showing a feedback loop for constant temperature output and (b) effect of load changes on the temperature of the water from the water heater.

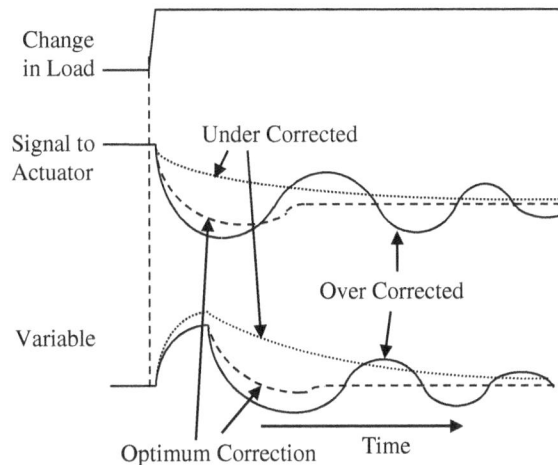

Figure 11.3 Effect of loop gain on correction time using proportional action with over correction and under correction.

This can, in turn, send a larger correction signal to the actuator, which will cause the system to oscillate or cause an excessively long settling or lag time. If the gain in the system is too low the correction signal is too small and the correction will never be fully completed, or again an excessive amount of time is taken for the output to reach the set reference level.

This effect is shown in Fig. 11.3, as can be seen in comparing the over corrected (excessive gain) and the under corrected (too little gain) to the optimum gain case (with just a little overshoot). The variable takes a much longer time for the correction to be implemented than in the optimum case. In many processes, this long delay or lag time is unacceptable.

11.3.4   Derivative Action

Proportional plus derivative (PD) action was developed in an attempt to reduce the correction time that would have occurred using proportional action alone. Derivative action senses the rate of change of the measured variable and applies a correction signal that is proportional to the rate of change only (this is also called rate action or anticipatory action). Figure 11.4a shows some examples of derivative action. As can be seen in this example, a derivative output is obtained only when the load is changing. The derivative of a positive slope is a positive signal and the derivative of a negative slope is a negative signal; zero slopes give zero signals as shown. An in-depth look at derivatives is outside the scope of this text.

Figure 11.4b shows the effect of PD action on the correction time. When a change in loading is sensed as shown, both the P and D signals are generated and added. The significance of combining these two signals is to produce a signal that speeds up the actuator's control signal. The faster reaction time of the control signal reduces the time to implement corrective action reducing the excursion of the measured variable and its settling time. The amplitudes of these signals must be adjusted for optimum operation, or over or under shoot can still occur.

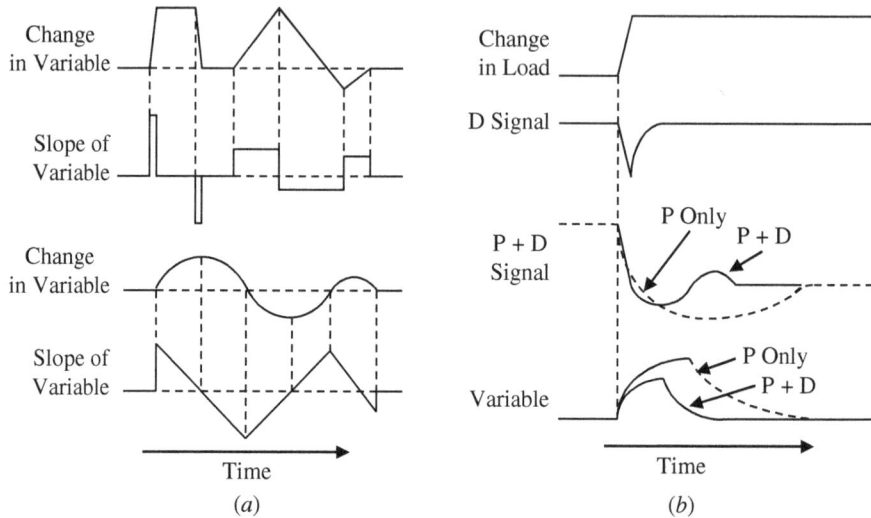

FIGURE 11.4 Proportional and derivative action (a) variable change with resulting slope and (b) effect of proportional and derivative action on a variable.

11.3.5 Integral Action

Proportional plus integral (PI) action, also known as reset action, was developed to correct for long-term loads and applies a correction proportional to the area under the change in the variable curve. Figure 11.5a gives some examples of the integration of a curve or the area under a curve. In the top example, the area under the square wave increases rapidly but remains constant when the square wave drops back to zero. In the triangular section, the area increases rapidly at the apex but increases slowly as the triangle approaches zero; when the triangle goes negative, the area reduces. In the lower example, the area increases more rapidly when the sine wave is at its maximum and slower as it approach the zero level. During the negative portion of the sine wave, the area is reduced. Proportional action gives a response to a change in the measured variable but does not fully correct the change in the measured variable due to its limited gain. For instance, if the gain in the proportional amplifier is 100, then when a change in load occurs 99 percent of the change is corrected. However, a 1 percent error signal is required for amplification to drive the actuator to change the manipulated variable. The 1 percent error signal is effectively an "offset" in the variable with respect to the reference. Integral action gives a slower response to changes in the measured variable to avoid overshoot, but has a high gain so that with long-term load changes it takes over control of the manipulated variable and applies the correction signal to the actuator. Because of the higher gain the measured variable error is reduced to close to zero. This also returns the proportional amplifier to its normal operating point, so that it can correct for other fluctuations in the measured variable. Note that these corrections are done at relatively high speeds. The older pneumatic systems are much slower and can take several seconds to make such a correction. Figure 11.5b shows the PI corrective action waveforms. When a change in loading occurs, the P signal responds to take corrective action to restore the measured variable to its set point; simultaneously, the integral signal starts to change

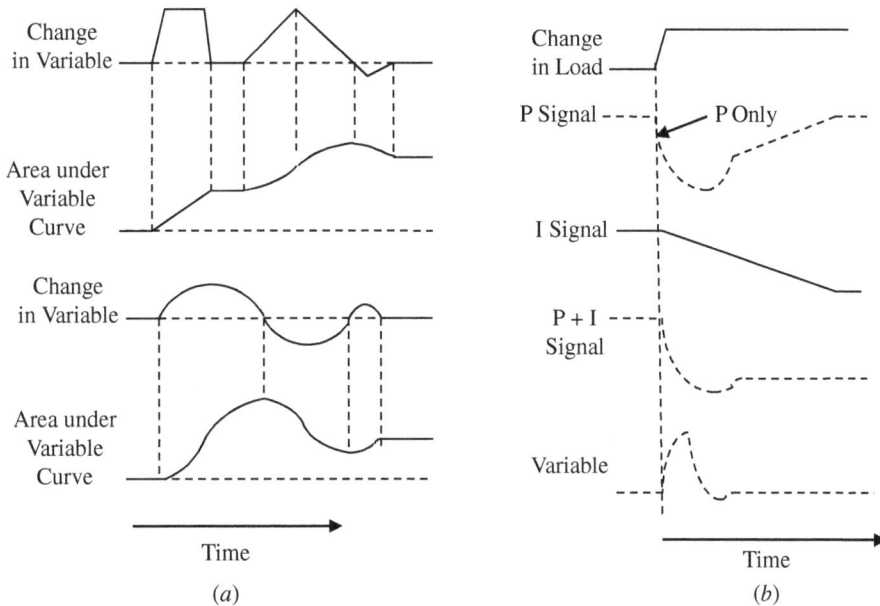

FIGURE 11.5 Proportional and integral action (a) variable change with area under the graph and (b) effect of proportional and integral action on a variable.

linearly to supply the long-term correction, thus allowing the proportional signal to return to its normal operating point as is shown. Here again integral action can become complex and further discussion is considered to be outside the scope of this text.

11.3.6 PID Action

A combination of all three of the actions described earlier is more commonly referred to as proportional, integral, and derivative (PID) action. The waveforms of PID action are illustrated in Fig. 11.6. PID is the most often used corrective action for process control. There are however, many other types of control actions based on PID action. Understanding the fundamentals of PID action gives a good foundation for understanding other types of controllers. The waveforms used have been idealized for ease of the explanation and are only an example of what may be encountered in practice. Loading is a function of demand and is not affected by the control functions or actions; the control function is to ensure that the variables are within their specified limits.

To give an approximate indication of the use of PID controllers for different types of loops, the following are general rules that should be followed:

Pressure control requires proportional and integral; derivative is not normally required.

Level control uses proportional and sometimes integral; derivative is not normally required.

Flow control requires proportional and integral; derivative is not normally required.

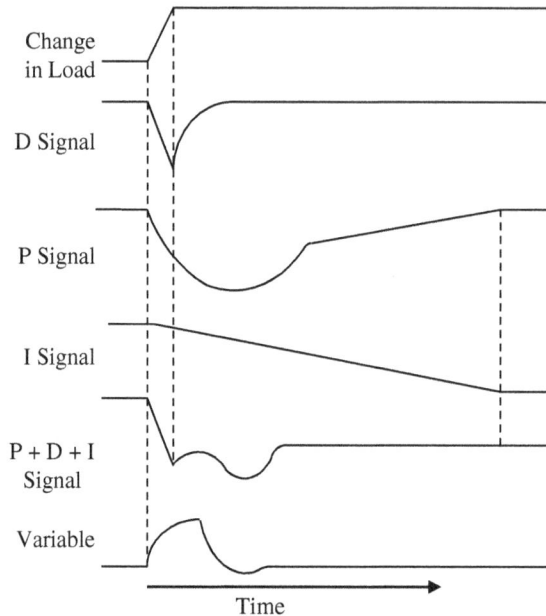

FIGURE 11.6 Waveforms for proportional plus integral action and waveforms for proportional plus derivative and integral action.

Temperature control uses proportional, integral; and derivative usually with integral set for a long time period.

However, the above are general rules and each application has its own requirements.

Typical feedback loops have been discussed. The reader should, however, be aware that there are other kinds of control loops used in process control such as cascade, ratio, and feed-forward.

11.4   Implementation of Control Loops

Implementation of the control loops can be achieved using pneumatic, analog, or digital electronics. The first process controllers were pneumatic. However, these have largely been replaced by electronic systems, because of improved reliability, less maintenance, easier installation, easier adjustment, higher accuracy, lower cost, use with multiple variables, and higher speed operation.

11.4.1   ON/OFF Action Pneumatic Controller

Figure 11.7 shows a pneumatic furnace control system using a pneumatic ON/OFF controller. In this case, the furnace temperature sensor moves a flapper that controls the air flow from a nozzle. When the temperature in the furnace reaches its set point the sensor moves the flapper toward the nozzle to stop the air flow and allow pressure to build up in the bellows. The bellows operates an air-control relay that shuts OFF the air flowing to the control valve turning OFF the fuel to the furnace. When the temperature in the furnace drops below a set level the flapper is opened by the sensor, reducing the

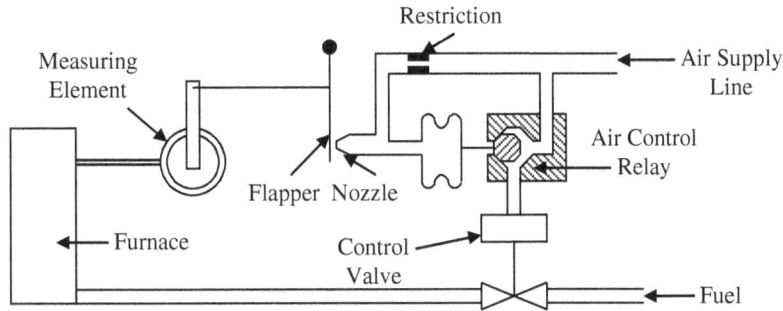

FIGURE 11.7 Pneumatic ON/OFF furnace controller.

air pressure in the bellows, which in turn opens the air-control valve allowing the air pressure to drop and the control valve to open, turning ON the fuel to the furnace.

11.4.2 ON/OFF Action Electrical Controller

An example of an ON/OFF action electrical room temperature controller is shown in Fig. 11.8. In this case, the room temperature is sensed by a bimetallic sensor. The sensor operates a mercury switch. As the temperature decreases the bimetallic element tilts the mercury switch down causing the mercury to flow to the end of the glass envelope and in so doing shorts the two mercury switch contacts together. The contact closure operates a low-voltage relay turning ON the blower motor and the heating element. When the room temperature rises to a predetermined set point the bimetallic strip tilts the mercury switch back causing the mercury to flow away from the contacts. The low-voltage electrical circuit is turned OFF, and the power to the heater and the blower motor is disconnected.

The ON/OFF controller action has many applications in industry; an example of some of these uses is shown in Fig. 11.9. In this case, cartons on a conveyer belt are being filled from a hopper. When a carton is full it is sensed by the level sensor, which sends a signal to the controller to turn OFF the material flowing from the hopper and to start the conveyer moving. As the next carton moves into the filling position it is sensed by the position sensor, which sends a signal to the controller to stop the conveyer belt and to start filling the carton. Once it is full the cycle repeats itself.

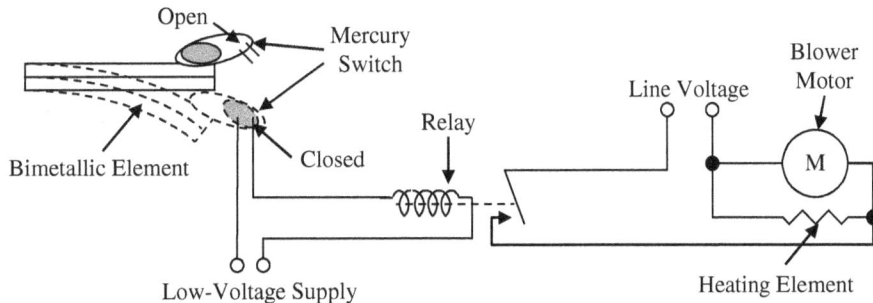

FIGURE 11.8 Simple ON/OFF room heating controller.

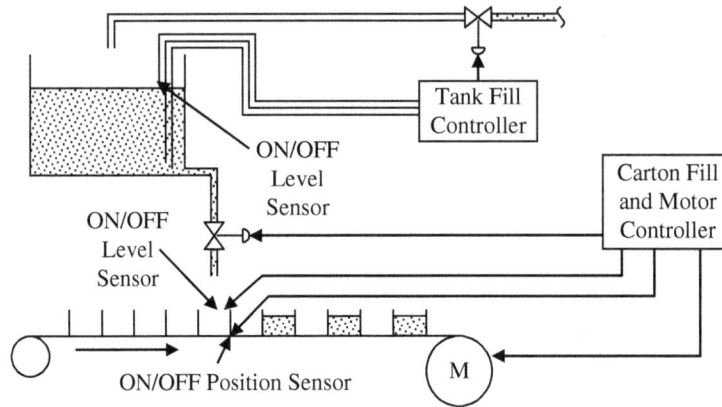

Figure 11.9   Example of the use of ON/OFF controls used for carton filling.

A level sensor in the hopper senses when the hopper is full and when it is almost empty. When empty, the sensor sends a signal to the controller to turn ON the feed valve to the hopper and when the hopper is full it is detected and a signal is sent to the controller to turn the feed to the hopper OFF.

Instead of direct feed from storage unit, a hopper is often used so that in the case of an accident only a limited amount of material is spilled. This is particularly important when dealing with toxic chemicals, inflammable liquids, or combustible material.

11.4.3   PID Action Pneumatic Controller

Many configurations for PID pneumatic controllers have been developed over the years, have served us well, and may still be in use in some older processing plants. But pneumatic controllers have, with the advent of the requirements of modern processing and the development of electronic controllers, achieved the distinction of becoming museum pieces and only a brief description is given. Figure 11.10 shows an example of a pneumatic PID controller. The pressure from the sensing device P_{in} is compared to a

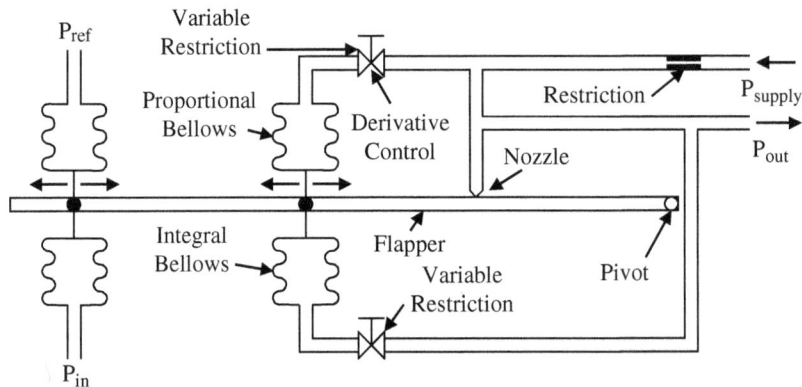

Figure 11.10   Pneumatic PID controller.

set or reference pressure P_{ref} to generate a differential force (error signal) on the flapper to move the flapper in relation to the nozzle giving an output pressure proportional to the difference between P_{in} and P_{ref}. If the derivative restriction is removed the output pressure is fed back to the flapper via the proportional bellows to oppose the error signal and to give proportional action. System gain is adjusted by moving the position of the bellows along the flapper arm; i.e., the closer the bellows is positioned to the pivot the greater the movement of the flapper arm.

By putting a variable restriction between the pressure supply and the proportional bellows, a change in P_{in} causes a large change in P_{out}, as the feedback from the proportional bellows is delayed by the derivative restriction. This gives a pressure transient on P_{out} before the proportional bellows can react, thus giving derivative action. The duration of the transient is set by the size of the bellows and the setting of the restriction.

Integral action is achieved by the addition of the integral bellows and restriction as shown. An increase in P_{in} moves the flapper toward the nozzle causing an increase in output pressure. This increase in output pressure is fed to the integral bellows via the restriction until the pressure in the integral bellows is sufficient to hold the flapper in the position set by the increase in P_{in}, creating integral action.

11.4.4   PID Action Control Circuits

PID action can be performed using either analog or digital electronic circuits. In order to understand how electronic circuits are used to perform these functions, the analog circuits used for the individual actions will be discussed. The circuit shown in Fig. 11.11a is used to compare the signal from the measured variable and the reference to generate the error signal. Proportional action is achieved as shown in Fig. 11.11b by amplifying the error signal V_{in}. The stage gain is the ratio of R_2/R_1; the gain can be adjusted using the potentiometer R_2 giving an inverted output.

The circuit for derivative action is shown in Fig. 11.12a. The feedback resistor can be replaced with a potentiometer to adjust the differentiation duration. The output signal is inverted which can be changed to a noninverted signal with an inverting amplifier stage if required. The waveforms of the differentiator are shown in Fig. 11.12b.

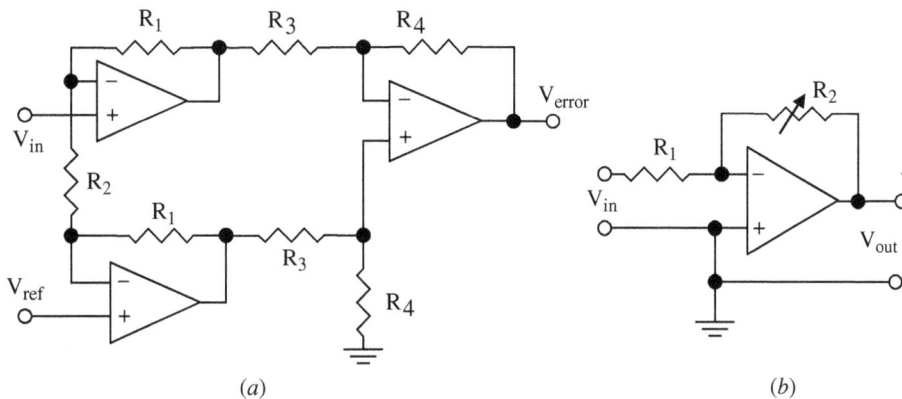

(a)   (b)

Figure 11.11   Circuits used in PID action: (a) error generating circuit and (b) proportional circuit.

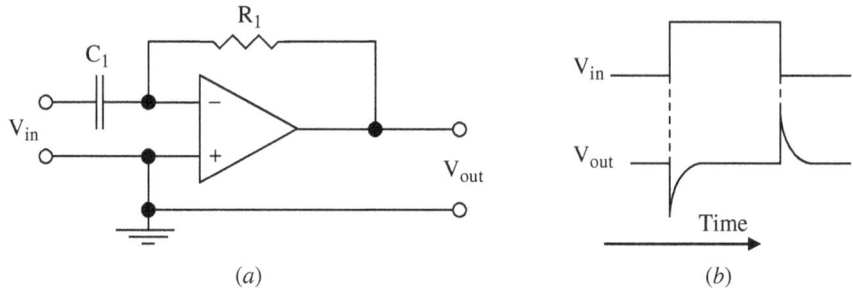

Figure 11.12 Derivative amplifier: (a) circuit and (b) waveforms.

Proportional and derivative action can be combined using the circuit shown in Fig. 11.13a. Derivative action is obtained by the input capacitor C_1 and proportional action by the ratio of the resistors R_1 and R_2. The inverted output signal is shown in Fig. 11.3b.

A circuit to perform integral action is shown in Fig. 11.14a. Capacitive feedback around the amplifier prevents the output from the amplifier from following the input change. The output changes slowly and linearly when there is a change in the measured variable as shown in the waveforms in Fig. 11.14b. The slope of the output waveform is set by the time constant of the feedback C_1 and the input resistance R_1. This is integral action and the

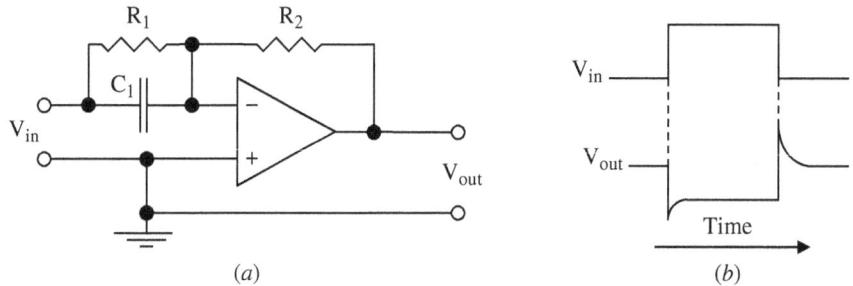

Figure 11.13 Proportional plus derivative amplifier: (a) circuit and (b) waveforms.

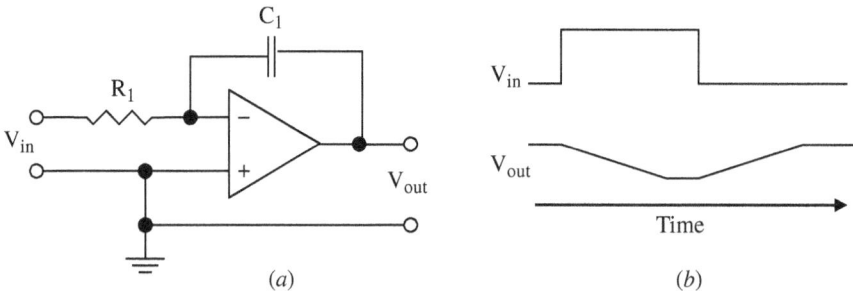

Figure 11.14 Integrating amplifier: (a) circuit and (b) waveforms.

output from the integrator is the area under the input waveform. This area can be adjusted by replacing R_1 with a potentiometer. This circuit gives an inverted output.

11.4.5 PID Electronic Controller

Figure 11.15 shows the block diagram of an analog PID controller. The measured variable from the sensor is compared to the set point in the first unity gain comparator; its output is the difference between the two signals or the error signal. This signal is fed to the integrator via an inverting unity gain buffer and to the proportional amplifier and differentiator via a second inverting unity gain comparator, which compares the error signal to the integrator output. Initially, with no error signal the output of the integrator is zero so that the zero error signal is also present at the output of the second comparator.

When there is a change in the measured variable, the error signal is passed through the second comparator to the proportional amplifier and the differentiator where it is amplified in the proportional amplifier, added to the differential signal in a summing circuit, and fed to the actuator to change the input variable. Although the integrator sees the error signal, it is slow to react so its output does not change immediately, but starts to integrate the error signal. If the error signal is present for an extended period of time, the integrator will supply the correction signal via the summing circuit to the actuator and inputs the correction signal to the second comparator to reduce the effective error signal to the proportional amplifier to zero, when the integrator is supplying the full correction signal to the actuator. Any new change in the error signal will still be passed through the second comparator as the integrator is only supplying an offset to correct for the first long-term error signal. The proportional and differential amplifiers can then correct for any new changes in the error signal.

The circuit implementation of the PID controller is shown in Fig. 11.16. This is a complex circuit because all of the amplifier blocks are shown doing a single function to give a direct comparison to the block diagram and is only used as an example. In practice, there are a number of circuit component combinations that can be used to produce PID action.

A single amplifier can also be used to perform several functions, which would greatly reduce the circuit complexity. Such a circuit is shown in Fig. 11.17, where feedback from the actuator position is used as the proportional band adjustment.

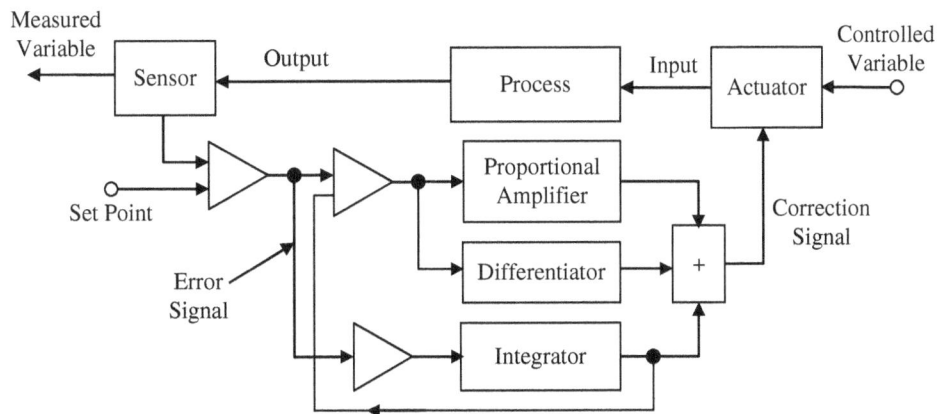

FIGURE 11.15 Block schematic of a PID electronic controller.

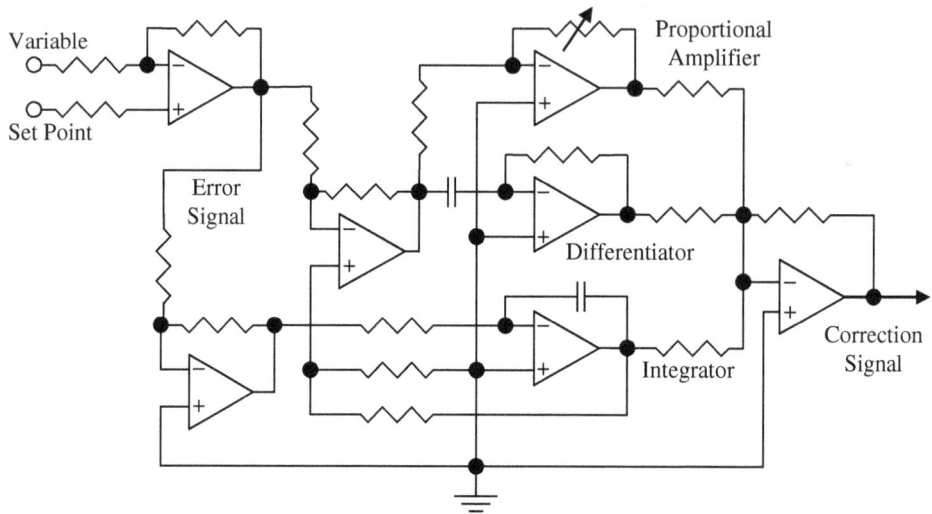

Figure 11.16 Circuit of a PID action electronic controller.

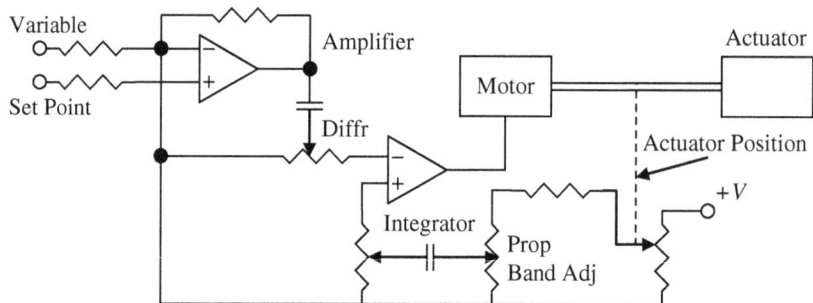

Figure 11.17 Circuit of a PID electronic controller with feedback from the actuator position.

In new designs, PLC processors can be used to replace the analog circuits to perform the PID functions using digital techniques.

11.5 Digital Controllers

Modern process facilities will use a computer or PLC processor as the heart of the control system. The system can control analog loops, digital loops, and will have a foundation fieldbus input/output for communication with smart sensors. All of these control functions may not be required in small process facilities but in large facilities are necessary. The individual control loops are not independent in a process but are interrelated and many measured variables may be monitored and manipulated variables controlled simultaneously. Several processors may also be connected to a mainframe computer for complex control functions. Figure 11.18 shows the block diagram of a processor controlling two digital loops. The analog output from the monitors is converted to a digital

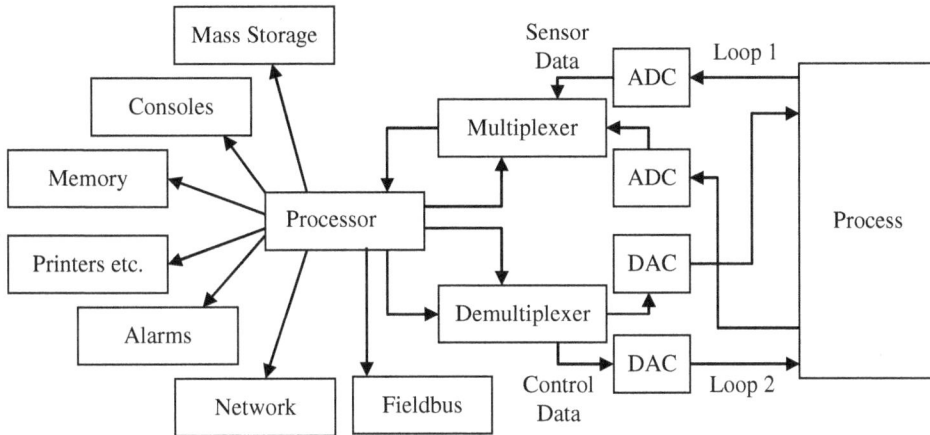

Figure 11.18 Computer-based digital controlled process.

signal in an analog-to-digital controller (ADC). The loop signal is then selected in a multiplexer and PID action is performed in the processor using software programs. The digital output signal is then fed to the actuator through a demultiplexer and a digital-to-analog controller (DAC). The processor will also have mass storage for storing process data for later use or making charts and graphs, and will also be able to control a number of peripheral units and monitors as shown.

Digital controllers will compare the digitized measured variable to the set point stored in memory to produce an error signal which it can amplify under program control and feed to an actuator via a DAC. The processor can measure the rate of change of the measured variable and produce a differential signal to add to the digital correction signal. In addition, the processor can measure the area under the measured variable signal which it will also add to the digital correction signal. All of these actions are under program control; the setting of the program parameters can be changed with a few key strokes, making the system much more versatile than the analog equivalent.

Summary

This chapter discussed process control and the various methods of implementation of the controller functions. Various controller modes and the methods of implementing the modes in pneumatic and electronic circuits are described. An understanding of these circuits will enable the reader to extend these principles to other methods of control.

The main points described in this chapter were as follows:

1. ON/OFF action and delayed ON/OFF action and their use in HVAC. A number of examples of ON/OFF action in process control were given.

2. Proportional, integral, and derivative action and their use in process control, the effects of gain setting in proportional control.

3. Circuits to perform proportional, integral, differential action, and methods of combining the various actions in a PID controller are described.

4. The operation of pneumatic controller actions using flappers, nozzles, and bellows combinations is given. A combination of the various pneumatic components is used to make a PID controller.

5. Digital controller concepts in modern processing facilities are given.

Problems

11.1   Describe controller ON/OFF action?

11.2   What is the difference between simple ON/OFF action and differential ON/OFF action?

11.3   What is proportional action?

11.4   What are integral actions?

11.5   What is derivative action?

11.6   Draw the derivative signal for the variable shown in Fig. 11.19a.

11.7   Draw the integral signal for the variable shown in Fig. 11.19a.

11.8   Draw the derivative signal for the variable shown in Fig. 11.19b.

11.9   Draw the integral signal for the variable shown in Fig. 11.19b.

11.10   Redraw the PID action controller in Fig. 11.16 as it would be if integral action was not required.

11.11   Give a list of applications for ON/OFF controller action.

11.12   Why is the gain setting critical in proportional action?

11.13   What is the difference between an error signal and a measured variable signal?

11.14   What is the difference between lag time and dead time?

11.15   What is the difference between offset and error signal?

11.16   What are some of the actions that can be taken to reduce correction time?

11.17   What is a dead-band?

11.18   What would be the effect of time constants on correction time?

11.19   What types of control do not normally require derivative action?

11.20   Why is ON/OFF action not normally suitable for control of a process?

(a)   (b)

FIGURE 11.19   Change in measured variable for Problems 11.6 through 11.9.

CHAPTER 12

Documentation and Symbol Standards

Chapter Objectives

This chapter will help you understand the need for documentation and become familiar with standard symbols and their use in process-control flow diagrams.

Topics discussed in this chapter are as follows:

- Alarm and trip systems documentation
- Documentation for safety procedures
- Programmable logic controller (PLC) and peripheral circuits documentation
- Interconnection symbols and flow line abbreviations used in piping diagrams
- Instrument Society of America (ISA) list of standard symbols
- Standard instrument symbols and identification letters
- Standard functional actuator symbols
- Control loop numbering system and pipe and identification diagrams (P and IDs)

12.1   Introduction

Documentation covers front-end engineering, detailed explanations of function and operation of all equipment, objectives of the process being controlled, acceptable limits of the product, and detailed engineering drawings. Of the overwhelming amount of documentation needed in a plant, the only documentation that will be introduced is limited to documentation an engineer or technician may encounter and need to use, such as alarm and trip systems, programmable logic controller (PLC) documentation, and pipe and identification diagrams (P and IDs). Of these the P and ID is the detailed documentation covering instruments, their location, process-control loops, and process flow details. Documentation standards and symbols have been set up and standardized by the Instrument Society of America (ISA) in conjunction with the American National Standards Institute (ANSI).

12.2   System Documentation

System modifications and changes are normally the easy part and supported by management. However, because documentation of earlier changes was not completed, the new changes could not be completed or many hours lost in correcting the documentation and product lost. Now, with the new changes, there is not enough time to update the documentation and other changes are pressing, but without the documentation updated the next changes could suffer the same fate or a wrong hastily made correction could cause tremendous damage.

12.2.1   Manuals

The first step in documentation is to have a complete set of vendor-supplied equipment and instrumentation manuals. These should contain maintenance instructions and schedules, calibration, wiring schematics, and parts list.

12.2.2   Alarm and Trip System Documentation

Alarm and trip system information and implementation is given in ANSI/ISA-84.01-1996—Application of Safety Instrumented Systems (SIS) for the Process Control Industry. The purpose of an alarm system is to protect operators, and maintenance personnel, the environment, product, and equipment.

Good up-to-date documentation is a must in alarm and trip systems. All SIS devices should be clearly marked and numbered. System drawings must show all SIS devices using standard symbols, their location, function, and set limits. Drawings must include interlock and logic diagrams.

The types of information required in alarm and trip documentation are as follows:

1. Safety requirement specifications
2. Logic diagram with functional description
3. Functional test procedures and required maintenance
4. Process monitoring points and trip levels
5. Description of SIS action if tripped
6. Action to be taken if SIS power is lost
7. Manual shutdown procedures
8. Restarting procedures after SIS shutdown

12.2.3   Safety Documentation

The documentation on safety should list all of possible hazards that can occur and hazard procedures. Valves will stick open, the SIS will take time to operate and spills will occur, and so on. When toxic or corrosive chemicals are involved first, treatment of personnel who come into contact with the chemicals and second, containment, neutralization, disposal of the chemicals and protective clothing that must be worn during containment and cleanup must be documented. If toxic gases escape or there is a fire personnel evacuation routs must be well documented. Breathing apparatus and venting procedures for toxic gas by the cleanup crew or containment and neutralization if the gas is an environmental hazard must be listed.

Among other safety documented cautions there should be warnings of the accidental generation of sparks if volatile chemicals or gases are present, and so on.

12.2.4 PLC Documentation

As with all technical devices detailed engineering records are essential. Without accurate drawings, changes and modifications needed for upgrading, diagnostics, and maintenance are extremely difficult or impossible. Every wire from the PLC to the monitoring and control equipment must be clearly marked at both ends and shown on the wiring diagram to facilitate wiring changes and diagnostics. The PLC must have complete up-to-date ladder diagrams (or other approved language) and every rung must be labeled with a complete description of its function.

The essential documents in a PLC package are as follows:

1. System overview and complete description of control operation
2. Block diagram of the units in the system
3. Complete list of every input and output, destination, and number
4. Wiring diagram of I/O modules, address identification for each I/O point, and rack location
5. Rung description, number, and function

12.2.5 Circuit Diagrams

In addition to a PLC, custom logic is often used to work with the PLC or in a local control loop, over time the logic may be modified which must be recorded. The logic diagrams must be accurately drawn using standard symbols and in compliance with ISA standards, the function of the circuit accurately described, and accurately maintained.

Accurate records must be kept of the proportional, integral, and derivative (PID) settings (see Chap. 11). These setting are critical in certain areas for process stabilization and fast settling times. If any changes are made to component values both the old and new settings should be recorded together with the reasons for making the change so that if any instability occurs or changes are needed in the future the record will be available to help in deciding on new component values necessary for making additional corrections.

12.2.6 Bussing Information

A critical area is the type of bussing used in the system. This information is required for servicing of the equipment by the engineer or technician, who needs to know if the system is a distributed control system (DCS) with the PLC as an integral part of the system or if the PLC is a standalone controller, what is the protocol being used for communication and is Foundation Field or Profibus being used. The type of communication between the PLC and sensors, smart sensors, and smart instruments can be a 4 to 20 mA analog signal using a twisted pair, or a highway addressable remote transducer (HART) scheme in a point-to-point mode where the PLC imposes a digital signal using frequency shift keying (FSK) on to the analog signal as a second means of communication to a smart sensor to send information such as trip point or set point and request temperature or status information from the sensor, or using the HART protocol. The PLC can be connected to a number of smart sensors or instruments with a single twisted pair and communicate with the end points using a serial data format. Each sensor will have a unique address. This information must be recorded together with the number of bytes being used in the address, bytes in the preamble, and data format. Information is required on the check sum (error code). Each piece of equipment is also assigned a HART identification code which must be recorded, and so on.

12.3   Pipe and Identification Diagrams

One of the most important documents in a process facility is the P and ID. It shows the plant layout with the location and the connection between equipment, piping, cabling, and instrumentation for the process in the plant. It is critical that this diagram is kept up-to-date whenever changes are made.

12.3.1   Standardization

The electronics industry has standard symbols to represent circuit components for use in circuit schematics. Similarly, the processing industry has developed standard symbols to represent the elements in a process-control system. Instead of a circuit schematic the processing industrial drawings are known as *pipe and identification diagrams* (P and IDs) (not to be confused with PID) and represent how the components and elements in the processing plant are interconnected. Symbols have been developed to represent all of the components used in industrial processing and have been standardized by ANSI and ISA. The P and ID document is the ANSI/ISA S5.1–1984 (R 1992)—Instrumentation Symbols and Identification Standards. An overview of the symbols used is given in this chapter but the list is not complete. The ISA should be contacted for a complete list of standard symbols.

P and IDs or engineering flow diagrams were developed for the detailed design of the processing plant. The diagrams show complete details of all the required piping, instruments and location, signal lines, control loops, control systems, and equipment in the facility. The process flow diagrams and plant-control requirements are generated by a team from process engineering and control engineering. Changes to the P and ID are normally the responsibility of process engineering and must be approved and signed off by the same. These engineering drawings must be correct, current, up-to-date, and rigorously maintained. Every P and ID change must be approved and recorded. If not, time is lost in maintenance, repair, and modifications, not to mention catastrophic errors that can be made by using obsolete drawings.

P and ID typically show the following types of information:

1. Plant equipment and vessels showing location, capacity, pressure, liquid-level operating range and usage, and so on

2. All interconnection lines distinguishing between the types of interconnection, i.e., gas or electrical and operating range of line

3. All motors giving voltage and power and other relevant information

4. Instrumentation showing location of instrument, its major function, process-control loop number, and range

5. Control valves giving type of control, type of valve, type of valve action, fail-safe features, and flow plus pressure information

6. The ranges for all safety valves, pressure regulators, temperatures, and operating ranges

7. All sensing devices, recorders, and transmitters with control loop numbers

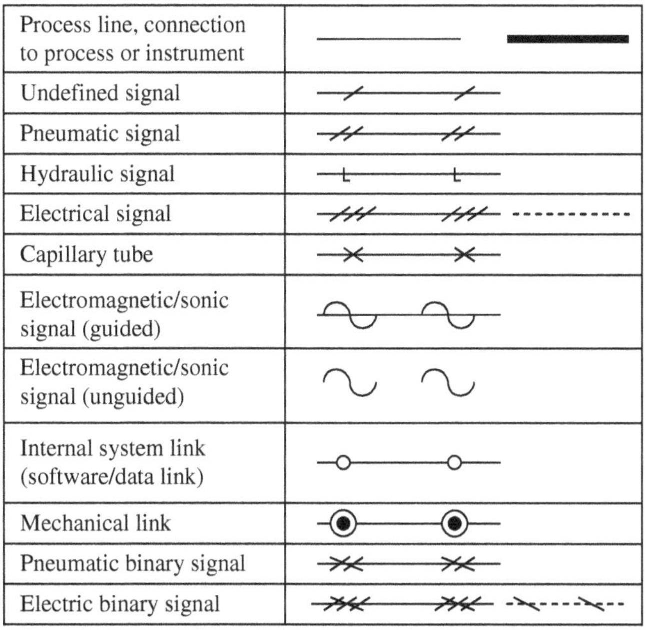

Process line, connection to process or instrument	
Undefined signal	
Pneumatic signal	
Hydraulic signal	
Electrical signal	
Capillary tube	
Electromagnetic/sonic signal (guided)	
Electromagnetic/sonic signal (unguided)	
Internal system link (software/data link)	
Mechanical link	
Pneumatic binary signal	
Electric binary signal	

FIGURE 12.1   Symbols for instrument line interconnection.

12.3.2   Interconnections

The standard on interconnections specifies the type of symbols to be used to represent the various types of connections in a processing plant (see Fig. 12.1). The solid bold lines are used to represent the primary lines used for process product flow and the plain solid lines are used to represent secondary flows such as steam for heating. Abbreviations for secondary flow lines are given in Table 12.1. The abbreviations are placed adjacent to the lines to indicate their function as shown in Fig. 12.2.

In the list of assigned symbols for interconnect lines given in Fig. 12.1, one symbol is undefined and can be assigned at the users discretion for a special connection not covered by any of the assigned interconnection symbols. The binary signals can be used for digital signals or pulses. It is also necessary to show on the P and ID the signal's content and range. For example, electrical interconnections can be either signal current or voltage and would be marked as 4 to 20 mA or 0 to 5 V. Examples of signal lines with the signals content and range marking is shown in Fig. 12.2.

AS	Air supply	NS	Nitrogen supply
HS	Hydraulic supply	GS	Gas supply
WS	Water supply	SS	Steam supply
ES	Electric supply		

TABLE 12.1   Abbreviations for Secondary Flow Lines

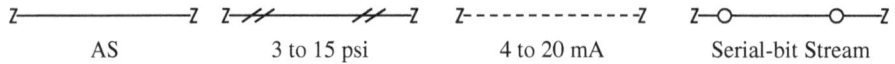

FIGURE **12.2**   Method of indicating the signal content of a line.

12.3.3   Instrument Symbols

Figure 12.3 shows the symbols designated for instruments. Discrete instruments are represented by circles, shared instruments by a circle in a rectangle, computer functions by hexagons, and PLC functions by a diamond in a rectangle.

12.3.4   Instrument Identification

A single horizontal line, no line, dashed line, or double line through the display is used to differentiate between location and accessibility to an operator; i.e., a line through an instrument may indicate the instrument is in a panel in the control room giving full access, no line could mean the instrument is in the process area and off limits to the operator, a double line has the possibility that the instrument is in a remote location but the operator can obtain access, whereas a dashed line means not available by virtue of being located in a totally inaccessible location. Instrument symbols should also contain letters and numbers. The letters are a shorthand way of giving the type of instrument, its use in the system, and the numbers identify the control loop. Usually two or three letters are used. The first letter identifies the measured or initiating variable, the following is a modifier, and the remaining letters identify the function. Table 12.2 shows some of the meaning of the assigned instrument letters.

FIGURE **12.3**   Standardized instrument symbols.

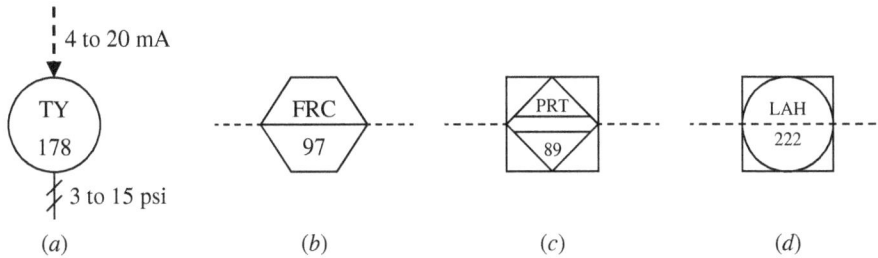

FIGURE 12.4   Examples of the letter and numbering codes.

Examples of instrument identification are shown in Fig. 12.4. By referring to Figs. 12.2, 12.3, and Table 12.2, the instrument identification can be determined as follows:

a. The first letter T indicates that the instrument is in temperature loop number 178. The second letter Y denotes conversion, which from the line description gives the conversion from a current of 4 to 20 mA to a pressure of 3 to 15 psi. The instrument is a discrete instrument located in the field.

b. The designation of F indicates flow, R is for recorder, and C is a controller indicating a recording flow controller in loop 97. This is an accessible computer function.

c. The letter P denotes pressure, R is recorder, and the third letter T is transmitter, giving a recording pressure transmitter in loop 89 which is located in a secondary accessible location and is a PLC function.

d. The first letter L stands for level, A indicates alarm, and H is high, which is an alarm for high liquid levels located in loop 222, and is not accessible.

12.4   Functional Symbols

A number of functional symbols or pictorial drawings are available for most P and ID elements. A few examples are given here to acquaint the student with these elements. They have been divided into actuators, primary elements, regulators, and math functions for clarity.

12.4.1   Actuators

The first row of examples and the last three drawings shown in Fig. 12.5 are the basic sections used in some of the actuator diagrams. The other drawings show how these basic sections can be combined to form families of actuators. For instance, the hand actuator and the pneumatic actuator are shown combined with the control valve symbol to give a representation of a hand-operated valve and a pneumatic-operated valve in the second row. Note should also be taken of the arrows to represent the state of the valve under system "fail" conditions.

12.4.2   Primary Elements

By far the largest numbers of elements used in P and ID are the primary elements; a sampling of these elements is given in Fig. 12.6. Lettering and numbers are included in the examples.

First Letter + Modifier		Succeeding Letters		
Initiating or Measured Variable	**Modifier**	**Readout, or Passive Function**	**Output Function**	**Modifier**
A Analysis		Alarm		
B Burner, combustion		User's choice	User's choice	User's choice
C User's choice			Control	
D User's choice	Differential			
E Voltage		Sensor		
F Flow rate	Ratio			
G User's choice		Glass, viewing device		
H Hand				High
I Current		Indicate		
J Power	Scan			
K Time	Time rate of change		Control station	
L Level		Light		Low
M User's choice	Momentary			Middle
N User's choice		User's choice	User's choice	User's choice
O User's choice		Orifice		
P Pressure		Test point		
Q Quantity	Integrate, totalize			
R Radiation		Record		
S Speed, frequency	Safety		Switch	
T Temperature			Transmit	
U Multivariable		Multifunction	Multifunction	Multifunction
V Vibration, mechanical analysis			Valve, damper louver	
W Weight, force		Well		
X Unclassified	x-axis	Unclassified	Unclassified	Unclassified
Y Event, state, or presence	y-axis		Ready, complete, covert	
Z Position, dimension	z-axis		Driver, actuator	

TABLE 12.2   Instrument Identification Letters

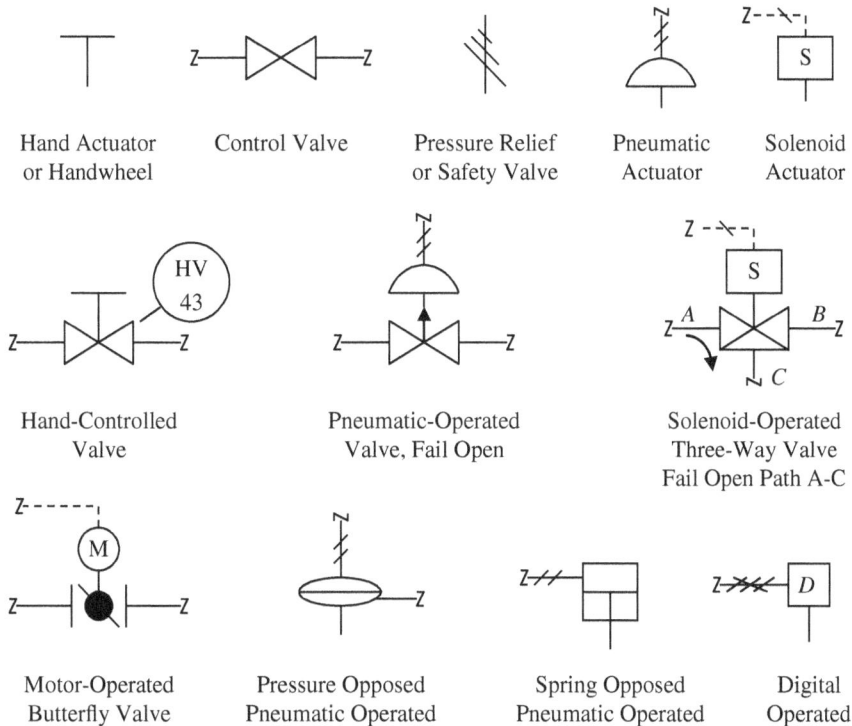

Figure 12.5 Examples of basic and actuator symbols.

12.4.3 Regulators

Typical examples of regulators and safety valves are shown in Fig. 12.7.

12.4.4 Math Functions

PLCs have a large number of math functions which can be implemented using software. If these math functions are incorporated into a P and ID they will probably be executed using hardware, e.g., use of a square root to convert a pressure measurement to flow data. These functions have been symbolized. An example of the math symbols is shown in Fig. 12.8.

12.5 P and ID Drawings

All processing facilities will have a set of drawings using the standardized ISA symbols to show the plumbing, material flow, instrumentation, and control lines. The drawings normally consist of one or more main drawing depicting the facility on a function basis with support drawings showing details of the individual functions. In a large processing plant, these could run into many tens of drawings. Each drawing should have a parts list, be numbered, have an area for revisions, notes, and approval signatures. It is imperative that these drawings are kept up-to-date; a few minutes taken to update a drawing can save many hours at a later date trying to figure out a problem on equipment

Orifice Plate with Vena
Contracta, Radius, or Pipe
Taps Connected to
Differential-Pressure
Flow Transmitter

Turbine-or Propeller-
Primary Element

Magnetic Flowmeter
with Integral
Transmitter

Venturi Tube

Level Indicator
Float Type

Counting Switch,
Photoelectric, with
Switch Action Based
on Cumulative Total

Bimetalic, Glass, or
Other Local
Thermometer
Temperature Indicator

Thermal-Radiation
Temperature Element

Speed Transmitter

Vibration Transmitter
for Motor

Weight Transmitter
Direct Connection

Roll-Thickness
Transmitter

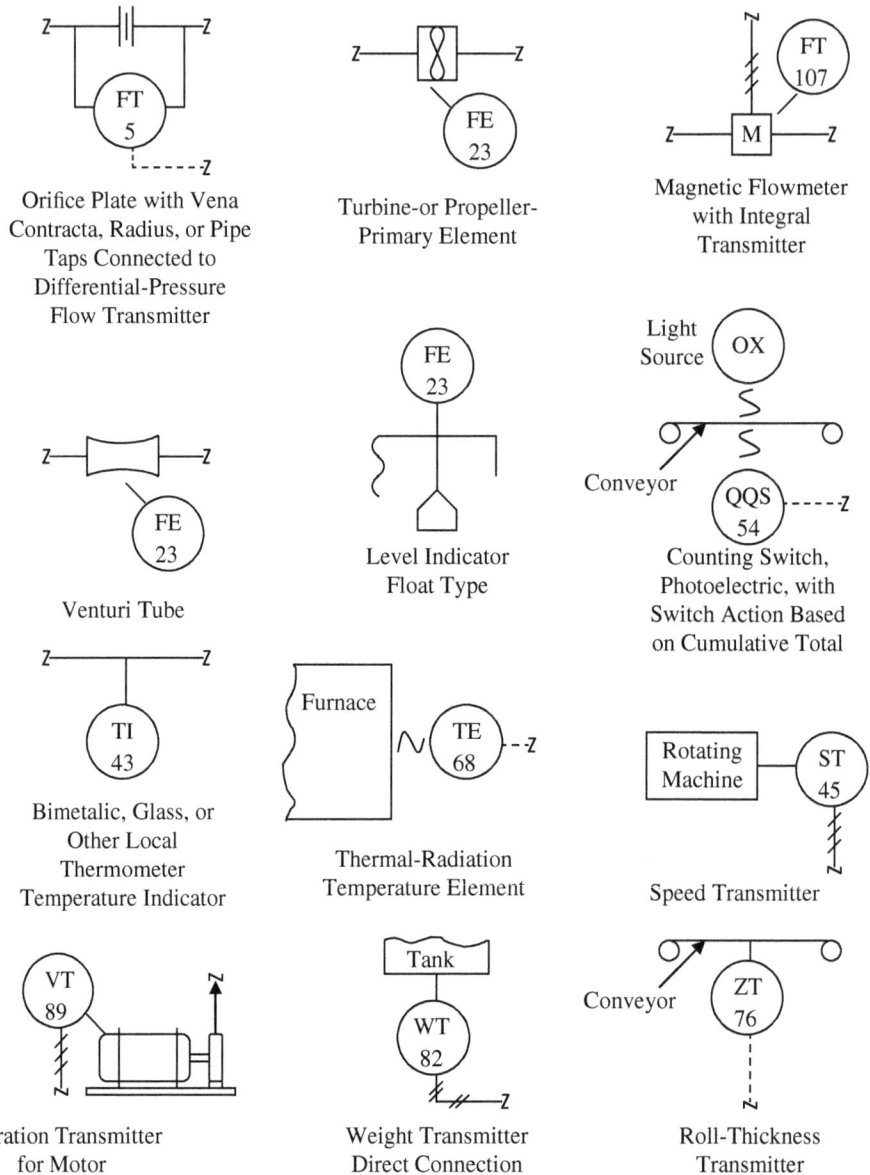

FIGURE **12.6** Examples of primary elements used in P and ID.

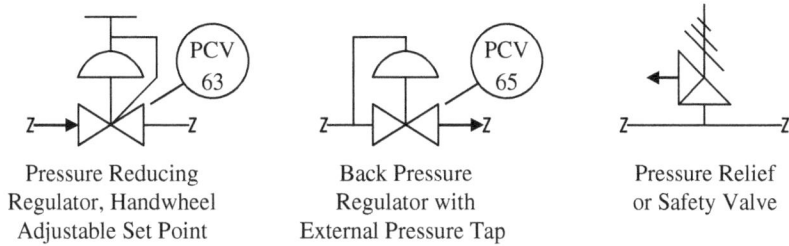

| Pressure Reducing Regulator, Handwheel Adjustable Set Point | Back Pressure Regulator with External Pressure Tap | Pressure Relief or Safety Valve |

FIGURE **12.7**   Examples of regulators and safety valve symbols used in P and ID.

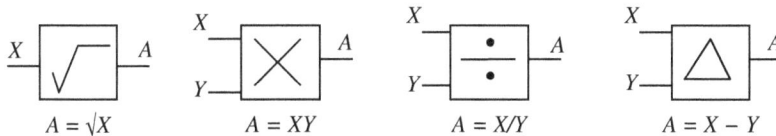

$A = \sqrt{X}$ $\qquad$ $A = XY$ $\qquad$ $A = X/Y$ $\qquad$ $A = X - Y$

FIGURE **12.8**   Examples of math symbols used in P and ID.

that has been modified but whose drawings have not been updated. Figure 12.9 shows an example of a function block. The interconnection lines and instruments are clearly marked, and control loops numbered. A materials list is attached with appropriate places for revisions and signatures.

Summary

This chapter introduced the documentation for alarm and trip systems, PLCs, P and IDs, and the standards developed for the symbols used in PID drawings.

The main points described in this chapter were as follows:

1. Alarm and trip systems and system documentation.

2. Documentation for safety procedures.

3. The documentation required for PLC systems and peripheral circuits.

4. The development of standards for process-control symbols and drawings by ISA. The standards cover interconnection, supply lines, and the line symbols to be used.

5. Symbols used for instruments, the identification, and functional letters used with instruments and their meaning.

6. Basic primary element symbols are shown and how they can be used make more complex elements.

7. Examples of P and ID facility drawings and the information that should be contained in the drawings.

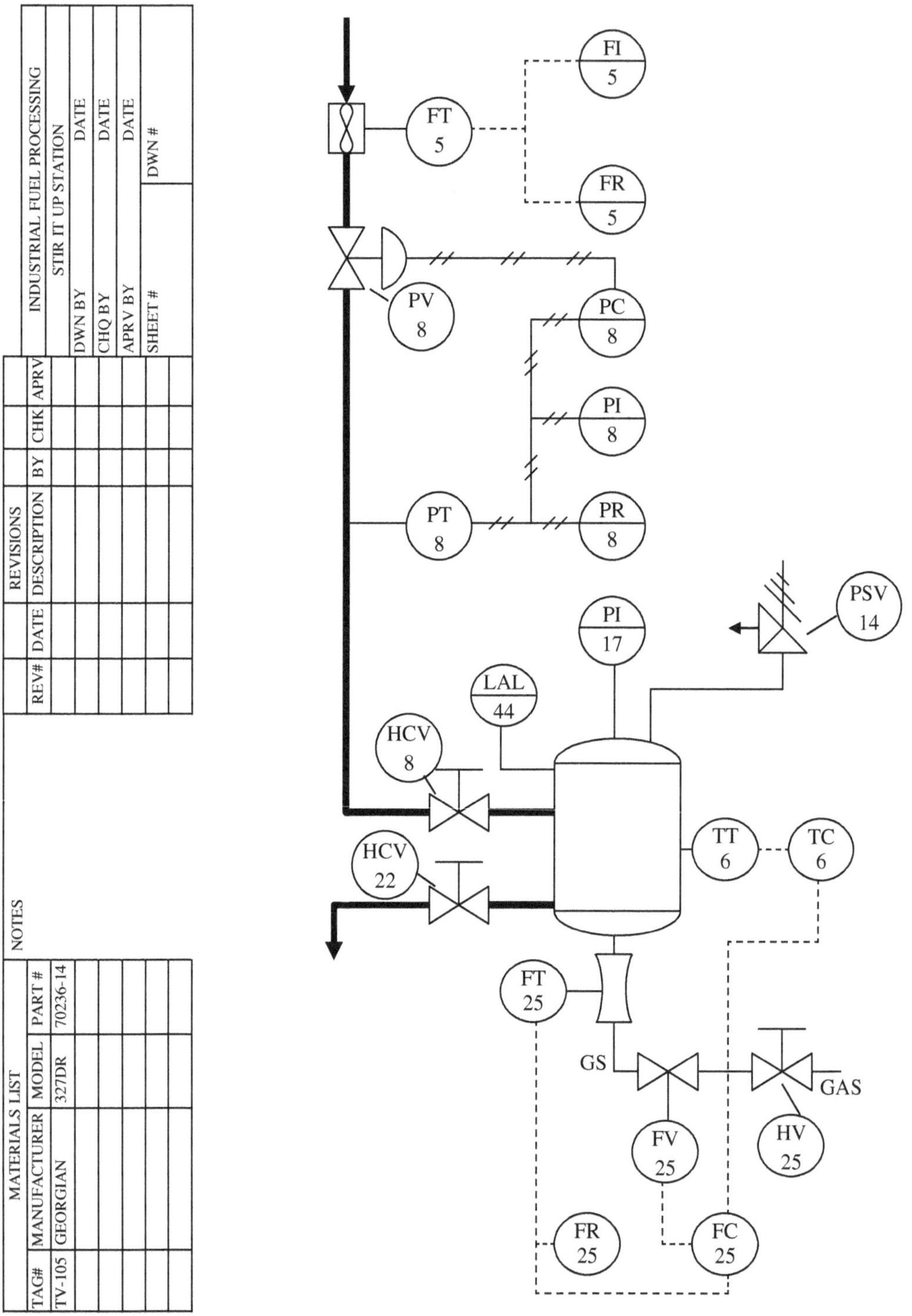

FIGURE 12.9   Illustration of a P and ID for a mixing station.

Problems

12.1   What does the drawing in Fig. 12.10*a* represent?

12.2   Draw a steam supply line and attach the line indicator.

12.3   What do you understand by the symbol shown in Fig. 12.10*b*?

12.4   Draw a speed recorder symbol as a computer function in the field location.

12.5   Describe the symbol shown in Fig. 12.10*c*.

12.6   Draw an electrically operated three-way valve.

12.7   What does the symbol in Fig. 12.10*d* represent?

12.8   Draw a solenoid-operated butterfly valve which is "open" in the fail mode.

12.9   What does the symbol in Fig. 12.10*e* represent?

12.10   What does the symbol in Fig. 12.10*f* represent?

12.11   Why should documentation be kept up-to-date?

12.12   Who normally has the responsibility for keeping P and ID up-to-date?

12.13   Who normally has the responsibility for developing P and ID?

12.14   List the information that should be contained in a P and ID.

12.15   List the information that should be contained in PLC documentation.

12.16   List the information that should be contained in alarm and trip documentation.

12.17   What is the purpose of the SIS?

12.18   What are the differences between the type of sensors used in SIS and process control?

12.19   Draw the symbol of an internal pressure-loaded regulator.

12.20   Draw the symbol of a pneumatic-operated butterfly valve.

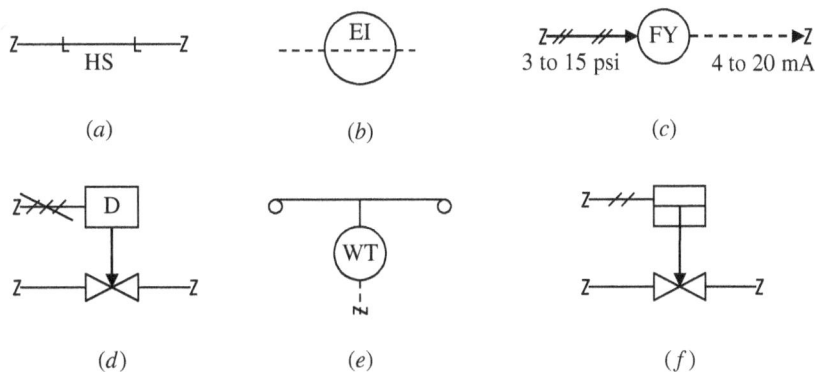

FIGURE 12.10   Diagrams and symbols for use with Problems 12.1 through 12.10.

CHAPTER 13

Signal Transmission

Chapter Objectives

This chapter will help you understand the prime modes of signal transmission and familiarize you with the various methods of signal transmission and where they are used.

Topics discussed in this chapter are as follows:

- Pneumatic signal transmission
- Types of analog electrical signal transmission
- Electrical to pneumatic signal converters
- Thermocouple and resistive type devices and temperature signal transmission
- The operation of the signal processor in signal transmission
- Smart sensors and Fieldbus
- Telemetry signal transmission

Measurement of variables are made by sensors, conditioned by transducers, and then transmitted to a central monitoring processor. In the case of process control, the accuracy of transmission of the value of the variable is very important; any errors introduced during transmission will be acted upon by the controller and degrades the accuracy of the signal. There are several methods of transmitting data. The chosen solution will depend on the sensor, application of the signal, the distance the signal needs to be sent, the accuracy requirements of the system, and cost. Unfortunately, the accuracy of the system can be degraded by poor transmission.

13.1 Introduction

The various methods of signal transmission are discussed in this chapter. Control signals can be transmitted pneumatically or electrically. Due to the needs of an air supply for pneumatic transmission, inflexible pluming, cost, slow reaction time, limited range of transmission, reliability, accuracy and the requirements of control systems, electrical transmission is now extensively used. Electrical signals can be transmitted in the form of voltages, currents, digital, or optical signals. Unfortunately, the terms transducer, converter, and transmitter are often confused and used interchangeably. These terms are defined in Chap. 9.

Transmitters are devices that accept low-level electrical signals and format them, so that they can be transmitted to a distant receiver. The transmitter is required to be able to transmit a signal with sufficient amplitude and power so that it can be reproduced at a distant receiver as a true representation of the input to the transmitter without loss of accuracy or information.

Offset refers to the low end of the operating range of a signal. When performing an offset adjustment, the output from the transducer is being set to give the minimum output (usually zero) when the input signal value is a minimum.

Span references the range of the signal, i.e., from zero to full-scale deflection. The span setting (or system gain) adjusts the upper limit of the transducer with maximum signal input. There is normally some interaction between offset and span; the offset should be adjusted first and then the span.

13.2 Pneumatic Transmission

Pneumatic signals were used for signal transmission, but are not in use in today's facilities except in applications where electrical signals or sparks could ignite combustible materials or to operate pneumatic actuators. Electrical signals (4 to 20 mA) can be converted to pneumatic signals using an *I* to *P* converter. Pneumatic transmission pressures were standardized into two ranges, i.e., 3 to 15 psi (20 to 100 kPa) and 6 to 30 psi (40 to 200 kPa); the 3 to 15 psi is now the standard range. Zero is not used for the minimum of the ranges as low pressures do not transmit well and the zero level can then be used to detect system failure.

13.3 Analog Transmission

13.3.1 Noise Considerations

Analog voltage or current signal lines are hard wired between the transmitter and the receiver. Compared to digital signals, these signals can be relatively slow to settle due to the time constant of the lead capacitance, inductance, and resistance, but are still very fast in terms of the speed of mechanical systems. Analog signals can loose accuracy if signal lines are long with high resistance; can be susceptible to ground offset, ground loops, noise and radio frequency (RF) pickup. Figure 13.1*a* shows the controller supplying dc power to the transmitter and the signal path from the transmitter to the controller.

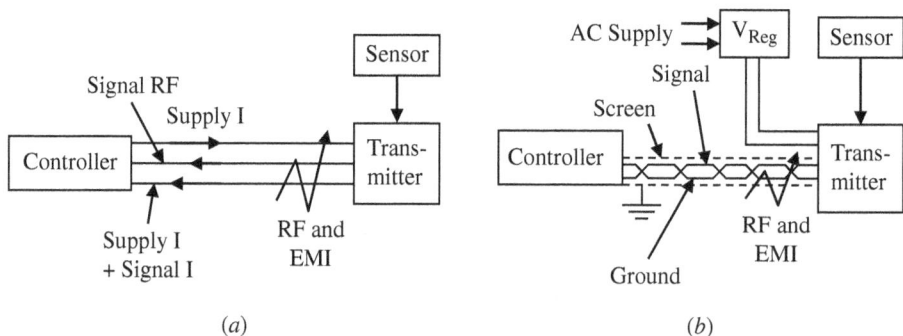

FIGURE 13.1 Supply and signal connections are shown between controller and transmitter using (a) straight leads and (b) a twisted pair.

The dc power for the sensors can be obtained from the controller to save the cost of deriving the power at the sensor as shown in Fig. 13.1b. However, the current flowing in the ground line (shown in Fig. 13.1a) from the supply will be much larger than the signal current and will produce a voltage drop across the resistance of the ground lead elevating the ground level of the transmitter which will give a signal offset error at the controller. The second problem with this type of hard wiring is that it is susceptible to RF and electromagnetic induction (EMI) noise pickup; i.e., the induced noise from RF transmitters, and motors, will produce error signals.

To reduce these problems, the setup shown in Fig. 13.1b can be used. This setup shows that the dc supply to the transmitter is generated from the ac line voltage via an isolation transformer and voltage regulator at the transmitter. The ground connection is used only for the signal return path. The signal and ground return leads are a screened twisted pair; i.e., the signal leads are screened by a grounded sheath. The RF and EMI pickup are reduced by the screen and the induced noise in both lines is greatly reduced. Because variations in the supply voltages can produce changes in the offset voltage and the gain of the sensor/transmitter, the supply voltage must be regulated.

An improved method of minimizing RF and EMI pickup is shown in Fig. 13.2. In this case, the transmitter sends a differential signal using a screened twisted pair. The pickup will affect both signals by the same amount and will cancel in the differential receiver in the controller as well as any ground offset voltages.

A differential output voltage signal can be generated using the circuit shown in Fig. 13.3. The output stages have unity gain to give low output impedance and equal and opposite phase signals. Op-amps are also commercially available with differential outputs, which can be used to drive buffer output stages.

13.3.2   Voltage Signals

Voltage signals are standardized in the voltage ranges 0 to 5 V, 0 to 10 V, and 0 to 12 V, with 0 to 5 V being the most common. The requirements of the transmitter are a low output impedance to enable the amplifier to drive a wide variety of loads without a change in the output voltage, low temperature drift, low offset drift, and low noise. Figure 13.4a shows a transmitter with a voltage output signal. A voltage change requires a settling time at the receiver due to line capacitance and the input voltage to the controller V_{in} can be less than the output voltage V_{out} from the transmitter due to resistance losses in the cables if the receiver is drawing any current, i.e.,

$$V_{in} = \frac{V_{out} \times \text{Internal } R}{\text{Internal } R + 2 \times \text{Wire } R} \qquad (13.1)$$

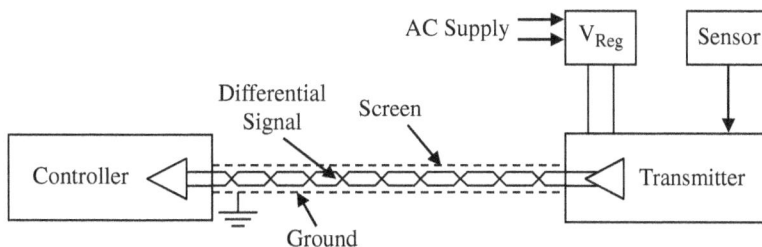

FIGURE 13.2   Screened differential signal connection between the controller and the transmitter.

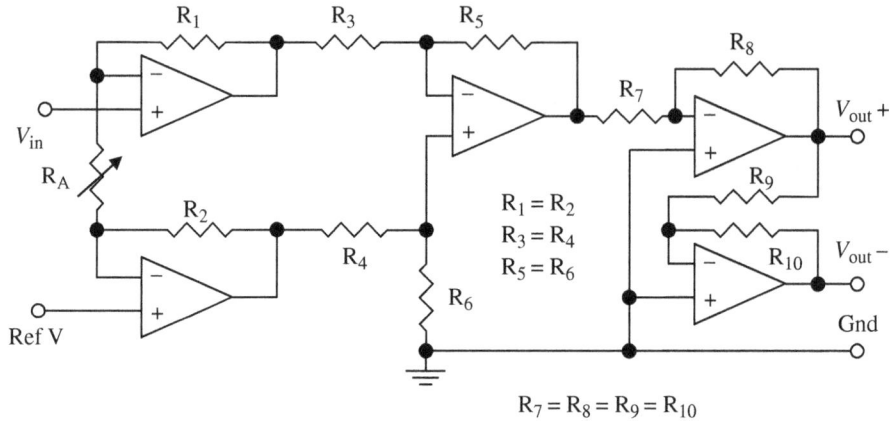

$R_1 = R_2$
$R_3 = R_4$
$R_5 = R_6$

$R_7 = R_8 = R_9 = R_{10}$

FIGURE 13.3   Differential amplifier with buffer outputs.

The internal R of the controller must be very high compared to the resistance of the wire and connections to minimize signal loss (which is normally the case). A differential signal, as shown in Fig. 13.2, will eliminate ground noise and offset problems.

13.3.3   Current Signals

Current signals were standardized into two ranges: these are 4 to 20 mA and 10 to 50 mA, where 0 mA is a fault condition. The latter range was the preferred standard, but has now been dropped, and the 4 to 20 mA range is the accepted standard. The requirements of the transmitter are high output impedance, so that the output current does not vary with load, i.e., changes in lead impedance due to length or temperature. Figure 13.4b shows a transmitter with a current output. The internal resistance of the controller receiver is low for current signals, i.e., a few hundred ohms, so that voltages with different signal levels are small and less affected by capacitance. The differential signal connection as shown in Fig. 13.2 minimizes noise and ground problems. Current signals are normally used for analog transmission.

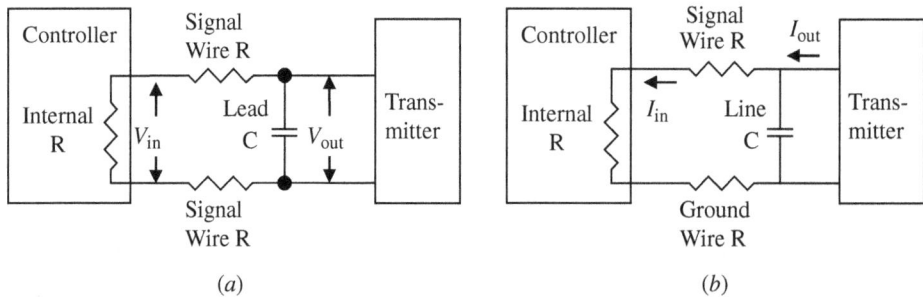

(a)

(b)

FIGURE 13.4   Effect of resistance and lead capacitance on (a) voltage signals and (b) current signals.

13.3.4 Signal Conversion

Signal conversion is required between the low-level signals transmitted and high-energy requirements of the control signals for actuator and motor control. Control signals can be either digital voltage, analog current, or pneumatic. It is often necessary to convert electrical signals to pneumatic signals for actuator control. Pneumatics is still the industry standard for the operation of most valves. Air-operated valves are also required in applications where electromagnetic (e/m) radiation could cause problems, or in a hazardous environment where sparks from electrical devices could cause volatile material to ignite.

A linear pneumatic amplifier or booster can be used to increase the pressure from a low-level pressure signal to a high-pressure signal to operate an actuator. Figure 13.5a shows a pressure amplifier. Gas from a high-pressure supply is controlled by a conical plug which in turn is controlled by a diaphragm whose position is set by a low-pressure signal. The gain of the system is set by the area of the diaphragm divided by the area of the base of the conical plug. The output pressure is inverted but linear with respect to the input pressure; the device shown is one of many different types. Pneumatic feedback can be used to improve characteristics of the amplifier.

One design of a current to pressure converter is shown in Fig. 13.5b. The spring tends to hold the flapper closed giving a high-pressure output (15 psi). When current is passed through the coil the flapper moves toward the coil opening the air gap at the nozzle reducing the output air pressure. The output air pressure is set to the maximum of 15 psi by the set zero adjustment when the current through the coil is 4 mA. The system gain and span is set by moving the nozzle along the flapper to give an output of 3 psi when the signal current is 20 mA. The output pressure is inverted with respect to the amplitude of the current in the setup as shown, but the converter could be set up to be noninverting. There is a linear relationship between current and pressure.

13.3.5 Thermocouples

Thermocouples have several advantages over other methods of measuring temperature, in that they are very small in size, have a low time response (10/20 ms compared to several seconds for some elements), are reliable, have good accuracy, a wide operating temperature range, and they convert temperature directly into electrical units. The disadvantages are the need for a reference and the low signal amplitude.

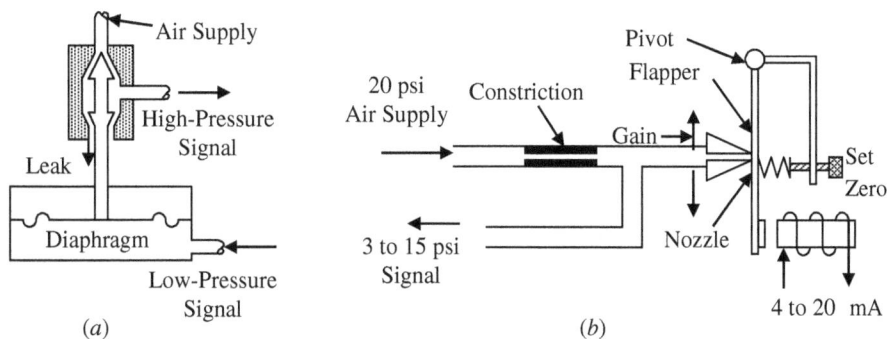

Figure 13.5 Signal conversion: (a) pressure amplifier and (b) current to pressure transducer.

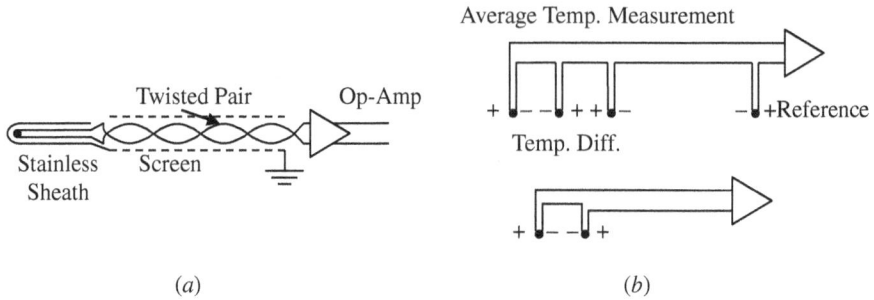

FIGURE 13.6   Different types of thermocouple connections to an op-amp: (a) direct using twisted pair to a reference and amplifier and (b) for average temperature measurement and differential temperature measurement.

Thermocouple signals can be amplified with a cold junction reference close to the amplifier and the signal transmitted in an analog or digital format to a controller, or the thermocouple can be connected directly to the controller for amplification and cold junction correction. This method is sometimes used to eliminate the cost of remote amplifiers and power supplies. Controller peripheral modules are available for amplification of several thermocouple inputs with cold junction correction. Figure 13.6a shows a differential connection between the amplifier and the thermocouple as a twisted pair of wires that is screened to minimize noise and the like. Figure 13.6b shows other configurations that can be used to connect thermocouples for temperature averaging, and differential temperature measurements.

13.3.6   Resistance Temperature Devices

Resistance temperature device (RTD) elements can be connected directly to the controller peripheral amplifiers using a two-, three-, or four-wire lead configuration; these are shown in Fig. 13.7. The RTD is driven from a constant current source I and the voltage drop across the RTD measured. The two-wire connection (a) is the simplest cheapest and least accurate, the three-wire connection (b) is a compromise between cost and accuracy, and the four-wire connection (c) is the most expensive but most accurate. The wires in all cases will be in screened cables. In the case of the two-wire connection, the voltage drop is measured across the lead wires as well as the RTD; the resistance in the two-lead wires can be significant giving a relatively high degree of error.

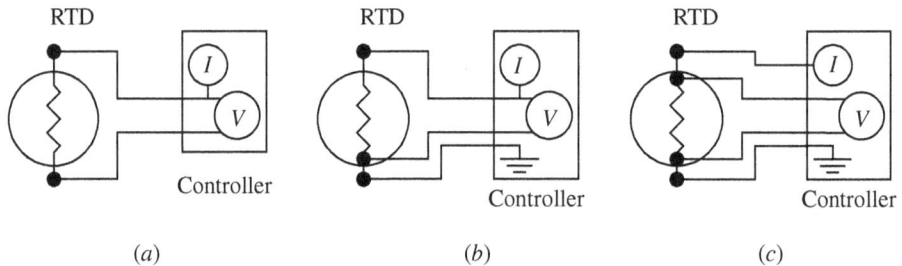

FIGURE 13.7   Alternative connection schemes between an RTD and a controller: (a) two lead, (b) three lead, and (c) four lead.

In the case of the three-wire connection, a direct return lead from the RTD to the voltmeter is added, as shown. The voltage drop δV between the ground connection and the lower RTD connection as well as the voltage drop V between the current source and the lower RTD connection can be measured. If the resistance in each supply lead to the RTD is assumed to be the same, the voltage across the RTD is $V - \delta V$ correcting for the error caused by the common lead wire. In most cases each lead wire will have about the same resistance, so this method is accurate enough for most applications. With the four-wire connection the voltmeter is connected directly to the RTD as shown in Fig. 13.7c and because no current flows in the leads to the voltmeter there is no voltage drop in the measuring leads and an accurate RTD voltage reading is obtained.

13.4 Digital Transmission

13.4.1 Transmission Standards

Digital signals can be transmitted via a hardwired parallel or serial bus, radio transmission or fiber optics without loss of integrity. Digital data can be sent faster than analog data due to higher speed transmission. Another advantage is digital transmitters and receivers require much less power than analog transmission devices.

Communication standards for digital transmission between computers and peripheral equipment are defined by the Institute of Electrical and Electronic Engineers (IEEE). The standards are the IEEE-488 or RS-232. However, several other standards have been developed and are now in use. The IEEE-488 standard specifies that a digital "1" level will be represented by a voltage of 2 V or greater and a digital "0" level shall be specified by 0.8 V or less, as well as the signal format to be used for logic circuits. The RS-232 standard specifies that a digital "1" level shall be represented by a voltage of between +3 V and +25 V and a digital "0" level shall be specified by a voltage of between –3 V and –25 V, and IEEE 802 defines signal levels used for Ethernet signals. Fiber optics are now also being extensively used to give very high-speed transmission over long distances, and are not affected by electromagnetic or RF pickup and dc isolation. Figure 13.8 shows a two-way fiber optic cable setup with light emitting diode (LED) drivers and photodiode receivers.

Digital signals can be transmitted without loss of accuracy and can contain error correction codes (Manchester encoded) for automatic error correction or to request data retransmission if an error in the number of bits received is detected. These networks are

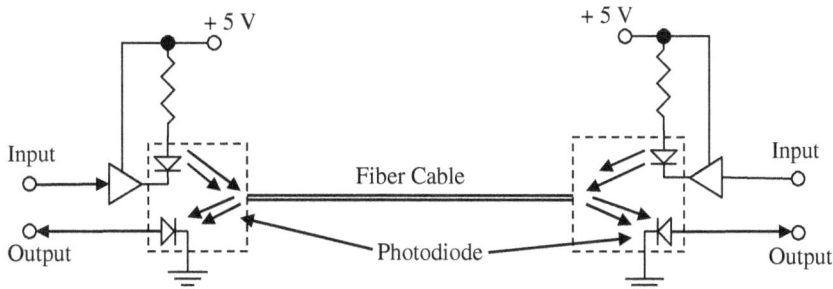

Figure 13.8 Fiber-optic bus.

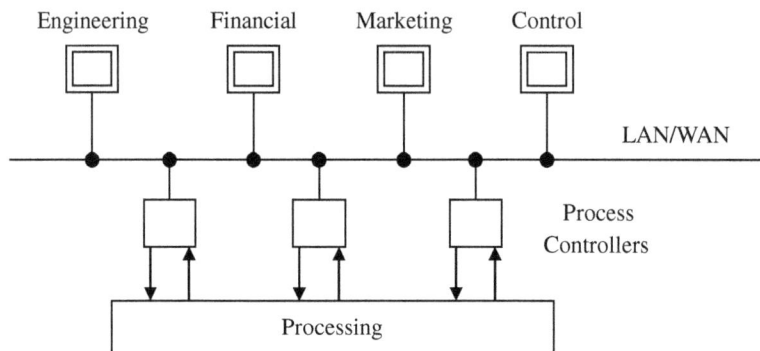

FIGURE **13.9** A LAN network.

known as local area networks (LANs) when used in a limited area such as a plant or wide area networks (WANs) when used as a global system. A typical LAN/WAN network is shown in Fig. 13.9. The network shown is a distributed control system (DCS) with a central processor communicating with the process controllers. Engineering, finance, and marketing are also in the loop and can monitor plant operations for such information as cost figures, and product delivery details over the network, directly from the control system.

Computer-based programmable logic control systems are flexible systems with a central processor and the ability to add a variety of analog and digital interface units. The interface units can be receivers for reception of analog and/or digital information from the monitoring sensors or transmitters for sending set point or control information for actuators. A typical receiver unit will contain analog amplifiers with analog-to-digital converters (ADCs) giving the unit the ability to interface with analog transmitting devices and change the data into a digital format to interface with the processor. Interface receiver units can contain thermocouple amplifiers or bridges for use with resistive sensors, logic level interfaces, serial bus receivers, and so on. The transmitter units will have the capability of transmitting analog, or digital actuator control information, serial bus information, and so on. This setup is shown in Fig. 13.10. Each input or output requires its own interconnect cable or bus resulting in a mass of wiring, which requires careful routing and identification marking.

13.4.2 Smart Sensors

Smart sensor is a name given to the integration of the sensor with an ADC, processor, and DAC for actuator control and the like: such a setup for furnace temperature control is shown in Fig. 13.11. The electronics in the smart sensor contains all the circuits necessary to interface to the sensor, amplify the signal, apply proportional, integral, and derivative (PID) control (see Chap. 11), sense temperature to correct for temperature variations in the process if required, correct for sensor nonlinearity, and so on. The ADC is used to convert the signal into a digital format for the internal processor, and the DAC to convert the signal back into an analog format for actuator control. The processor has a serial digital bus interface for interfacing via the fieldbus to a central computer. This enables the processor in the smart sensor to receive updated information on set points, gain, operating mode, and so on, and to send status information to the central computer.

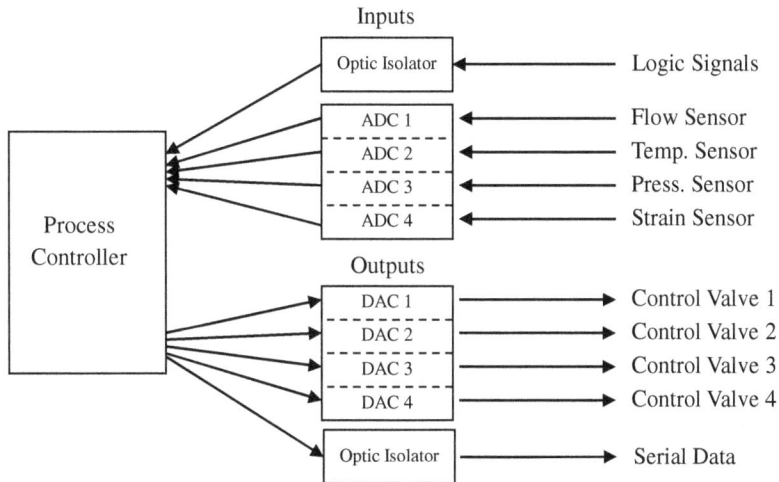

FIGURE 13.10 Process system with individual inputs and outputs for each variable.

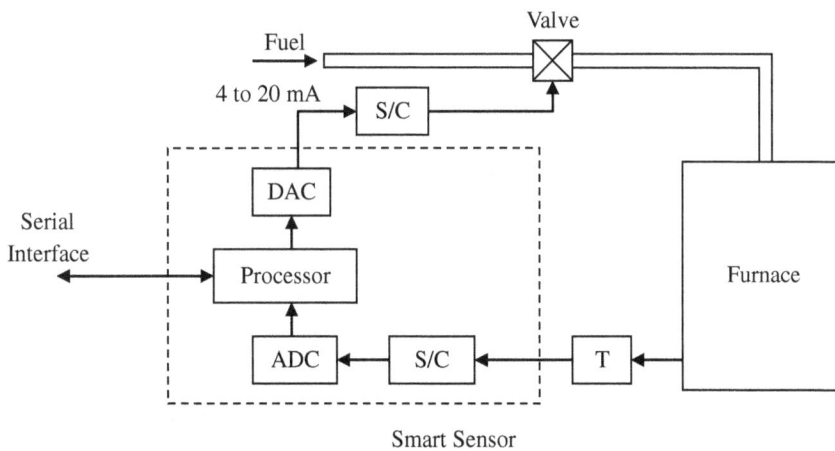

FIGURE 13.11 Smart sensor block diagram.

13.4.3 Foundation Fieldbus and Profibus

Foundation Fieldbus (FF) and Profibus are the two most universal serial data bus formats that have been developed for interfacing between a central processor and smart sensing devices in a process-control system. The FF is primarily used in the United States and the Profibus format is primarily used in Europe. Efforts are being made for a universal acceptance of one bus system.

At present, process-control equipment is being manufactured for one system or the other; global acceptance of equipment standards would be preferred. The HART (highway addressable remote transducer) protocol is an implementation of the Fieldbus. HART is the most widely used of the server protocols available. It communicates over the 4 to 20 mA analog serial data bus that uses a single pair of twisted copper wires by

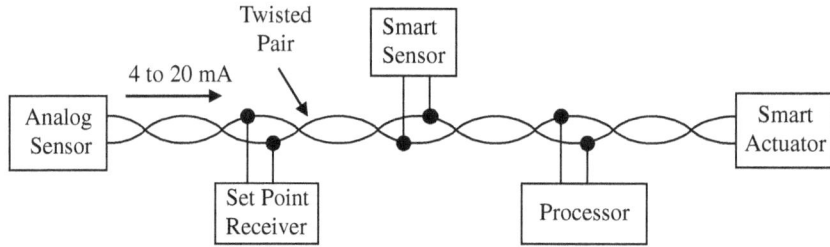

FIGURE **13.12**   Foundation Fieldbus cable connection.

superimposing a frequency-modulated (FM) signal on to the analog signal; the FM signal uses frequency shift keying (FSK) for digital data. The system can operate in a point-to-point mode to obtain additional information from or send information to the sensing point such as temperature, set point, status of sensing unit, and so on, or in a multidrop mode to enable communication between the central processing computer and several sensors, smart sensors, and actuators. This is shown in Fig. 13.12. An advantage of the serial bus data is that it uses the same bus that is already in use for 4 to 20 mA analog data. At the receiver the digital data are filtered from the analog data. This system can replace the individual leads to all the monitoring points by using the analog data bus. New units can be added to the bus with no extra wiring. A plug and play feature is provided, giving faster control, and programming is the same for all systems. Higher accuracies are obtained than using analog and more powerful diagnostics are available. The bus system uses time division multiplexing. The serial data word from the central processor contains the address of the peripheral unit being addressed in a given time slot and the data being sent plus the data bits are exclusively "ORED" to give a check sum. Information on FF is given in the ISA 50.02 standards.

One disadvantage of the FF is that a failure of the bus such as a broken wire can shut the whole process down, whereas, with the direct connection method only one sensor is disabled. This disadvantage can be overcome by the use of a redundant or a backup bus in parallel to the first bus, so that if one bus malfunctions the backup bus can be used.

A comparison of the characteristics of the serial data buses is given in Table 13.1. The original FF was designated H1, a new generation of the H1 is the HSE, which will use an

	Fieldbus (H1)	Profibus	Fieldbus (HSE)
Bus type	Twisted pair copper	Twisted pair copper or fiber optic cable	Twisted pair copper and fiber optic cable
Number of devices	240 per segment 65,000 segments	127 per segment 65,000 segments	Unlimited
Length	1900 m	100 m copper plus 24 km fiber	100 m copper plus 24 km fiber
Max speed	31.25 kb/s	12 Mb/s	100 Mb/s
Cycle time (milliseconds)	<600	<2	<5

TABLE **13.1**   Comparison of Bus Characteristics

Ethernet LAN bus to provide operation under the TCP/IP protocol used for the internet. The advantages are increased speed, unlimited addresses, and standardization.

13.5 Digital Signal Converters

Instrumentation variables are analog signals that cannot be readily transmitted over long distances or the analog value stored. Process-control systems are computer based and require digital signals. The analog signals are converted to digital signals for transmission, processing, and storage. Actuators normally require analog signals for control; hence, digital-to-analog converters are also required.

13.5.1 Analog-to-Digital Conversion

The amplitude of an analog signal can be represented by a digital number, e.g., an 8-bit word can represent numbers up to 256, giving 255 steps so that it can represent an analog voltage or current with an accuracy of 1 in 255 (assuming the conversion is accurate to 1 bit) or 0.4 percent accuracy. Similarly, 10- and 12-bit words can represent analog signals to accuracies of 0.1 percent and 0.025 percent, respectively.

Commercial integrated A/D converters are readily available for instrumentation applications. Several techniques are used for the conversion of analog-to-digital signals. These are:

> *Flash converters* which are very fast and expensive with limited accuracy, i.e., 6-bit output with a conversion time of 33 ns. The device can sample an analog voltage 30 million times per second.

> *Successive approximation* is a high-speed, medium-cost technique with good accuracy; i.e., the most expensive device can convert an analog voltage to 12 bits in 20 µs, and a less expensive device can convert an analog signal to 8 bits in 30 µs.

> *Resistor ladder networks* are used in low-speed, medium-cost converters. They have a 12-bit conversion time of about 5 ms.

> *Dual slope converters* are low-cost, low-speed devices but have good accuracy and are very tolerant to high noise levels in the analog signal. A 12-bit conversion takes about 20 ms.

Analog signals are constantly changing, so that for a converter to make a measurement a sample and hold technique is used to capture the voltage level at a specific instant in time. Such a circuit is shown in Fig. 13.13a, with the waveforms shown in Fig. 13.13b. The N-channel field effect transistor (FET) in the sample-and-hold circuit has a low impedance when turned "ON" and a very high impedance when "OFF." The voltage across capacitor C follows the input analog voltage when the FET is "ON," and holds the DC level of the analog voltage when the FET is turned "OFF."

During the "OFF" period the ADC measures the DC level of the analog voltage and converts it into a digital signal. As the sampling frequency of the ADC is much higher than the frequency of the analog signal, the varying amplitude of the analog signal can be represented in a digital format during each sample period and stored in memory. The analog signal can be regenerated from the digital signal using a DAC.

Figure 13.14a shows the block diagram of the ADC0804, a commercial 8-bit ADC. The analog input is converted to a byte of digital information every few milliseconds.

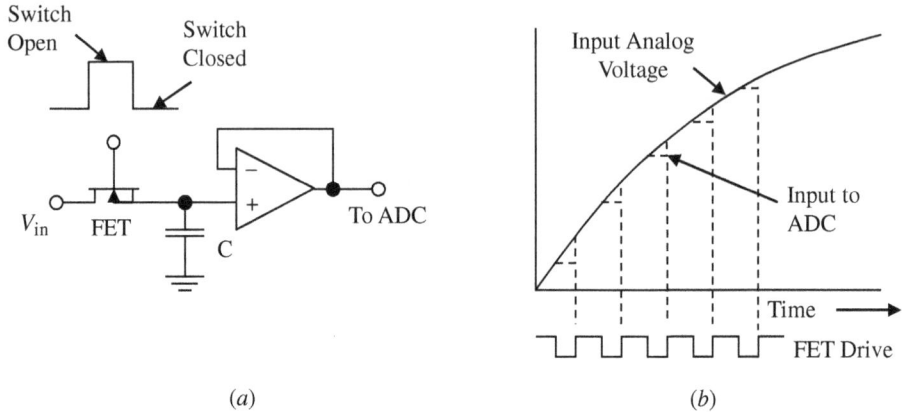

FIGURE **13.13** Sample-and-hold (a) circuit and (b) waveforms for the circuit.

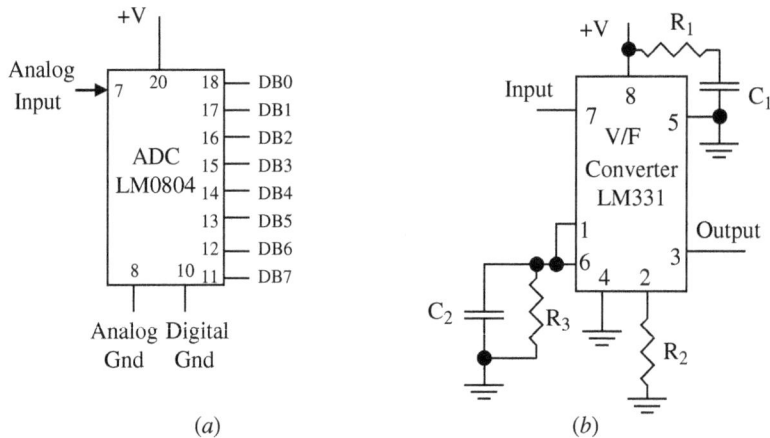

FIGURE **13.14** Different types of converters: (a) LM 0804 ADC and (b) LM 331 V/F converter.

An alternative to the ADC is the voltage to frequency converter; in this case the analog voltage is converted to a frequency. Commercial units such as the LM331 shown in Fig. 13.14b are available for this conversion. These devices have a linear relation between voltage and frequency. The operating characteristics of the devices are given in the manufacturers' data sheets.

13.5.2 Digital-to-Analog Conversion

There are two methods of converting digital signals to analog signals. These are digital-to-analog converters, which are normally used to generate a voltage reference or low-power voltage signals, and pulse width modulation which is used in high power circuits, i.e., actuator and motor control, and so on.

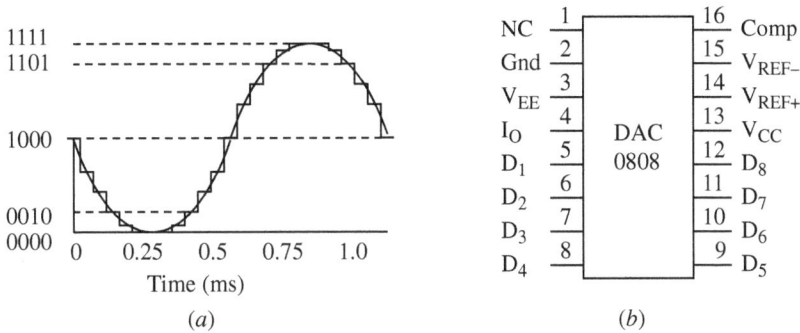

FIGURE 13.15   Illustration of (a) 1-kHz sine waveform reproduced from a DAC and (b) commercial 8-bit DAC.

Digital-to-analog converters (DACs) decode digital information into analog using a resistor network or similar method. The analog signals are normally used for low-power applications but can be amplified and used for control. Figure 13.15a shows the generation of a 1-kHz sine wave. In the example shown, the digital signal is converted to a voltage every 0.042 ms giving the step waveform shown. In practice, the conversion rate could be higher approximating the steps to a complete sine wave. Shown also is the binary code from the DAC (4 bits only), the step waveform can be smoothed by a simple RC filter to get a pure sine wave. The example is only to give the basic conversion idea. Commercial DACs, such as the DAC 0808 shown in Fig. 13.15b are readily available. The DAC 0808 is an 8-bit converter that will give an output resolution or accuracy of 1 in $(2^8 - 1)$ (-1 is because the first number is zero leaving 255 steps) or an accuracy of ± 0.39 percent. For higher accuracy analog signals, a 12-bit commercial DAC would be used (± 0.025 percent accuracy).

Pulse width modulation (PWM) changes the duration for which the voltage is applied to reproduce an analog signal, and is shown in Fig. 13.16. The width of the output pulses shown is modulated going from narrow to wide and back to narrow. If the voltage pulses shown are averaged, the width modulation shown will give a half-sine wave. The other half of the sine wave is generated using the same modulation, but with a negative supply, or with the use of a bridge circuit to reverse the current flow.

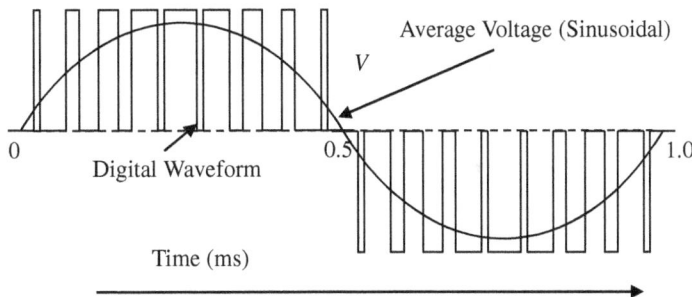

FIGURE 13.16   PWM signal to give a 1-kHz sine wave using positive and negative supplies.

The current is limited by the load. This type of width modulation is normally used for power drivers for ac motor control from a dc supply. The output devices such as insulated gate bipolar transistors (IGBTs) are used as switches, i.e., they are "ON" or "OFF" and can control over 100 kW of power. This method of conversion gives low internal dissipation with high efficiency, which can be as high as 95 percent of the power going into the load. Whereas, analog power drivers are only 50 percent efficient at best and have high internal power dissipation.

13.6   Telemetry

Telemetry is the wireless transmission of measurement data from a remote location to a central location for processing and/or storage. This type of transmission is used for sending data over long distances such as from weather stations and the like, and data from rotating machinery where cabling is not feasible. Wireless communication can be used to eliminate cabling or to give flexibility in moving the positioning of temporary monitoring equipment. Broadcast information is a wireless transmission using amplitude modulation (AM) or frequency modulation (FM) techniques. These methods are not accurate enough for the transmission of instrumentation data, as reception quality varies and the original signal cannot be accurately reproduced. In telemetry, transmitters transmit signals over long distances using a form of FM or a variable width amplitude modulated signal. When transmitting from battery or solar cell operated equipment it is necessary to obtain the maximum transmitted power for the minimum power consumption. FM transmits signals at a constant power level, whereas, AM transmits at varying amplitudes and pulsing techniques can transmit only the pulse information needed which conserves battery power; hence, for the transmission of telemetry data pulse AM is preferred.

13.6.1   Width Modulation

Width coded signals or PWM are blocks of RF energy whose width is proportional to the amplitude of the instrumentation data. Upon reception the width can be accurately measured and the amplitude of the instrumentation signal reconstituted. Figure 13.17*a* shows the relation between the voltage amplitude of the instrumentation signal and the width of the transmitted pulses, when transmitting a series of 1 V signals and a series of voltages through 10 V.

For further power saving, PWM can be modified to pulse position modulation (PPM). Figure 13.17*b* shows a typical PWM modulation and the equivalent PPM signal. The PWM signal shows an OFF period for synchronization of transmitter and receiver. The receiver then synchronizes on the rising edge of the transmitted zero; the first three pulses of the transmission are calibration pulses followed by a stream of width modulated data pulses. In the case of the PPM, narrow synchronization pulses are sent, and then only a pulse corresponding to the lagging edge of the width modulated data is sent. Once synchronized the receiver knows the position of the rising edge of the data pulses, so that information on the lagging edge is all that is required for the receiver to be able to regenerate the data. This form of transmission has the advantage of greatly reducing power consumption and extending battery life.

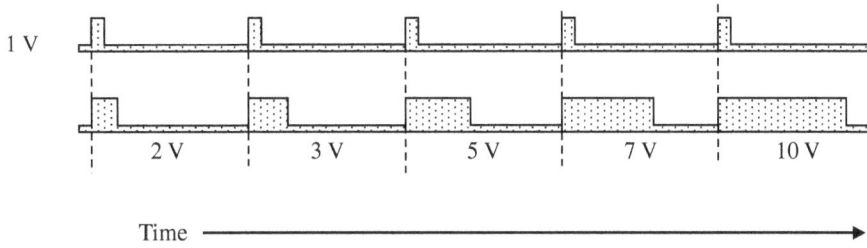

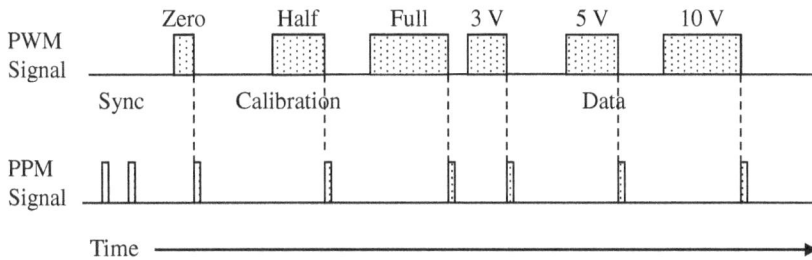

FIGURE 13.17 (a) An amplitude modulated waveform where the width of the modulations correspond to voltage levels and (b) PWM and PPM waveforms compared.

13.6.2   Frequency Shift Modulation

One type of *frequency shift modulation* is FSK (frequency shift keying). It is the transmission of binary data using FM where the binary "0" and the binary "1" are represented by two different frequencies as shown in Fig. 13.18. The unmodulated carrier has fixed amplitude and frequency.

FSK has been further developed into multiple frequency shift keying (MFSK). This is shown in Fig. 13.19. The carrier shown in Fig. 13.19a is modulated into different frequencies to represent different numbers as shown in Fig. 13.19b. Upon reception the

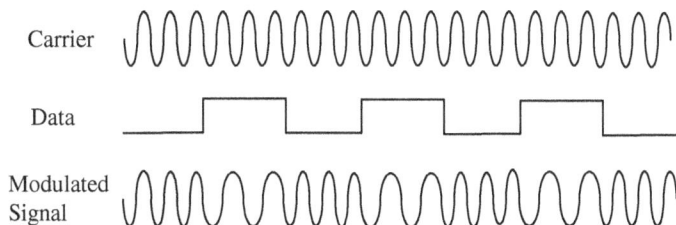

FIGURE 13.18 FSK waveforms.

1 V 2 V 3 V 4 V

Fixed
Amplitude

Time ⟶

Frequency

(a)   (b)

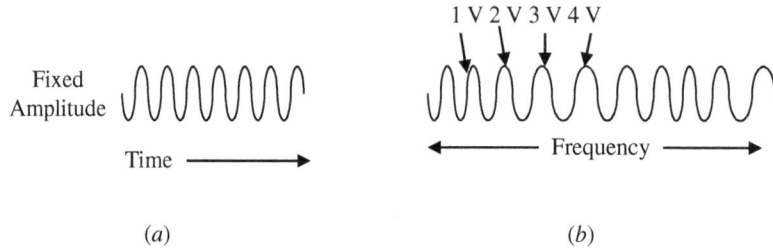

FIGURE 13.19 Frequency modulation: (a) unmodulated waveform and (b) data modulated waveform.

base frequency of the transmission is subtracted from the received signal leaving the frequency of modulation that can then be reconstituted to determine the original data.

Summary

This chapter discusses the various types of data used in signal transmission and their advantages and disadvantages and the methods of interconnecting multiple devices in a process-control system and controller operation.

The main points described in this chapter were as follows:

1. The pros and cons of pneumatic signal transmission compared to electrical transmission in new and old systems.

2. Electrical analog signal transmission can use voltage or current but current transmission is preferred when trying to minimize errors.

3. Electrical signals can be converted to pneumatic signals for driving actuators for controlling hazardous material, where electrical sparks could ignite the material and so forth.

4. Thermocouples can be set up and interconnected to measure differential and average temperatures or connected in series to increase their sensitivity as in a thermopile.

5. RTDs can be interconnected using two, three, or four wires. Two-wire connection is the lowest cost method, but is less accurate than the more expensive four-wire connection.

6. Digital signal transmission has higher signal integrity and is faster than analog signal transmission and can use error correction codes to correct for any errors in transmission.

7. Smart sensors and the interconnection schemes using FF in the United States and Profibus in Europe.

8. Transmission of telemetry signals and methods of reducing the power required by using PPM.

Problems

13.1 Name various methods of data transmission.

13.2 What types of signal transmissions are used in telemetry?

13.3 What are the various types of connections used for RTD elements and what are their advantages and disadvantages?

13.4 What are the standard ranges used in the transmission of pneumatic and electrical analog signals?

13.5 What conversion techniques are used to convert digital into analog signals?

13.6 Define offset and span.

13.7 What are the digital transmission standards?

13.8 Describe where a fiber cable is used and its advantages and disadvantages.

13.9 What are the advantages of digital over analog transmission?

13.10 What does FSK stand for?

13.11 What transmission speeds are used in the FF system?

13.12 What is a smart sensor?

13.13 How are FSK serial signals separated from analog signals?

13.14 Why is a twisted pair of wires used in preference to a single signal wire?

13.15 What are the advantages of AM over FM?

13.16 What are the advantages and disadvantages of current over voltage signal transmission?

13.17 Why are pneumatic signals used in electrical signal transmission?

13.18 Name the modes used in a controller.

13.19 How many steps are there in a 12-bit DAC and what is the percent resolution?

13.20 How many devices can be connected to the Fieldbus and Profibus?

CHAPTER 14

Logic Gates

Chapter Objectives

This chapter will help you understand the relation between the decimal system and the binary system, and the design criterion for various types of logic gates, gate symbols, and logic gate use for computer building blocks.

Topics discussed in this chapter are as follows:

- Understand the relation between decimal and binary numbers
- Digital building blocks
- Why digital communication is used in an analog world
- Boolean equations
- Gate descriptions and truth tables
- Gate symbols
- Basic logic circuits

14.1   Introduction

In electronics, it is necessary to be able to convert decimal numbers to binary, octal, and hexadecimal codes, and understand how logic gates are used to control digital signals. Other codes that may be encountered are specialized codes, such as the Gray code, binary coded decimal (BCD), and ASC II code.

Digital logic and gates are the building blocks of logic control circuits and the processor and computer that have given us the power to accurately control extremely complex processes and complex mathematical functions. Sensors and instrumentation measurements are analog in nature. However, analog signals are hard to transmit accurately and analog values cannot be stored, but can be accurately converted to digital signals using commercially available analog-to-digital converters (ADCs). It may be necessary in some cases to implement the combination of several digital signals in a remote area before transmission to a local processor. An understanding of logic circuits will help in understanding the design and function of processor building blocks.

Some of the advantages of digital signals are as follows:

Low power requirements

Ease of programing logic systems

Can transmit signals over long distances without loss of accuracy

Can operate in a relative noisy electrical environment

High-speed signal transmission

Ease of information storage

Compatibility with controllers and alpha numeric displays

14.2 Digital Numbers

Digital signals are either high- or low-voltage levels. Most digital circuits are designed for use with a 5 V supply. The logic low (binary 0) level is from 0 to 1 V, and logic high (binary 1) level is from 2.5 to 5 V; 1 to 2.5 V is an undefined region (no man's land). Technology has made it possible to design gates that draw little or no power and can operate in an integrated environment at clocking speeds greater than 1 GHz.

In our everyday system of mathematics, we use the decimal system that contains the numbers 0 through 9. However, in the world of electronics only two numbers "0" and "1" are available to us. So to understand electronics computing, it is necessary that technicians, engineers, and programmers understand the power of numbers in both the binary system and the decimal system and be able to convert between the numbering systems.

We use the decimal system (base 10) for mathematical functions, whereas electronics use the binary system (base 2) to perform the same functions. The rules are the same when performing calculations using either numbering system (to the base 10 or 2). Table 14.1 gives a comparison between counting in the decimal and binary systems.

Decimal	Binary	Decimal	Binary
0000	0000000	0021	0010101
0001	0000001	—	—
0002	0000010	0031	0011111
0003	0000011	0032	0100000
0004	0000100	—	—
0005	0000101	0063	0111111
0006	0000110	0064	1000000
0007	0000111	—	—
0008	0001000	0099	1100011
0009	0001001	0100	1100100
0010	0001010	0101	1100101
0011	0001011	—	—
0012	0001100	0999	1111100111
—	—	1000	1111101000
0015	0001111	1001	1111101001
0016	0010000	1002	1111101010
—	—	—	—
0020	0010100	1024	10000000000

TABLE 14.1 Decimal and Binary Equivalents

The least significant bit (LSB) or unit number is the right-hand bit. In the decimal system, when the unit numbers are used we go to the tens, i.e., 9 goes to 10, and when the tens are used we go to the hundreds, i.e., 99 goes to 100, and so on. The binary system is the same when the 0 and 1 are used in the LSB position, then we go to the next position, and so on, i.e., 1 goes to 10, 11 goes to 100, and 111 goes to 1000, and so on. The only difference is that to represent a number it requires more digits when using the binary system than in the decimal system.

14.2.1 Converting Binary Numbers to Decimal Numbers

Binary numbers can be easily converted to decimal numbers by using the power of numbers. Table 14.2 gives the power value of binary numbers versus their location from the LSB and their decimal equivalent.

To convert a binary number to a decimal number start at the LSB and work to the right to the most significant bit (MSB) recording the decimal value of each "1" from Table 14.2, and adding the decimal numbers to get the decimal equivalent. The binary number 1101001 equals decimal 105. This determination is given below:

Binary number	1	1	0	1	0	0	1
Decimal equivalent	64 +	32 +		8 +			1
Sum	= 105						

Example 14.1 What is the decimal number equivalent of the binary number 101100101?
The values and decimal equivalent are given in Table 14.3.

Solution When programming digital numbers four bits (nibble) are normally used as a block where each nibble is represented by a decade number plus six letters as shown in Table 14.4. This system is known as the hexadecimal system.

To convert binary numbers to hexadecimal code the binary number must be broken down into groups of four bits (nibble) from the LSB end and Table 14.4 is used to convert the groups of binary bits to hexadecimal code. The binary number 10110111110 can be written as 5 B E in hexadecimal code as shown below:

0101	1011	1110
5	B	E

Location	8	7	6	5	4	3	2	1	0
Power Value	2^8	2^7	2^6	2^5	2^4	2^3	2^2	2^1	2^0
Decimal Number	256	128	64	32	16	8	4	2	1

TABLE 14.2 Power Value of Binary Numbers

Binary Number	1	0	1	1	0	0	1	0	1
Location	8		6	5			2		0
Power Value	2^8		2^6	2^5			2^2		2^0
Decimal Number	256	+	64+	32	+		4	+	1
Decimal Total		=	357						

TABLE 14.3 Equivalent Power Values

Binary Number	Hexadecimal Code	Binary Number	Hexadecimal Code
0000	0	1000	8
0001	1	1001	9
0010	2	1010	A
0011	3	1011	B
0100	4	1100	C
0101	5	1101	D
0110	6	1110	E
0111	7	1111	F

TABLE 14.4 Numbering Equivalent in the Hexadecimal (H) System

The MSB in this case has only three binary bits but higher order bits are understood as 0. Hexadecimal code or octal numbers can be converted to decimal numbers using the power of numbers as with binary to decimal conversion; see Examples 14.2 and 14.3.

Example 14.2 What is the hexadecimal code for the binary word 1101001110110111? What is the decimal value?

Solution The binary word is broken down into groups of four bits (bytes) starting from the LSB and going to the MSB.

	MSB ——— LSB
Binary number	= 1101001110110111
Groups	= 1101 – 0011 – 1011 – 0111
Hexadecimal code	= D 3 B 7
Conversion to decimal	

	MSB ——————— LSB			
Power value	16^3	16^2	16^1	16^0
Hexadecimal code	D	3	B	7
Decimal equivalent	4096	256	16	1
	$\times 13 \times 3 \times 11 \times 7$			

Decimal number	53248 + 768 + 176 + 7
Decimal total	54,199

Example 14.3 Convert the octal number (base 8) 5 0 7 2 to decimal.

Solution Conversion to decimal

	MSB			LSB
Power value	8^3	8^2	8^1	8^0
Octal number	5	0	7	2
Decimal equivalent	512	0	8	1
	$\times 5 \times 64 \times 7 \times 2$			

Decimal number	2560	0	56	2
Decimal total	2618			

14.2.2 Converting from Decimal to Binary

It is often necessary to convert a decimal number to another base number. One method of converting the number is the repeated division of the binary number by the base number. The remainder at each step is the number of the new base with the first remainder being the LSB of the new number. For instance, to convert decimal 25 to a binary number, divide 25 repeatedly by 2 as shown:

First divide	$25 \div 2 = 12 + \text{remainder} = 1$	LSB
Second divide	$12 \div 2 = 6 + \text{remainder} = 0$	↓
Third divide	$6 \div 2 = 3 + \text{remainder} = 0$	↓
Fourth divide	$3 \div 2 = 1 + \text{remainder} = 1$	↓
Fifth divide	$1 \div 2 = 0 + \text{remainder} = 1$	MSB

$$\text{MSB} \text{——} \text{LSB}$$
Binary number $= 11001$

Example 14.4 What is the octal equivalent of the decimal number 2618?

Solution

First divide	$2618 \div 8 = 327 + \text{remainder} = 2$ LSB
Second divide	$327 \div 8 = 40 + \text{remainder} = 7$ ↓
Third divide	$40 \div 8 = 5 + \text{remainder} = 0$ ↓
Fourth divide	$5 \div 8 = 0 + \text{remainder} = 5$ MSB
Octal number	$= 5\,0\,7\,2$

Compare this answer with the numbers in Example 14.3.

14.3 Digital Logic Gates

The basic building blocks used in digital circuits are called gates. The types of gates are buffer, NOT, AND, NAND, OR, NOR, XOR, and XNOR. These basic blocks are interconnected to build functional blocks, such as encoders, decoders, adders, counters, and so on. The functional blocks are then interconnected to make systems, i.e., calculators, computers, microprocessors, to name a few. The equivalent switch circuit of each gate is shown.

14.3.1 Buffer Gate

The buffer gate shown in Fig. 14.1 is the simplest gate. The output follows the input. The gate is used as a signal buffer or to reduce loading on a previous stage.

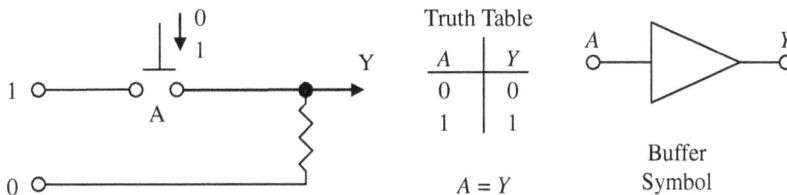

Truth Table

A	Y
0	0
1	1

$A = Y$

Buffer Symbol

FIGURE **14.1** Buffer gate.

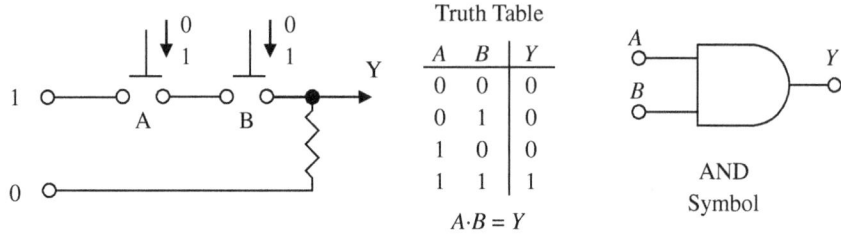

FIGURE **14.2** Basic AND gate and truth table.

In this configuration

$$Y = A \qquad (14.1)$$

14.3.2 AND Gate

The basic AND gate is configured so that the inputs are serially connected and all switches must be closed to obtain an output. The truth table and switch circuit for this configuration are shown in Fig. 14.2. The truth table shows the output Y is equal to $A \times B$, giving $0 \times 0 = 0$, $1 \times 0 = 0$, and $1 \times 1 = 1$.

In the AND configuration

$$Y = A \cdot B \qquad (14.2)$$

14.3.3 OR Gate

The basic OR gate is configured so that the inputs are connected in parallel and an output obtained if any of the switches are closed. The truth table and switch circuit for this configuration are shown in Fig. 14.3. The truth table shows that the output Y is equal to $A + B$, giving $0 + 0 = 0$, $0 + 1 = 1$, and $1 + 1 = 1$.

$$Y = A + B \qquad (14.3)$$

Similarly, the configuration for AND gates with more than two inputs is that all inputs are in series so that for a four input AND gate (inputs A, B, C, and D), we get

$$Y = A \cdot B \cdot C \cdot D \qquad (14.4)$$

and with a four input OR gate, we get

$$Y = A + B + C + D \qquad (14.5)$$

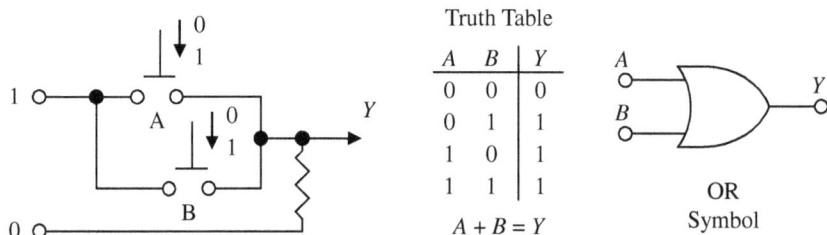

FIGURE **14.3** Basic OR gate and truth table.

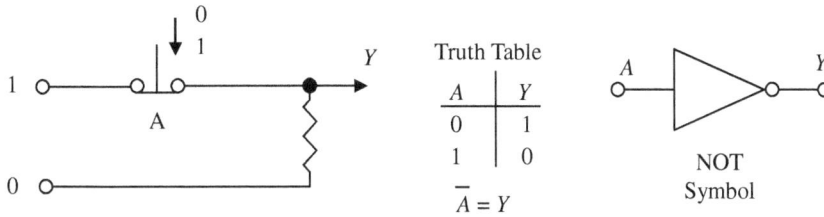

FIGURE 14.4 Basic NOT gate and truth table.

14.3.4 NOT Gate

In many cases, it is necessary to invert a signal, i.e., change a "1" to a "0" or a "0" to a "1." This change is made by a NOT gate. Note the bar over the A to denote signal inversion and the inversion "o" symbol in Fig. 14.4.

14.3.5 Signal Inversion

Figure 14.5 shows the signal being inverted twice by two NOT gates so that the output signal is the same as the initial input signal. The initial signal A is inverted by the first NOT gate to become $\bar{A}$ which is inverted by the second NOT gate to become A double bars showing that a double inversion bar is the same and can be replaced by the original signal.

Figure 14.6 shows the conversion of the AND and OR gates into NAND and NOR gates by the addition of a NOT gate. The AND + NOT gate and the OR + NOT gate can

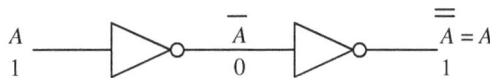

FIGURE 14.5 Double inversion.

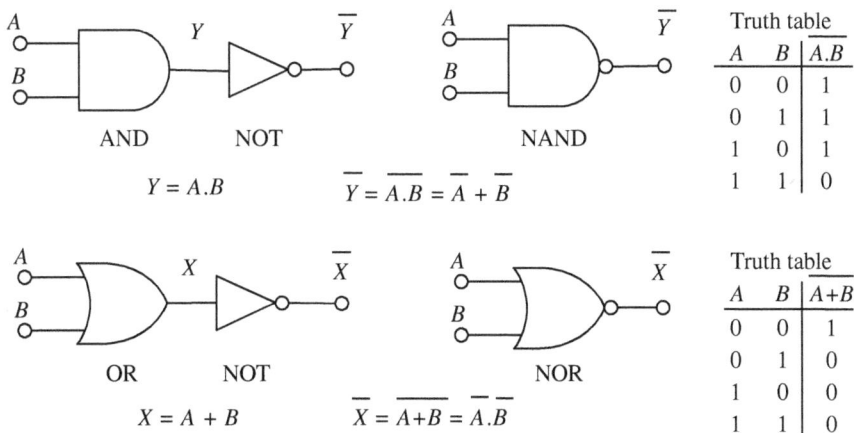

FIGURE 14.6 NAND and NOR gates.

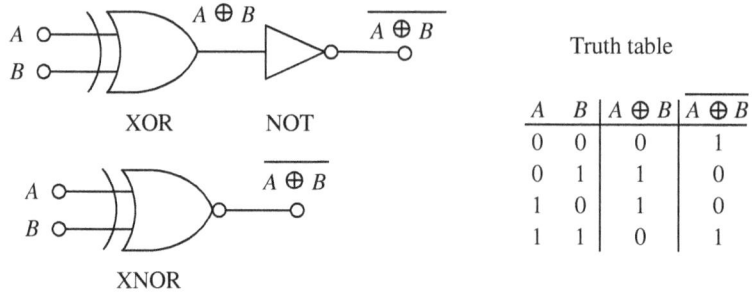

FIGURE **14.7** XOR and XNOR gates.

be replaced by a NAND and NOR gate symbol as shown. The circle "o" at the outputs of the gates indicates signal inversion.

14.3.6 XOR Gates

Figure 14.7 shows the exclusive OR or XOR gate and conversion to the XNOR using a NOT gate. The XOR gate has a high output "1" when the inputs have an odd number of "1"s or in the case of the XNOR the output is "0" when the inputs contain an odd number of "1"s as shown. This criterion is the same if there are more than two inputs. Note the XOR and XNOR symbols.

The input signals to a gate can also be inverted with the use of NOT gates. An example of this is shown in Fig. 14.8 with the truth table. The inversion is shown by a circle "o" at the inputs to the gate.

14.3.7 Logic Symbols

Figures 14.1 through 14.7 show the traditional logic symbols used together with the Boolean equation describing the gate functions first used by the military and now covered by ANSI/IEEE Std. 91-1984. The American National Standards Institute (ANSI) and the IEEE have developed an alternative set of standard symbols for gates given in ANSI/IEEE Std. 91a-1991, which they are pushing very hard for acceptance. These symbols are given in Fig. 14.9. Either set of logic symbols may be encountered in practice, so it is necessary to be familiar with both sets of logic symbol configurations. A lesser known third set of logic symbols was developed by the National Electrical Manufactures Association (NEMA). Alternative symbols can be found in Europe.

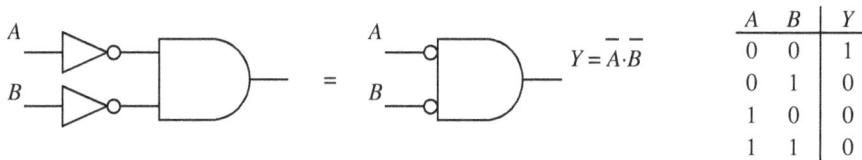

FIGURE **14.8** Gate input signal inversion.

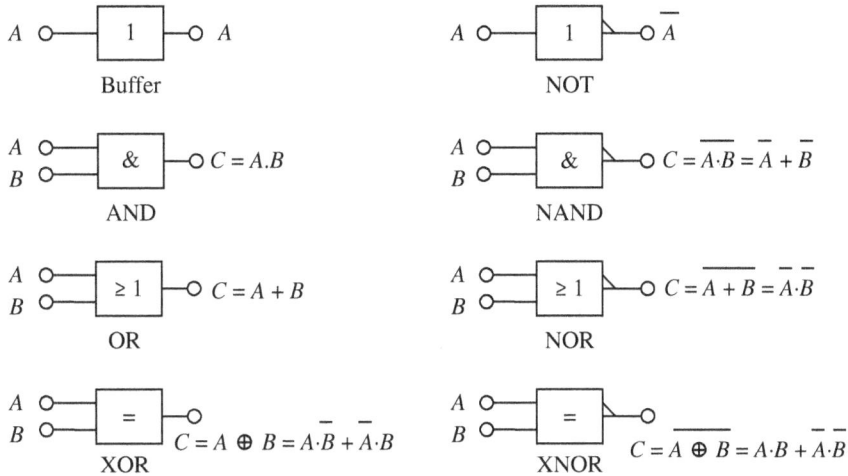

FIGURE 14.9   ANSI/IEEE standard logic symbols.

14.4   Boolean Algebra

Boolean algebra rules and theorems are used in logic designs to reduce complex logic problems to a much simpler and easier to understand format. Complex logic equations can be reduced to the minimum number of logic gates required to perform what seemed to be an unmanageable system design by the use of minterms and maxterms, Karnaugh mapping, or the use of a logic simulation program. Complete sets of computer building blocks for mathematical functions and detailed instructions for minimizing gate count in functional expressions can be found in any book on digital logic.

A summary of Boolean rules are given in Table 14.5.

Boolean theorems can easily be proven using truth tables. Table 14.6 shows the truth table for the theorem $A + \bar{A} \cdot B = A + B$ as can be seen for all values of A and B the equation holds.

Basic Boolean Rules	DeMorgan's Theorems	Other Theorems
$A \cdot A = 1$	$\bar{\bar{A}} = A$	$A + \bar{A} \cdot B = A + B$
$A \cdot 1 = A$	$\overline{A \cdot B} = \bar{A} + \bar{B}$	$A \cdot B + \bar{A} \cdot C + B \cdot C = A \cdot B + \bar{A} \cdot C$
$A \cdot 0 = 0$	$\overline{A + B} = \bar{A} \cdot \bar{B}$	
$A + A = A$	$\overline{\bar{A} \cdot \bar{B}} = A + B$	
$A + 1 = 1$	$\overline{\bar{A} + \bar{B}} = A \cdot B$	
$A + 0 = A$		
$\bar{A} \cdot A = 0$		
$\bar{A} + A = 1$		

TABLE 14.5   Summary of Boolean Rules and Theorems

A	$\bar{A}$	B	$\bar{A} \cdot B$	$A + \bar{A} \cdot B$	A + B
0	1	0	0	0	0
1	0	0	0	1	1
0	1	1	1	1	1
1	0	1	0	1	1

TABLE 14.6   Truth Table for $A + \bar{A} \cdot B = A + B$

The truth table for $A \cdot B + \bar{A} \cdot C + B \cdot C = A \cdot C + \bar{A} \cdot C$ is given in Table 14.7.

To understand how a logic problem is analyzed, it is necessary to obtain a Boolean expression for the variables in the system. Consider the system shown in Figure 14.10. In a chemical process a storage tank contains a mixture. It is necessary to sound an alarm when certain parameters exceed specification. The variables are labeled as follows: temperature is denoted by A, the pressure by B, and the level by E. Set points have been established for these variables, and depending on weather the variables are above or below the set points a "1" or "0" is assigned to each variable in order to develop a Boolean expression for the system.

A	$\bar{A}$	B	C	$A \cdot B$	$\bar{A} \cdot C$	$B \cdot C$	$A \cdot B + \bar{A} \cdot C + B \cdot C$	$A \cdot B + \bar{A} \cdot C$
0	1	0	0	0	0	0	0	0
1	0	0	0	0	0	0	0	0
0	1	1	0	0	0	0	0	0
1	0	1	0	1	0	0	1	1
0	1	0	1	0	1	0	1	1
1	0	0	1	0	0	0	0	0
0	1	1	1	0	1	1	1	1
1	0	1	1	1	0	1	1	1

TABLE 14.7   Truth Table for $A \cdot B + \bar{A} \cdot C + B \cdot C = A \cdot C + \bar{A} \cdot C$

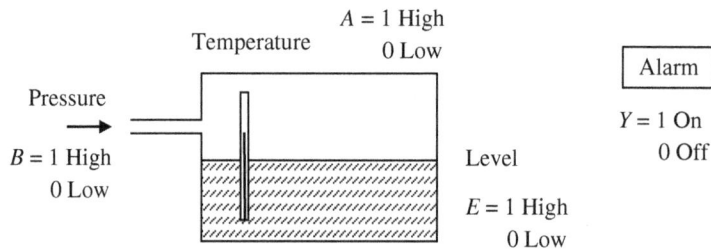

FIGURE 14.10   Setup to illustrate the development of a Boolean equation.

The conditions for turning on the alarm are:

- High temperature with low liquid level
- High temperature with high pressure
- Low temperature with high pressure and low level

If Y represents the alarm states, where in the "1" state the alarm will be activated, a Boolean expression can be developed to meet the above conditions.

The equations for the above alarm conditions are as follows:

- $Y = 1 = A \cdot \overline{E}$ condition 1
- $Y = 1 = A \cdot B$ condition 2
- $Y = 1 = \overline{A} \cdot B \cdot \overline{E}$ condition 3

These three equations can now be combined with the "OR" function, so that if any of these conditions exist the alarm will be activated. This gives:

$$Y = A \cdot \overline{E} + A \cdot B + \overline{A} \cdot B \cdot \overline{E} \tag{14.6}$$

This equation can now be used to define the digital logic required to activate the alarm system.

Example 14.5 It is required to develop a digital circuit to implement the alarm conditions discussed for Fig. 14.10 using the traditional gate symbols.

Solution The starting point is the Boolean expression in Eq. (14.6).

$$Y = A \cdot \overline{E} + A \cdot B + \overline{A} \cdot B \cdot \overline{E}$$

There are many gate combinations that can be used to implement this logic equation. One possible solution using AND and OR gates is given in Fig. 14.11. The three expressions $A \cdot \overline{E}$, $A \cdot B$, and $\overline{A} \cdot B \cdot \overline{E}$ are produced using AND gates and then combined in an OR gate to give Eq. (14.6). As can be seen if any of the alarm conditions occur, Y will go from "0" to "1."

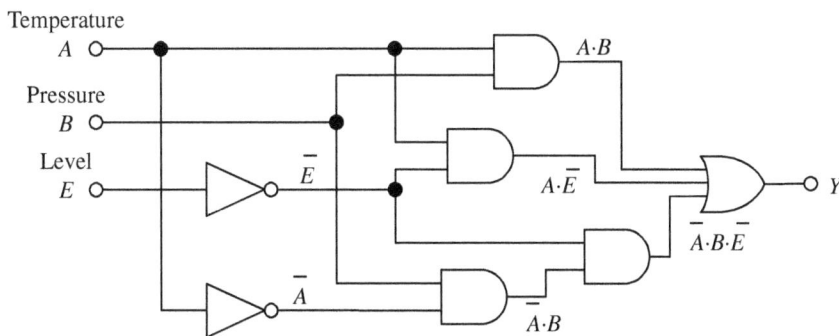

FIGURE 14.11 Possible solution for Example 14.5.

14.5   Functional Building Blocks

Basic logic gates can be interconnected to build functional blocks for custom logic functions or computer logic functions. Two examples of computer building blocks are given in Example 14.6, which gives the logic circuit for a full adder, and Example 14.7, which gives the logic circuit for a two bit decoder.

Example 14.6   Draw and describe the logic circuit for a full adder.

Solution   The logic circuit using NAND and XOR gates of a full adder is shown in Fig. 14.12*a* and the truth table in Fig. 14.12*b*. In an adder, it is necessary to sum the carry from the previous stage and the bits (*A* and *B*) from the two numbers being added to produce the sum and carry bit for the next stage. The input bits are summed in the XOR. If *A* and *B* are both "1"s or "0"s the output of the gate will be a "1." If the "carry in" is a "1" the sum from the second XOR is a "1," but if it is a "0" the output sum is a "0," giving a sum of "1" when there are an odd number of input "1"s and a carry out "1" when there are two or more input "1"s as shown in the truth table.

Example 14.7   Draw and describe the logic circuit for a two bit decoder using AND gates.

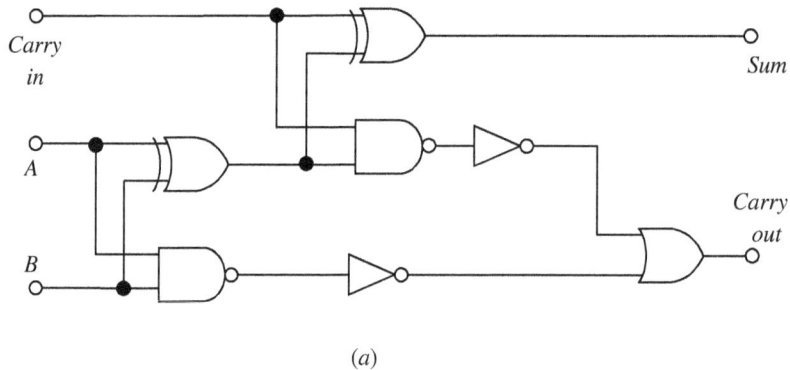

(*a*)

INPUTS			OUTPUTS	
Carry in	A	B	Sum	Carry out
0	0	0	0	0
0	1	0	1	0
0	0	1	1	0
0	1	1	0	1
1	0	0	1	0
1	1	0	0	1
1	0	1	0	1
1	1	1	1	1

(*b*)

FIGURE 14.12   (a) Adder and (b) truth table.

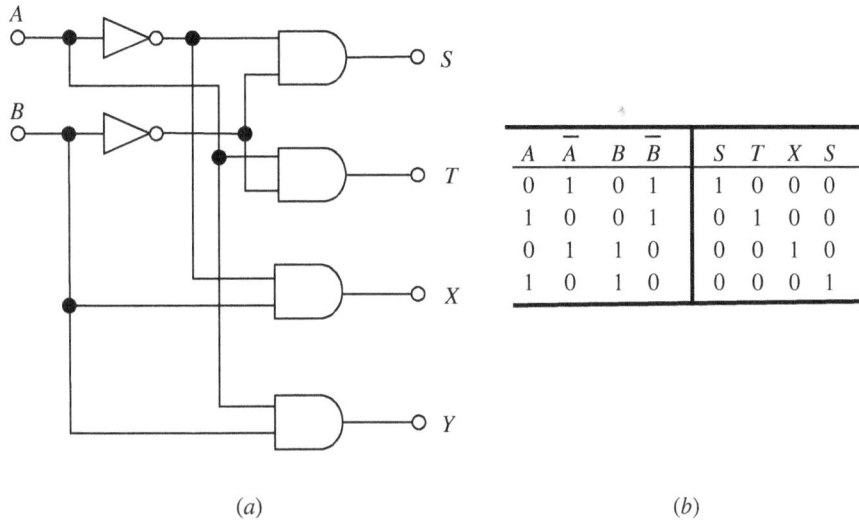

A	$\overline{A}$	B	$\overline{B}$	S	T	X	S
0	1	0	1	1	0	0	0
1	0	0	1	0	1	0	0
0	1	1	0	0	0	1	0
1	0	1	0	0	0	0	1

(a)   (b)

FIGURE **14.13** (a) Two bit decoder and (b) decoder truth table.

Solution The logic circuit of a two bit decoder using AND gates is shown in Fig. 14.13a. The truth table is shown in Fig. 14.13b. A two bit decoder will select one of four possible outputs. The two input bits A and B are inverted and these together with the inverted bits are alternately fed to four two input AND gates as shown in Fig. 14.13a that will give a single "1" at the outputs. The truth table is given in Fig. 14.13b.

Summary

This chapter discussed the various types of basic logic gates and their use as building blocks in the design of digital controllers and the construction of processors and computing systems.

The main points described in this chapter were as follows:

1. Buffer, NOT, AND, NAND, OR, NOR, XOR, and XNOR gates.
2. The truth table for each gate.
3. The powers of numbers and relation between the decimal and binary systems.
4. The advantages of using digital communication in an analog world.
5. Gate symbols in use.
6. Boolean rules and theorems.
7. Proof of theorems using truth tables.

Problems

14.1 Give the truth table for a three input NAND gate.

14.2 Give the truth table for a three input NOR gate.

14.3 Give the truth table for a three input XNOR gate.

14.4 Prove the equation in Fig. 14.14a using truth tables.

14.5 Prove the equation in Fig. 14.14b using truth tables.

14.6 Prove the equation in Fig. 14.14c using truth tables.

14.7 Prove the equation in Fig. 14.14d using truth tables.

14.8 What is the decimal equivalent of 1011001?

14.9 What is the binary equivalent of 0037?

14.10 What is the hexadecimal equivalent of 011010011100?

14.11 What is the hexadecimal equivalent of 111000111010?

14.12 Name the types of gates used in logic circuits.

14.13 What is the output "Y' for the gate shown in Fig. 14.14e?

14.14 What is the output "Y' for the gate shown in Fig. 14.14f?

14.15 Draw a logic circuit for equation given in Fig. 14.15a.

14.16 Draw a logic circuit for equation given in Fig. 14.15b.

14.17 Draw a logic circuit for equation given in Fig. 14.15c.

14.18 Draw a logic circuit for equation given in Fig. 14.15d.

14.19 Draw a logic circuit for equation given in Fig. 14.15e.

14.20 Draw a logic circuit for equation given in Fig. 14.15f.

(a) $\overline{A \cdot B} = \overline{A} + \overline{B}$

(b) $\overline{A + B} = \overline{A} \cdot \overline{B}$

(c) $\overline{\overline{A \cdot B}} = A + B$

(d) $\overline{\overline{A + B}} = A \cdot B$

(e)

(f)

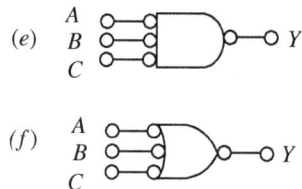

FIGURE 14.14 DeMorgan's equations.

(a) $Y = \overline{A} \cdot \overline{B} \cdot \overline{C}$

(b) $Y = \overline{A} + \overline{B} \cdot \overline{C}$

(c) $Y = \overline{A} + \overline{B} + \overline{C}$

(d) $Y = (A + B) \cdot (\overline{A} + \overline{B})$

(e) $Y = A \cdot \overline{B} + \overline{A} \cdot B$

(f) $Y = A \cdot \overline{C} + B(\overline{A} + C)$

FIGURE 14.15 Equations for Problems 14.15 through 14.20.

CHAPTER 15

Programmable Logic Controllers

Chapter Objectives

The programmable logic controller is used in virtually all process-control applications. In this chapter the basics of the programmable logic controller are introduced to familiarize you with its operation and how it is used in process control.

Topics discussed in this chapter are as follows:

- Controller operation
- Input/output modules
- Ladder diagrams
- Input/output symbols
- Ladder logic
- Scan modes

15.1 Introduction

Due to the complexity and large number of variables in many process-control systems, microprocessor-based programmable logic controllers (PLCs) are used for decision-making. The PLC can be configured to receive a small number of inputs (both analog and digital) and control a small number of outputs or the system can be expanded with plug-in modules to receive a large number of signals and simultaneously control a large number of actuators, displays, or other types of devices. In very complex systems, PLCs have the ability to communicate with each other on a global basis and to send operational data to, and be controlled from, a central computer terminal. The systems are designed not only for continuous monitoring and adjustment of process variables but also for sequential control, which is an event-based process, and alarm functions. This chapter introduces the PLC, its operation, and use for sequential, continuous control, and alarm functions.

15.2 Programmable Controller

The PLC is a rack mounted microprocessor-based system. The system contains memory for data recording and storage and has its own internal programmable memory for operation, control information, and parameter settings. The racks contain slots for input

Material Flow

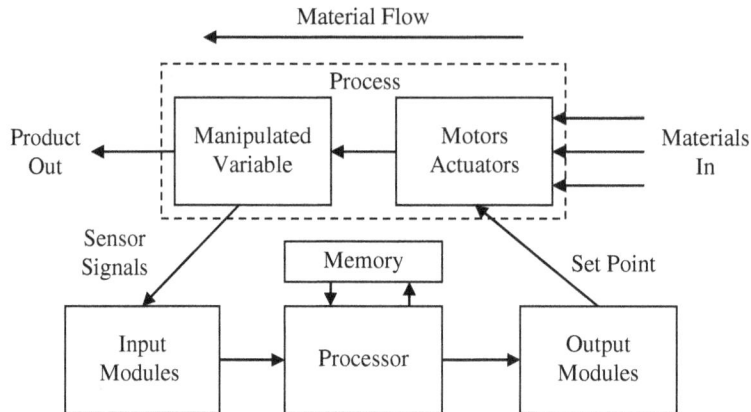

FIGURE **15.1**   Block diagram of a control loop.

and output modules for system expansion on as needed basis. Power supplies are also available for the modules. The basic setup is shown in Fig. 15.1.

The processor in a PLC system has software that is easily programmable and flexible making the initial program, updates, modifications, and changes easy to implement. Because of the complexity and large number of variables in many process-control systems, microprocessor-based PLCs are used for decision-making. The PLC can be configured to receive a number of inputs from a variety of input modules that are available for interfacing from ON/OFF, digital, and analog signals to proportional, integral, and derivative (PID) functions that are received and multiplexed on to the processor's data bus for use by the processor. The PLC can then multiplex new information onto the data bus for the output modules that are used for actuator control, indicators, alarm outputs, ON/OFF signals, and timing functions. As the modules are rack mounted only, the required modules can be used leaving rack space for expansion allowing the system to be expanded with plug-in modules to receive a large number of signals and simultaneously control a large number of actuators, displays, or other types of devices. PLCs are categorized into low-end, mid-range, and high-end, where low-end is from 64 expandable to 256 input/outputs, mid-range is expandable up to 2048 I/O's, and high-end is expandable up to 8192 I/O's. PLCs have the ability to communicate with each other on a local area network (LAN) or a wide area network (WAN), and send operational data to, and be controlled from, a central computer terminal. Figure 15.1 shows a typical controller setup for monitoring logic inputs sequentially. A decision can then be made by the PLC on any changes needed and the appropriate control signal sent via output modules to the actuators or motors in the process-control system.

The memory can be divided into ROM and RAM for system operation and EEPROM and Flash EEPROM (nonvolatile memory) for storing set-point information, look-up tables, and so on, with additional memory for data storage.

The processor not only controls the process but must be able to communicate to the outside world as well as use the Foundation Fieldbus (see Chap. 13, Sec. 13.4.3) for communication to smart sensors and other devices. All of these control functions may not be required in a small process facility, but are necessary in large facilities. The individual

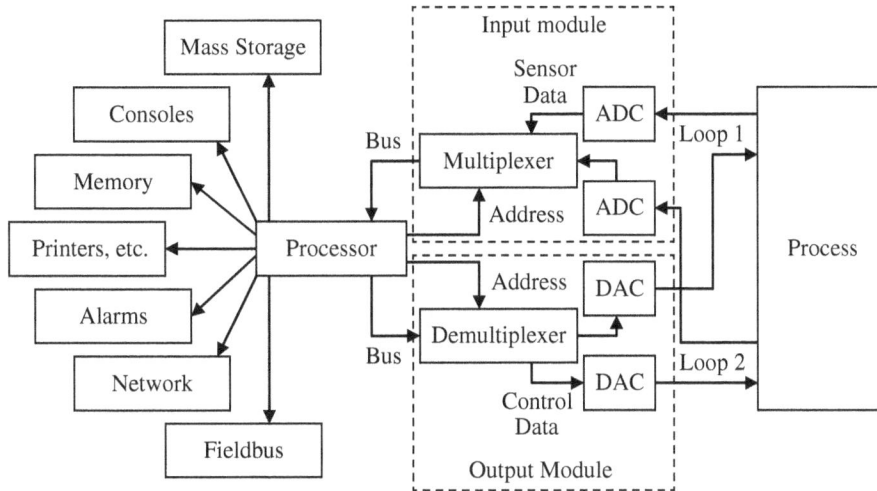

Figure 15.2   Control of analog loops.

control loops are not independent in a process, but are interrelated and many measured variables may be monitored and manipulated variables controlled simultaneously. Several PLCs may also be connected to a mainframe computer for complex control functions. Figure 15.2 shows the block diagram of a processor controlling two analog loops. The analog output from the process sensors is converted to a digital signal in an analog-to-digital converter (ADC). The digital signal is selected in a multiplexer and put into memory by the process to await evaluation and further action. After processing, the digital output signal is fed to the actuator through a demultiplexer and converted to an analog signal by a digital-to-analog converter (DAC). The processor will also have mass storage for storing process data for later use or making charts and graphs. The process will also be able to control a number of peripheral units and sensors as shown, and perform alarm functions.

15.3   Controller Operation

The central processing unit can be divided into the processor, memory, and input and output units or modules as shown in Fig. 15.3. The units are interconnected by a data bus, address bus, and an enable bus. The operation cycle in the PLC is made up of two separate modes; these are the I/O scan mode followed by the execution mode.

I/O scan mode can be divided into two operations. First is the period when the processor updates the output control signals based on the information received from the previous I/O scan cycle after its evaluation of the input signals. Second is the period when the processor scans the inputs in a serial mode and updates its internal memory as to the status of each of the inputs.

Execution mode follows the I/O scan mode. In this mode, the processor evaluates the input data stored in memory against the data programmed into the CPU. The programs are usually set up using ladder networks, where each rung of the ladder is an instruction

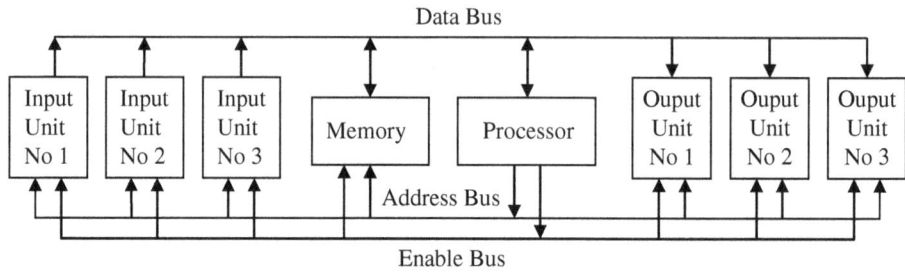

Figure 15.3 Block diagram of processing unit.

for the action to be taken for each given input data level. The rung instructions are sequentially scanned and the input data evaluated. The processor can then determine the actions to be taken by the output modules and puts the data into memory for transfer to the output modules during the next I/O scan mode.

Scan time is the time required for the PLC to complete one I/O scan plus the execution cycle. This time depends on the number of input and output channels, the length of the ladder instruction sets, and the speed of the processor. A typical scan time is between 5 and 20 ms. In addition to data evaluation, the PLC can also generate accurate time delays, store and record data for future use, and produce data in chart or graph form.

Watchdog timers are used to monitor scan times to ensure the PLC is working correctly and executing commands in a timely fashion and not hung up or stopped scanning.

15.4   Input and Output Modules

Input and output modules are used to act as the signal interface between the monitoring sensors, control actuators, and the processor. Modules provide electrical isolation, convert the input signals into a digital format suitable for evaluation by the processor if necessary, memory storage, and to format the output signals for displays and control functions. Modules fall into three categories—those for use with discrete ON/OFF levels, those with analog signal levels, and those that have intelligence to process the input signals before they are used by the controller. Input/output modules will typically have 16 inputs or outputs, but can range from 4 to 32. Modules that have both input and output ports are also available, as well as modules that have HART (highway addressable remote transducer) protocol interface capability.

15.4.1   Discrete Input Modules

Discrete input modules serve as ON/OFF signal receivers for the processor. The basic function of the input module is to determine the presence or absence of a signal. The inputs from peripheral devices to the input modules can be ac or dc signals. The voltage ratings for input modules can vary from 24 to 240 V ac or dc as well as 5 and 12 V TTL levels. The various types of applications that can be used with the discrete input modules are given in Table 15.1.

Normally a discrete input module will have 16 inputs which can be segmented into groups of 4, 8, or 16. Switches can be connected in blocks of four with either ac or dc

Type of Input	Application
Discrete ON/OFF	Push button, switch, relay contacts, starter contacts, float switch, temperature switch, level switch
Logic level	TTL levels, Ethernet levels
DC or AC	General purpose high, medium, or low level
Discrete parallel	Thumbwheel switches, bar codes, weigh scales, position encoders, ADC, BCD/parallel data devices

Note: ADC – analog to digital converter, BCD – binary coded decimal.

TABLE 15.1 Discrete Input Applications

power supplies. An open switch gives a "0" level input and a closed switch gives a "1" level input. The inputs can also be configured to accept transistor logic levels.

The input stage of a dc or ac module is used to detect presence or absence of a voltage and to convert the input voltage to a logic 5 V level. Figure 15.4 shows the block diagram of a discrete input module. The front ends of both the dc and ac modules are shown. With a high dc input voltage, the voltage is stepped down to a low voltage which goes through a debounce circuit with noise filter, and threshold detector for "1" or "0" detection followed by optical isolation, so that the signal can be referenced to the signal ground of the processor. The ac module input uses a bridge rectifier to convert the ac to dc and then uses the same circuit blocks as in the dc module. The LED is used to indicate the input logic level of the input signal. The input level LED indicators are normally located above the tag strip.

15.4.2 Analog Input Modules

Analog input modules are used to convert analog signals to digital values or words. Analog signals are derived from temperature, pressure, flow, position, or rate measurements. Analog input voltage and current are normally the standard levels of 0 to 5 V or 0 to 10 V, and current ranges from 4 to 20 mA. However, modules are available for thermocouple inputs and interfacing with RTD (resistance temperature device) for temperature

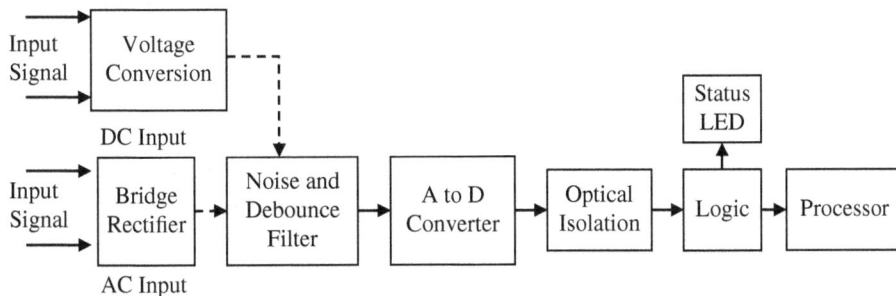

FIGURE 15.4 Block diagram of an input module with ac or dc input.

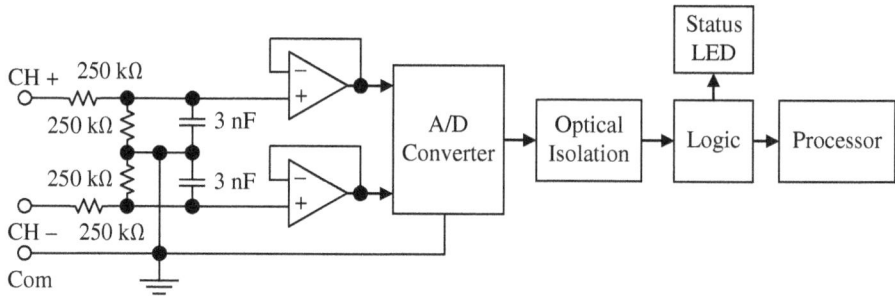

Figure 15.5   Analog input module block diagram.

measurements. Figure 15.5 shows the block diagram of a differential analog voltage receiver. The signal is converted into a digital word. The resolution of the ADC is normally 305.176 μV when using voltage input, and 1.2207 μA when using a current input. The ADC converter words are 14 to 16 bits in length. Because the inputs are referenced to the signal ground, the ADC uses an optoisolator to reference the digital signal to the ground level of the processor.

15.4.3   Special Function Input Modules

To satisfy some special cases, a variety of discrete modules are available to process or interface special signals. A list of these functions is given in Table 15.2.

The interrupt function module is used to interrupt the processors scan sequence to perform a task that requires immediate attention.

The voltage comparator module is used to compare the amplitude of the input to an internally generated voltage or an externally derived voltage.

To detect fast transients of a few microseconds that would normally be missed by standard input modules, the latching input module is used to set a latch.

The fast input performs a similar function to the latching module, but does not latch the transient; it only holds the information for a scan cycle so that it can be detected and recorded.

The rapid response module is similar to the latching input module but can immediately enable an output without having to wait for a scan cycle.

Module Type	Application
Interrupt input	Immediate response to signal changes
Voltage comparator unit	Analog set point comparison
Latching input	Detection of short-duration signals
Fast input	Fast response to dc level changes
Rapid response I/O	Provides fast input/output response
Relay contact output	High current switching and signal multiplexing
Wire fault input	Wire break and short-circuit detection

Table 15.2   Special Function Input Modules

The relay output module has isolated relay contacts to handle high currents and to multiplex signals.

The fault input module is used to detect wire faults.

15.4.4 Discrete Output Modules

Discrete output modules are used to transfer output information from the processor to peripheral control units. They provide electrical isolation and data in a suitable format for use by the external units. The output from the modules can be either discrete ac or dc outputs or relay contacts. The output voltage can be 12 to 230 V ac, dc, or TTL and Ethernet levels. Table 15.3 shows a list of discrete output applications.

Figure 15.6 shows the block diagram of solid-state discrete output drivers using TRIACs. Only two drivers are shown; normally the drivers would be in groups of four or eight in a module. The outputs have filtering and surge suppression to protect the drivers against transients and inductive spikes. The output is fused for protection against overloads. A LED located above the tag strip is used to indicate the logic state of the output.

15.4.5 Analog Output Modules

Analog outputs from the PLC are used to drive analog meters, chart recorders, proportional valves, variable speed drive controllers, and for current or voltage to pneumatic transducers. The voltage and current output ranges are normally the same as the

Types of Output	Application
Discrete outputs	Motor starters, solenoids, alarms, horns, buzzers, fans
Logic outputs	TTL logic and Ethernet devices
AC/DC outputs	General purpose high, medium, or low ac and dc loads
Parallel outputs	Seven segment displays, BCD-controlled message displays, DAC, BCD/parallel data input devices

Table **15.3**   Discrete Output Applications

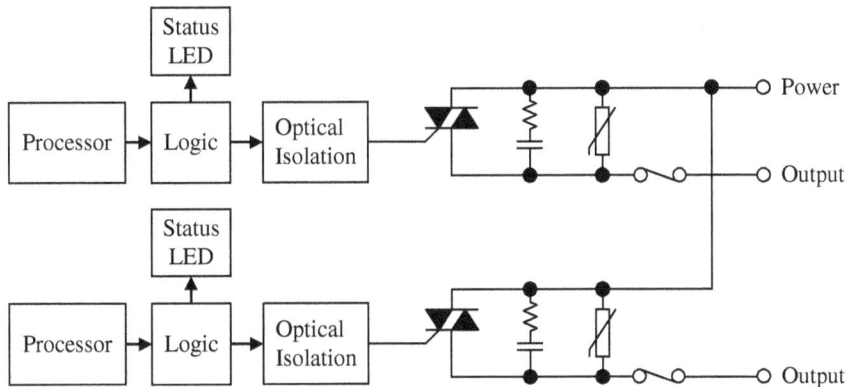

Figure **15.6**   Discrete AC output module.

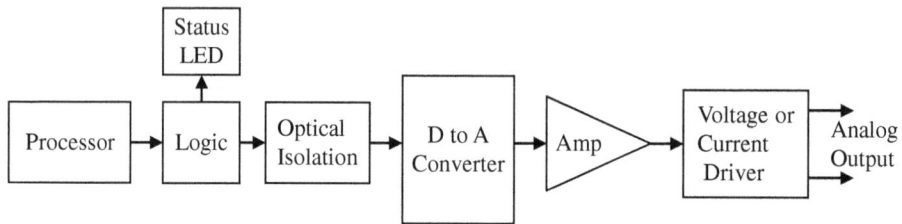

FIGURE 15.7 Analog output block diagram.

standard input ranges. Figure 15.7 shows the block diagram of an analog output stage. The digital output from the controller is fed via an optical isolator to an ADC converter to reference the signal to the ground of the peripheral device. The analog output of the converter is amplified and fed to a voltage or current driver which can have a single-ended output or a differential output.

15.4.6 Smart Input and Output Modules

A number of specialized modules have been developed to interface to the processor that normally contain their own processor and memory, and can be programmed to perform operations independent of the central processor. These modules are categorized in Table 15.4.

Serial and network modules are used for data communication. The serial modules are used to communicate between other PLCs, message displays, operator terminals, and intelligent devices. The network modules are used for LAN and WAN. The manufacturing automation protocol (MAP) is used for communication to robotic devices and a variety of computers of differing manufacture that can support MAP.

The coprocessor modules are basically used for housekeeping functions, such as mathematical functions, algorithms, and data manipulation, outputting reports, outputs

Intelligent Device Types	Module Function
Serial and network	ASC II communications module, serial communication
Communication	Module, loop controller interface module, proprietary LAN network module, MAP network module
Computer coprocessor	PC/AT computer module, basic language module, I/O logic processor module
Closed loop	PID control module, temperature control module
Position and motion	High-speed counter module, encoder input module, stepper positioning module, servo positioning module
Process specific	Precision control module, injection molding module, press controller module
Artificial intelligence	Voice output module, vision input module
HART protocol	Digital plus analog signal module

TABLE 15.4 Intelligent Input/Output Module Categories

to printer, displays, and mass data storage using basic programming language functions that would be hard for the main processor to perform with ladder logic.

Intelligent modules performing closed-loop control algorithms are required for PID functions, such as maintaining temperature, pressure, flow, and level at set values. However, with the introduction of smart sensors, these functions can be performed by the smart sensors reducing the load on the processor. Communication to the processor in this case is via the Fieldbus.

The internal control function of the PID module using an analog loop is shown in Fig. 15.8. The analog temperature signal from the furnace is the input to the module, where it can be converted to a digital signal and recorded in the computer memory, fed to the PID controller via a gain control, and to an analog differencing circuit where it is compared to the set point signal from the processor. The furnace signal and set point are subtracted giving an analog error voltage. The error voltage is fed to the PID controller that produces an analog control voltage/current to adjust the valve controlling the fuel flow to the furnace. The PID module could also use digital techniques; the analog input signal would be converted into a digital signal, compared to the set point in a digital comparator to generate the error signal. The controller will use the error signal to generate a pulse width modulation (PWM) signal to control the furnace fuel actuator.

The temperature control module is normally used to control 8 to 16 temperature zones. The module is configured for two position control (heat ON/OFF) or three position control (heat ON/OFF/cool). The set points are stored in the processor. A typical application is large building heating ventilation and air conditioning (HVAC), or to control the zone temperatures required in plastic injection molding machines.

Position and motion modules enable PLCs to control stepper and servo motors in feedback loops, measure and control rotation speeds and acceleration, and are used for precision tool control. This category of devices uses high-speed counting, rotational and linear position decoders, and open and closed-loop control techniques for the measurement of axis rotation and linear speed and position. Typical applications of position and motion modules are given in Table 15.5.

Process-specific modules are intelligent modules designed to perform specific control functions or a specific series of operations. Many machines built by different vendors perform similar functions and are similar in operation, but use a different interface. These modules were developed to interface with such machines; the operations they

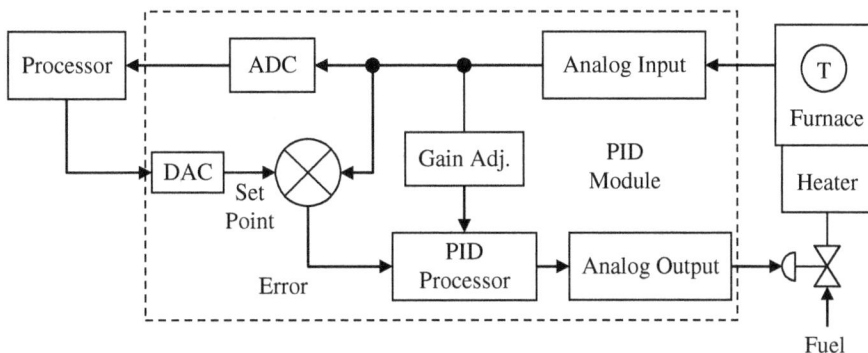

FIGURE 15.8 Supervisory PID control using an analog control loop.

Module Type	Application
High-speed counter module	Up/down counting, generate input for set count, generate gating, generate delays
Encoder input module	Absolute position tracker, incremental position tracker
Stepper positioning module	Open-loop position, setting dwell times, define motion speed, motion acceleration
Servo-positioning module	Transfer and assembly lines, material handling, machine tool setting, automatic parts insertion, table positioning, precision parts placement

TABLE 15.5   Applications of Position and Motion Modules

perform are normally repetitive, requiring precise measurements, and complex numerical algorithms. Typical applications are profiling and control in plastic molding and injection systems, and so on.

Artificial intelligence modules have a number of industrial applications in voice recognition, synthesized speech, and visual inspection. The sound module can be used to give alarm announcements, voice recognition, and echo evaluation when using sound waves for flaw detection. The video module can be used for dimension gauging, visual inspection and flaw and defect detection, position analysis, and product sorting.

HART is a hybrid protocol because it combines a single 4 to 20 mA analog signal with a low-level FM (frequency modulated) carrier superimposed on the analog signal. The FM signal contains digital information using a technique called FSK (frequency shift keying). Each peripheral unit on the communications bus is given a unique digital address. The digital information sent on the bus is in the form of serial words, with each word containing the address of a peripheral unit and data for that unit or requesting information from that unit allowing two-way communication with any number of peripheral units. The receiver filters the FM signal from the analog signal keeping the integrity of the analog signal.

15.5   Ladder Diagrams

The ladder diagram is universally used as a symbolic and schematic way to represent the interconnections between the elements in a PLC and as a tool for programming the operation of the PLC. The elements are interconnected between the supply lines for each step in the control process, giving the appearance of the rungs in a ladder. A number of programming languages are in common use for controllers; they are as follows:

Ladder instruction list

Boolean flowcharts

Functional blocks

Sequential function charts

High-level languages (ANSI, C, structured text)

15.5.1 Input and Output Symbols

Figure 15.9 shows some of the typical symbols used to represent external ON/OFF circuit elements in a ladder diagram. A number of momentary action switches are shown. These are a push to close (normally open NO) and push to open (normally closed NC). These switches are the normal momentary action panel mounted operator switches.

Position limit switches are used to sense the position of an object and set to close or open when a desired position is reached. Pressure, temperature, and level switches are used to set limits and can be designed to open or close when the set limits are reached.

The symbol for timer relay has the designation TR with a number and its associated NO and NC contacts will be named and numbered. A control relay is a circle with the designation CR and is shown in Fig. 15.9.

External control device symbols used in ladder diagrams are shown in Fig. 15.10. A motor is represented by a circle with the letter M and an appropriate number, an indicator is represented by a circle with radiating arms and a letter to indicate its color, i.e., R = red, B = blue, O = Orange, G = green, and so on. A solenoid is as shown and other elements are represented by boxes with the name of the element and a number.

15.5.2 Ladder Layout

In a ladder diagram, the supply lines are represented by the verticals forming the sides of the ladder, with the elements connected serially between the supply lines as shown

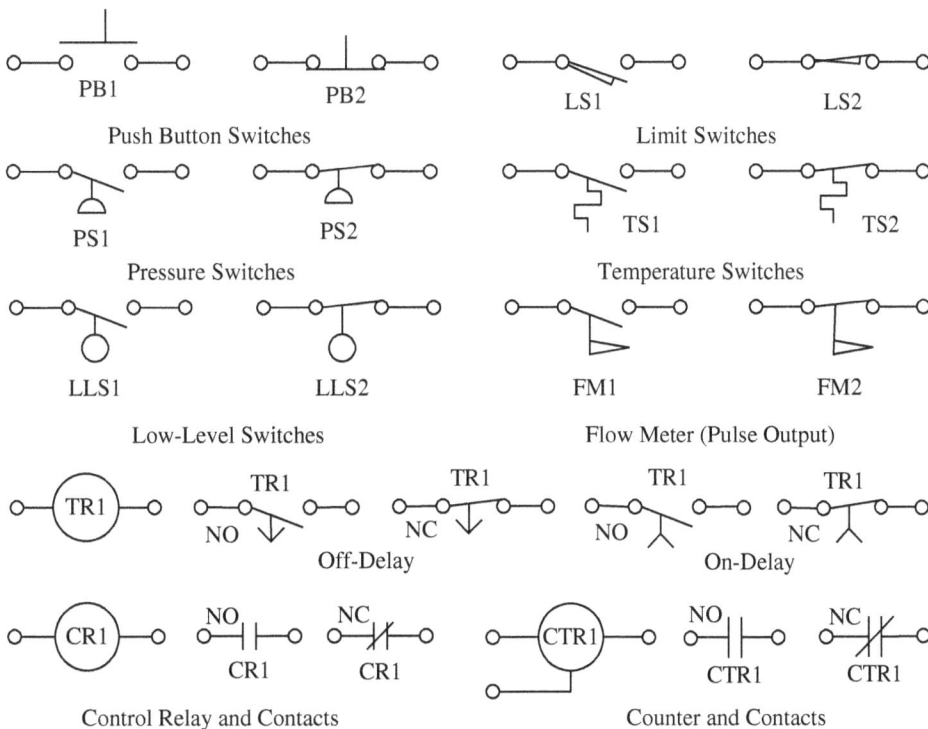

FIGURE 15.9 Symbols in use for ladder diagrams.

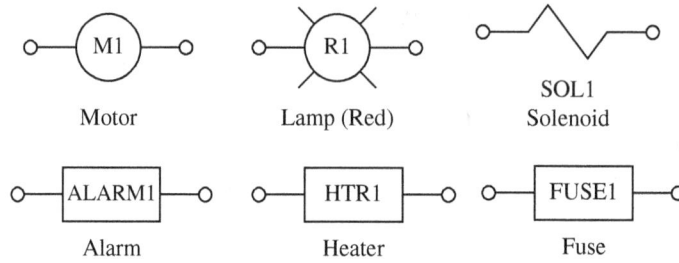

FIGURE 15.10 Ladder diagram control device symbols.

in Fig. 15.11. Each ladder rung is numbered using the hexadecimal numbering system with a note describing the function of the rung (see Fig. 15.14). The notes are required for debugging and to assist in future fault finding, upgrading, and modification. Figure 15.11 shows an electrical component wiring diagram and the equivalent ladder diagram. The components are represented using the open and closed contact symbols in the ladder diagram.

15.5.3 Ladder Gate Equivalent

Ladder diagrams can be made from logic diagrams, or Boolean expressions as well as a component electrical wiring diagram as shown in Fig. 15.11.

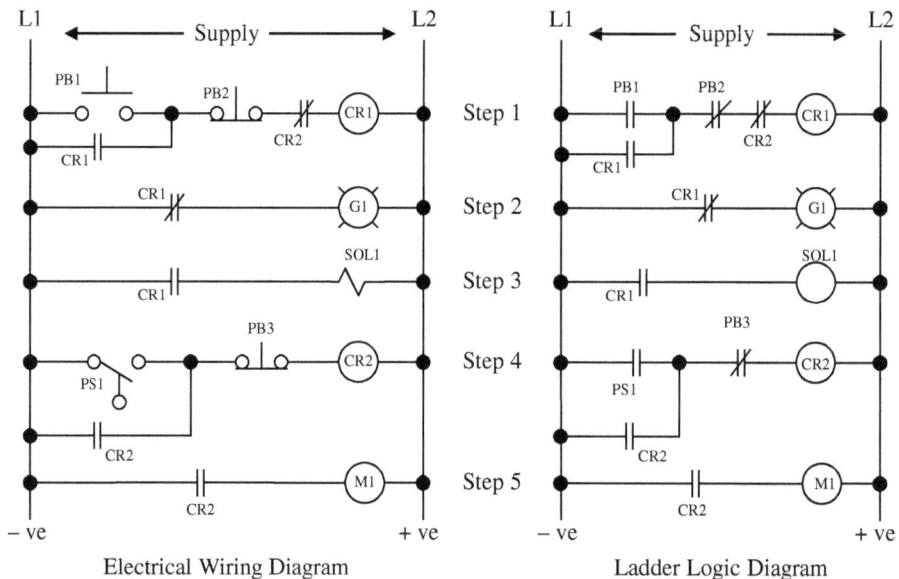

FIGURE 15.11 Comparison of component electrical wiring diagram and ladder logic diagram.

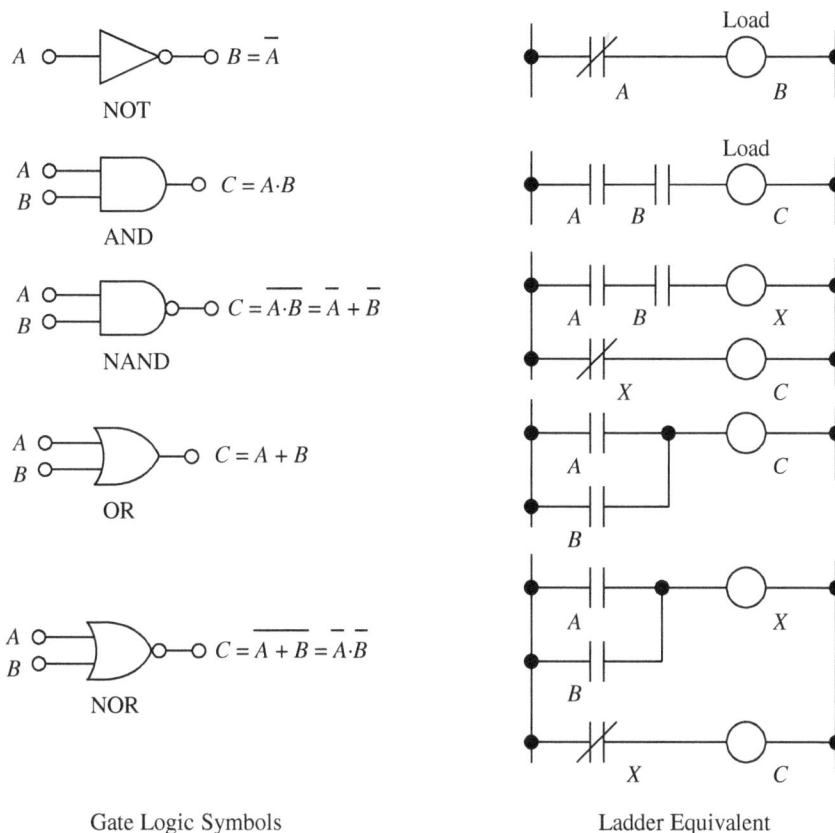

NOT $\quad B = \bar{A}$

AND $\quad C = A \cdot B$

NAND $\quad C = \overline{A \cdot B} = \bar{A} + \bar{B}$

OR $\quad C = A + B$

NOR $\quad C = \overline{A + B} = \bar{A} \cdot \bar{B}$

Gate Logic Symbols   Ladder Equivalent

FIGURE 15.12   Logic ladder equivalent of electronic logic gates.

15.5.4   Ladder Applications

When using more complex logic diagrams the logic can be optimized for component count using Boolean algebra to minimize the switch count and the number of operations required by the PLC. Figure 15.12 shows how the switch contacts in the ladder diagram are arranged to give the same function as the gate logic. The ladder equivalent of the inverter, AND, NAND, OR, and NOR gate logic are shown.

Example 15.1   Figure 15.13 shows a basic chocolate dispenser. Under PLC control, a proximity sensor (LS1) is used to sense a cup is in place under the dispenser, the water heater can then be filled via SOL1 and the water heated (HTR1) when a start button (PB1) is pushed. Buttons are also provided on the control panel for dispensing chocolate (PB2) and sweetener (PB3) and indicator lights to show each step in the operation of the dispenser; red indicates cup in position, blue water heating, cyan chocolate dispensed, orange sweetener dispensed, and green indicates cup full. Design a ladder diagram for a PLC to perform the dispenser control function, number, and add notes to the rungs.

Figure 15.14 shows a possible solution to Example 15.1 using a ladder diagram for PLC control. Placement of the cup in the dispenser is sensed by the close proximity switch LS1 which operates relay CR1 (rung S00) and the red indicator light (rung S01)

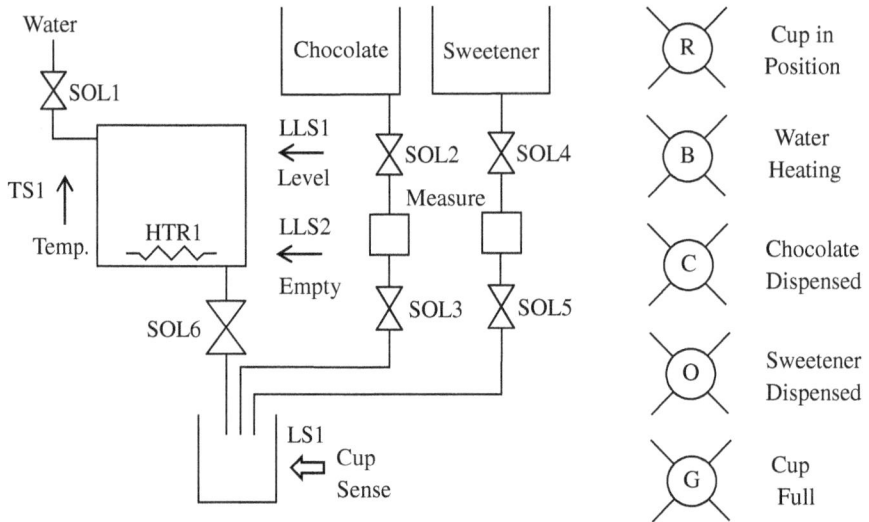

FIGURE **15.13** Chocolate dispenser.

and enables operation of the dispenser. Removal of the cup will stop the dispenser from operating. The ladder rungs will perform the following operations:

S02 Closure of CR1 enables PB1 to energize CR2; CR2 NO contact will keep CR2 energized after the release of PB1.

S03 CR2 operates the blue light to indicate water is filling the heater container and being heated.

S04 CR2 operates valve SOL1 to fill container.

S05 CR2 closure enables PB2 to energize CR3; CR3 NO contact will keep CR3 energized after the release of PB2.

S06 Closure of CR2 and CR3 energizes cyan light to indicator chocolate dispensed.

S07 Closure of CR2 and CR3 operates SOL3 to dispense chocolate.

S08 CR2 closure enables PB3 to energize CR4; CR4 NO contact will keep CR4 energized after the release of PB3.

S09 Closure of CR2 and CR4 energizes orange light to indicator sweetener dispensed.

S0A Closure of CR2 and CR4 operates SOL5 to dispense sweetener.

S0B CR2 closure lets float switch LLS1 energize CR5 when container is full, CR5 NO contact will keep CR5 energized after LLS1 opens.

S0C Opening of CR2 deenergizes SOL2 to prevent chocolate refilling measure during operation.

S0D Opening of CR2 deenergizes SOL4 to prevent sweetener refilling measure during operation.

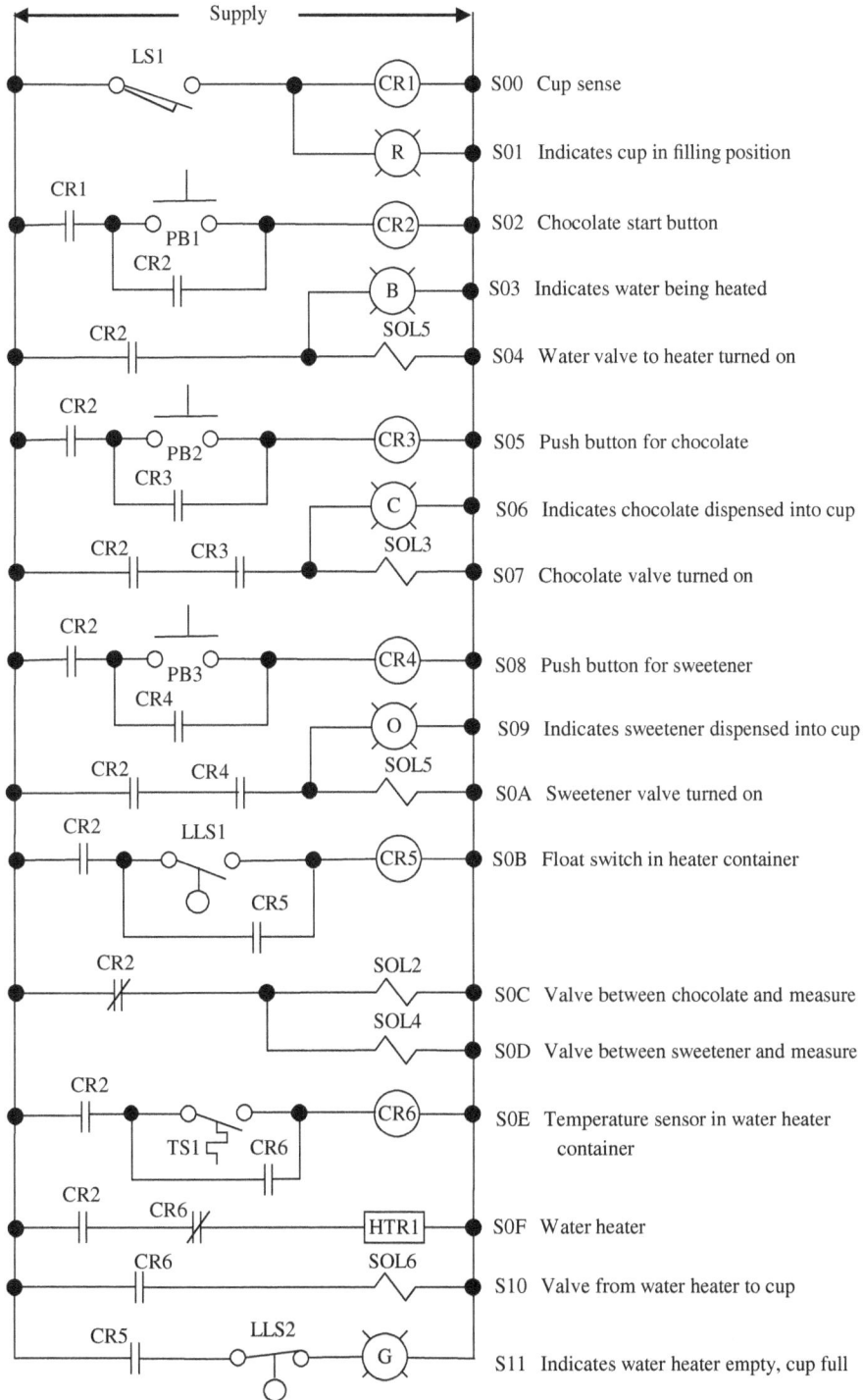

FIGURE 15.14 Dispenser ladder diagram.

S0E CR2 closure lets temperature switch TS1 energize CR6 when water is hot, CR6 NO contact will keep CR6 energized after TS1 opens.

S0F CR6 NC contacts deenergize the water heater when the water is hot.

S10 CR6 energizes SOL6 to fill cup when the water is hot.

S11 Float switch LLS2 closes when water container is empty energizing the green light to indicate end of cycle and cup is full.

Removal of cup will cause proximity switch LS1 to open deenergizing CR1, CR2, and resetting dispenser relays for the next cycle.

Summary

This chapter introduces the PLC and its use in sequential logic and continuous system control. The main points described in this chapter were as follows:

1. Internal operation of the controller.
2. The modular design of the PLC for ease of expansion from small to large system operation.
3. Description of interface modules for discrete and analog signal inputs.
4. Modules available for actuator control.
5. Intelligent plug in modules for special functions such as motion control and artificial intelligence.
6. Programming of the PLC using ladder networks, instruction lists, and so on.
7. The common symbols used to represent switch functions in ladder networks.
8. Example of a sequential logic system and the control ladder diagram.

Problems

15.1 Describe the modes of operation of a PLC.

15.2 What is the scan time of a PLC?

15.3 What is a ladder diagram?

15.4 What are the three general categories of input modules?

15.5 What is the HART protocol?

15.6 Name fives types of discrete inputs.

15.7 How is electrical isolation obtained between input signals and the processor?

15.8 What is the resolution of an input voltage ADC?

15.9 What is the resolution of an input current ADC?

15.10 Name three analog inputs.

15.11 Name four programming languages for ladder diagrams.

15.12   Name six types of input switches used in ladder diagrams.

15.13   How many drivers are present in an output module?

15.14   What is a watchdog timer?

15.15   Name three types of memory used in a PLC.

15.16   What are the main blocks in an ac input module?

15.17   What are main blocks in an ac discrete output module?

15.18   How many inputs are in an input module?

15.19   Name four types of intelligent input/output modules.

15.20   Name four types of position and motion modules.

CHAPTER 16

Motor Control

Chapter Objectives

The electric motor is extensively used in process control for driving conveyer belts, pumps, actuators, and so on. Hence, it is necessary to have knowledge of the different type of motors, their characteristics, operation, and applications.

Topics discussed in this chapter are as follows:

- DC motors
- AC motors
- Stepper motors
- Servo and synchro motors
- Motor characteristics
- Motor control
- Motor applications

16.1   Introduction

Many types of ac and dc motors are available with widely varying characteristics such as power requirements, starting torque, starting current, running current, speed control, and so on, which need to be evaluated when selecting a motor for a specific application.

16.2   Motor Classification

Electric motors are typically classified into three broad categories as shown in Fig. 16.1. These are motors that are designed to operate from a dc supply, motors that are designed to operate from a single ac or multiphase supply, and universal motors that are designed to operate from either an ac or dc supply.

DC motors can be subdivided into series, shunt, compound, brushless, and stepper motors; universal motors can be subdivided into compensated and noncompensated; and ac motors can be divided into induction and synchronous. The induction type of ac motor is a motor that normally has no connection to the rotor; current in the rotor conductors is induced by the magnetic field of the stator windings. A specialized group of induction motors, however, has its rotor current carrying conductors connected to slip rings so that resistance can be connected in series with the conductors to reduce

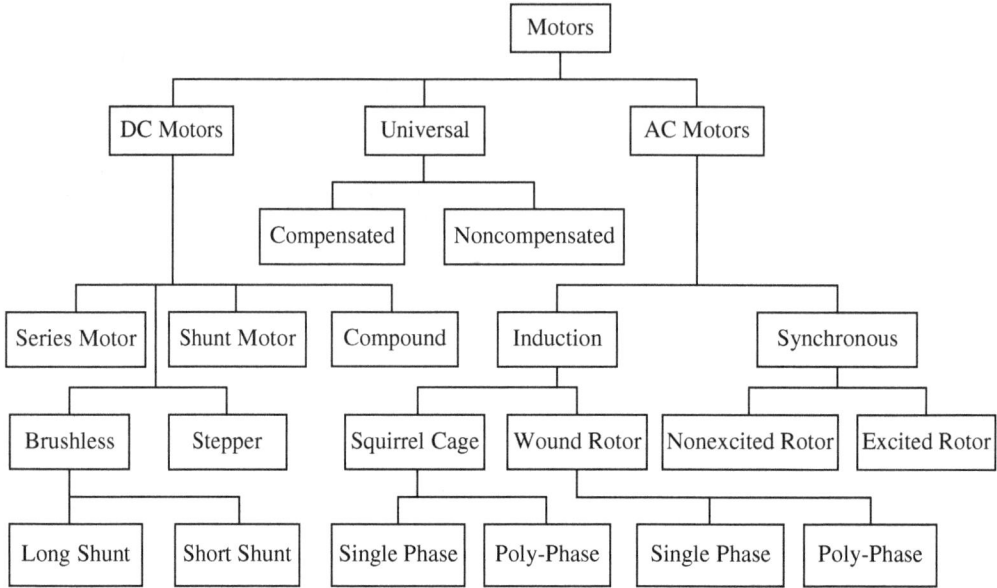

FIGURE 16.1 Motor classification.

the current on startup giving a high starting torque and speed control. Hence, the induction motor can have a nonconnected squirrel cage or wound rotor and both can operate from a single-phase or a poly-phase supply. The synchronous motor can be divided into a nonexcited or an excited rotor classification.

Motors can also be classified by their characteristics and application; ac motor speeds are dependent on the supply frequency and can have internal gears to give more than one output speed or to increase output torque. Motor stators can be wired so that the speed can be changed by switching coil connections, as opposed to variable speed controlled by varying the supply frequency. In the case of dc motors, the speed can be varied by changing the supply voltage.

Motors can also be subdivided by their power rating into integral hose-power (ihp) which is greater than 1 hp, fractional horse-power (fhp) 0.05 to 1 hp, and subfractional horse-power (sfhp) which is less than 0.05 hp or less than 50 millihorse-power (mhp).

16.3 Motor Operation

New permanent magnetic materials are being developed and used in specialized motor applications to replace motor windings to give higher torque, lower current requirements, and higher efficiency.

16.3.1 DC Motors

DC motors are used where high torque and variable speeds are required. Four types of dc motors in common use are the series, parallel, the compound, and brushless. The type of motor used is determined by the requirements and characteristics of the load. The motor connections shown in Fig. 16.2a and b are the series (S) and shunt (F) windings

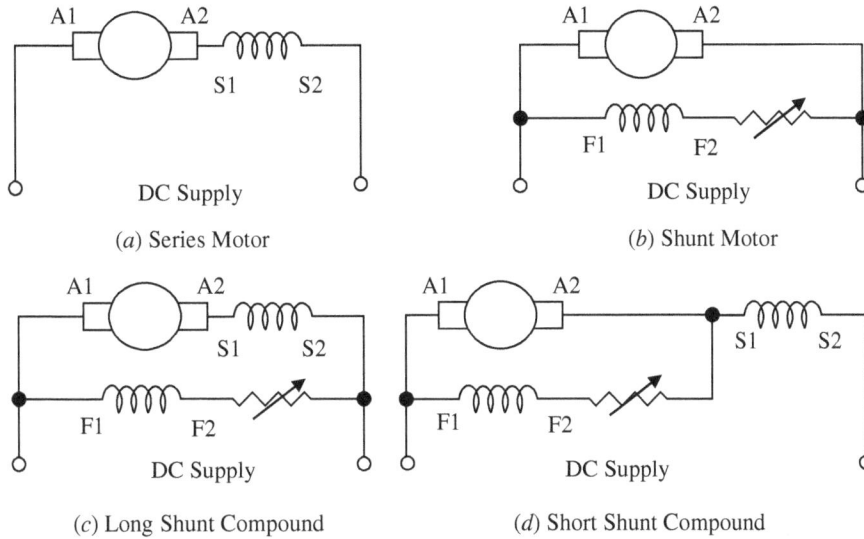

FIGURE 16.2 Various dc motor connections.

connection to the motor armature, and in *c* and *d* are the long shunt and the short shunt compound connection to the motor armature. The cross section of a dc motor is shown in Fig. 16.3*a*. The field shunt and series windings are shown on the pole pieces, with the armature coils connected via slip rings, so that the field windings can be connected in any of the configurations shown in Figure 16.2. When both are used they can be wound to enhance each other's magnetic field, or wound so that their magnetic fields oppose each other. The opposing field configuration is rarely used.

The potentiometers shown in Fig. 16.2*b*, *c*, and *d*, are used to vary the voltage to the shunt field windings for speed control. Figure 16.3*b* shows the cross section of a brushless

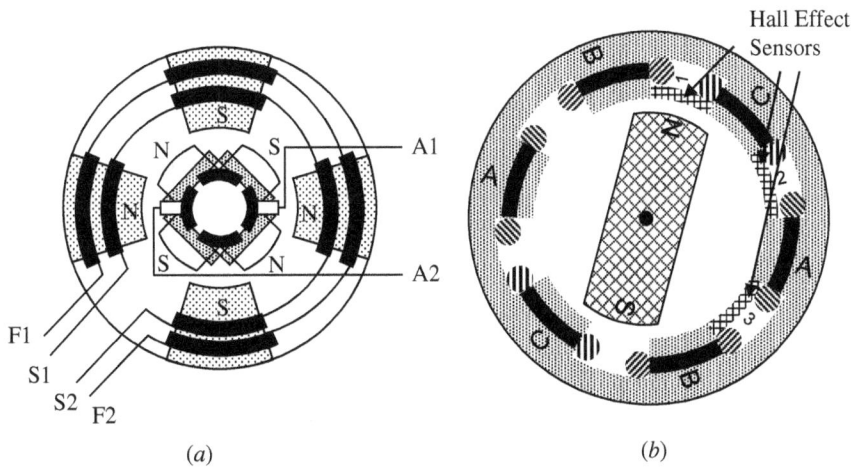

FIGURE 16.3 Cross section of (a) dc motor (b) a brushless motor.

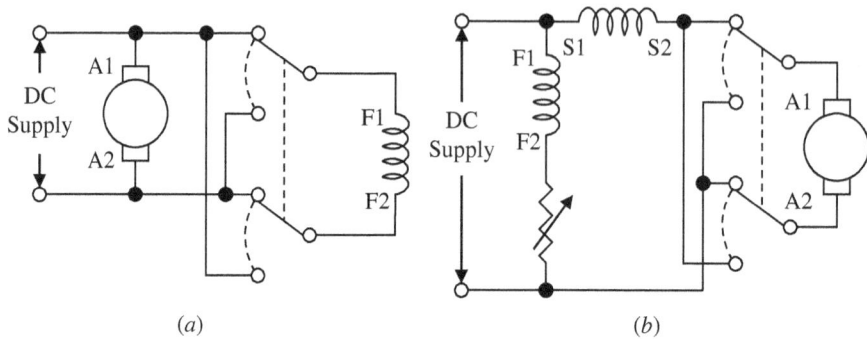

FIGURE 16.4 Double-pole double-throw switch to reverse (a) shunt motor and (b) compound motor.

motor. The armature is a permanent magnetic and its position is sensed by three Hall effect sensors that control currents through the three sets of field windings (A, B, and C) to produce a rotating magnetic field giving rotation to the armature. The advantages of the brushless motor are no brushes to wear, low maintenance, and safe operation in a volatile environment without being fully encased.

In Fig. 16.4, the fields are shown configured for clockwise rotation of the armature. By reversing the current through either the field winds or armature winding the armature will rotate in an anticlockwise direction. A double-pole double-throw switch is used to change the direction of rotation in a shunt wound motor by changing the direction of current through the shunt winding shown in Fig. 16.4a and to change the direction of rotation in a compound motor the current through the armature can be reversed as shown in Fig. 16.4b.

The characteristics of dc motors are given in Table 16.1.

16.3.2 AC Motors

AC motors can be divided into two groups—the induction which can be subdivided into single and three-phase motors, and the synchronous motor. In the induction motor, the rotor's magnetic field is induced by transformer action (Squirrel cage motor), see Fig. 16.1.

The rotor in a squirrel cage motor has low-resistance copper or aluminum conducting bars running the length of the silicon steel rotor that are shorted at the ends of the rotor as shown in Fig 16.5a such that the rotor acts as a single-turn secondary and the stator acts as the primary of a transformer. The squirrel cage is so called as the framework of the conducting bars looks like the exercise wheel in a squirrel cage.

Motor	Series	Shunt	Compound	Brushless
Stating torque	Very high	Medium	High	Medium
Variable speed	No	Yes	No	Yes
Speed regulation	Very poor	Good	Medium	Very good

TABLE 16.1 DC Motor Characteristics

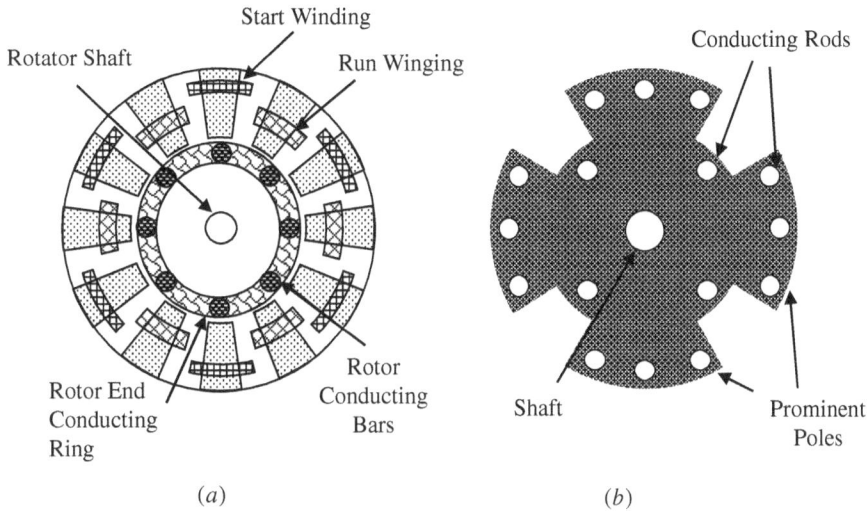

FIGURE **16.5** Representation of (a) stator windings in a single-phase squirrel cage motor and (b) a rotor with prominent poles.

The transformer action between the stator and the rotor produces a high current in the rotor conduction bars that produces a magnetic field in the rotor.

16.3.3 Single-Phase Motor

To produce a rotating magnetic field and rotation of the rotor in a single-phase ac motor it is necessary to produce an out-of-phase current so that there is an out-of-phase magnetic field. The out-of-phase fields can be produced by the combination of resistance and inductance, resistance and capacitance, or inductance and capacitance circuits. In Fig. 16.5a, the supply voltage is connected to the run windings and to the start windings via the phase shift network to give the required rotating magnetic fields to start the rotor turning. Once the rotor is up to speed, the out-of-phase current is normally disconnected.

The following are some of the methods used in single-phase motors to produce an out-of-phase current and magnetic field:

Split-phase internal resistance: The two stator windings in the motor have a different number of turns and are wound with different gauge wire to give different inductance and series resistance which in turn will produce two out-of-phase currents.

Capacitor start: A capacitor is connected in series with the start winding to obtain the phase shift between the run and start windings for a rotating magnetic field. When the motor is up to speed the start winding is disconnected by a centrifugal switch.

Permanent-split capacitor start: Uses the same technique as the capacitor start but does not switch the capacitor or start windings off once the rotor is up to speed. The capacitor in this case must be rated for continuous duty. The direction of rotation in this motor can be reversed by connecting the capacitor in series with the run winding instead of the start winding.

Two-capacitor start: Uses a second high-value capacitor in parallel with the capacitor in the permanent-split capacitor motor to give a high starting torque and low start current. This capacitor is switched out when the motor is up to speed.

Shaped pole: A number of motors use shaped stator poles to give different reluctance across the pole. One method is to use a shorted winding (shading) on a part of the pole. This produces a rotating magnetic field. These motors have very low torque, are inefficient with a low power factor but cheap to manufacture and as such are only used for small fractional horsepower motors.

Dual-speed voltage motors: These are sometimes designed for starting at a low voltage and then connecting the run and start windings in series when up to speed for higher running voltage and lower speed.

Synchronous motors: There are many types of synchronous motors. The above motors can be configured for synchronous operation by reshaping the rotor to have prominent poles as shown in Fig. 16.5b. These poles will almost align themselves with the stator magnetic field flux. The alignment will be off by several degrees due to the reluctance effect caused by rotational torque.

16.3.4 Three-Phase Motors

The three-phase motor is shown in Fig. 16.6a. The motor has different windings for each phase giving it a rotating magnetic field. The three-phase motor is designed in many configurations from the number of poles in use, brushless to wound rotor, and delta to wye operation. The three-phase motor has constant shaft torque, generates less noise, and has higher efficiency than the single-phase motor. The power of the single-phase motor is limited to about 10 hp whereas the three-phase motor can be designed for several hundred horsepower and is today's workhorse. Figure 16.6b shows the difference between the delta and wye connections. Some three-phase motors are started in the wye configuration in applications where low starting torque and current are required and then switched to the delta configuration when up to speed. The running speed is

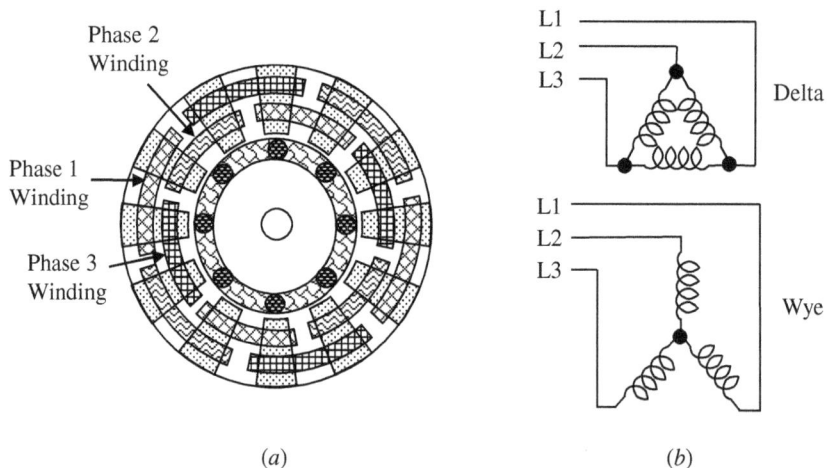

(a)   (b)

FIGURE 16.6 Rotor with prominent poles.

Motor Type	Split-Phase	Capacitor			Shaded Pole	Wound Armature	Three Phase
		Starting	Permanent	Two Value			
Starting current	High	Medium	Low	Medium	Very low	Low	Medium
Running current	Medium	Medium	Low	Low	High	Medium	Medium
Starting torque	Medium	Very high	Low	Very high	Very low	High	Medium
Efficiency	Medium	Medium	High	High	Very low	High	High
Size (hp)	Low	Medium	Low	Medium	Medium	Medium	Very high
Variable speed	No	No	Yes	No	Yes	Yes	Yes

TABLE 16.2 AC Motor Characteristics

about the same in the wye or delta connection. Three-phase motors are easily reversible by switching one of the phases. AC motor characteristics are given in Table 16.2.

16.3.5 Universal Motors

The universal motor is designed to operate from a dc or ac supply. The motor is similar in construction to a series dc motor and has similar characteristics. The major difference is that the laminates in both the armature and field are thinner than in the dc motor to reduce losses due to eddy currents when used with an ac supply.

16.3.6 Stepping Motors

Stepping motors differ from dc or ac motors as their rotor rotates a specific number of degrees each time the motor receives an input pulse. As an example, Fig. 16.7a shows a basic four-pole stepper motor that is powered by four clocked inputs shown in Fig. 16.7b. At each input pulse the permanent magnet rotor moves 90°; rotation in the example shown is anticlockwise, but the rotor can rotate in either direction depending on the pulse sequence.

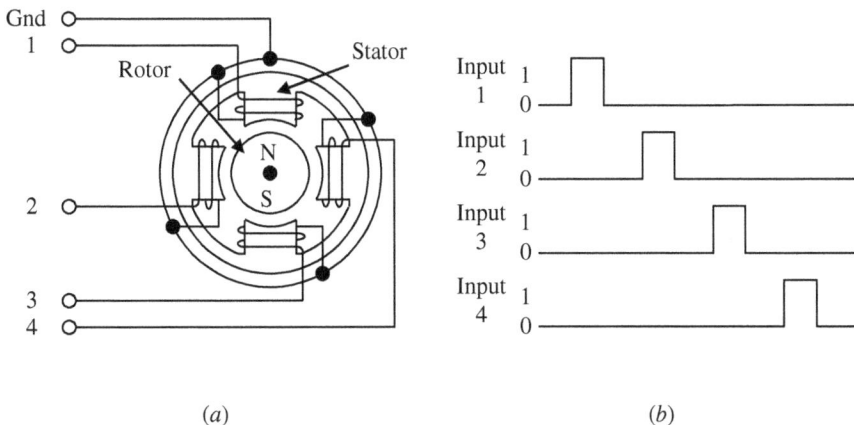

FIGURE 16.7 Illustration of (a) cross section of a stepper motor and (b) inputs for anticlockwise rotation.

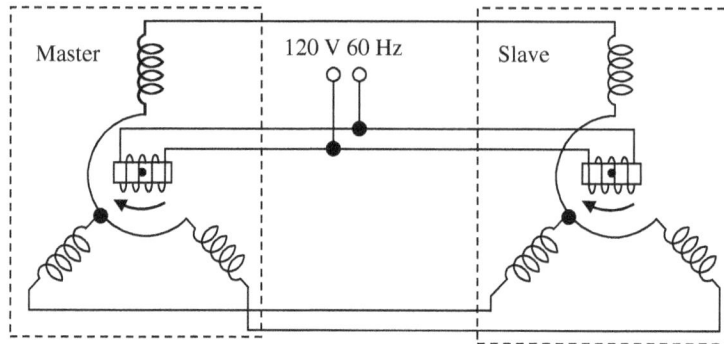

Figure 16.8 Synchro motors showing master and slave.

Basically, a stepper motor can be considered as a poly-phase synchronous motor. This principle can be applied to many stepper motor configurations. The stator and rotor can use a toothed configuration and achieve an angular rotation of only 1.8° per step. The stepper has high starting torque without high current requirements, has low maintenance, long life expectancy, low cost, rugged, and simple construction.

16.3.7   Servomotors and Synchro Motors

Servomotors are rotary or linear actuators used in precise control-loop applications. Optical rotary encoders are used to monitor position and are used in conjunction with a PID system for control of speed, direction, and position. The term is sometimes applied to motors that are suitable for use in servo systems. Typical applications are robotics and automated manufacturing.

Synchro or selsyn motors operate in a master–slave configuration. The angular position or rotation of the master is duplicated by the slave. These motors operate on a principle similar to that of a transformer. Figure 16.8 shows the diagram of a master and slave synchro. Both rotors are supplied from a common ac source. The magnetic field produced by the rotor in the master motor induces currents into the three stator windings. These current amplitudes are directly related to the position of the rotor. The induced currents are fed directly to or amplified and fed to the stator windings of the slave synchro where they will duplicate the magnetic field in the master synchro. Because the rotor in the slave is fed from the same supply as the rotor in the master, its magnetic field will align it with the magnetic field produced by the slave's stator windings to give it the same angular position as the master rotor. The accuracy of the slave rotor can be within 1° of the master. Synchro motors are used in instrumentation and rotary displays.

16.4   Motor Ratings

The parameter standard for motor ratings are set by the National Electrical Manufacturers Association (NEMA) and cover voltage, current, power, speed, temperature, frequency, torque, duty cycle, service factor, and efficiency. (The supply voltage to ac motors is assumed to be 60 Hz ±5 percent in North America.) These ratings are given on the motor's nameplate similar to that shown in Fig. 16.9.

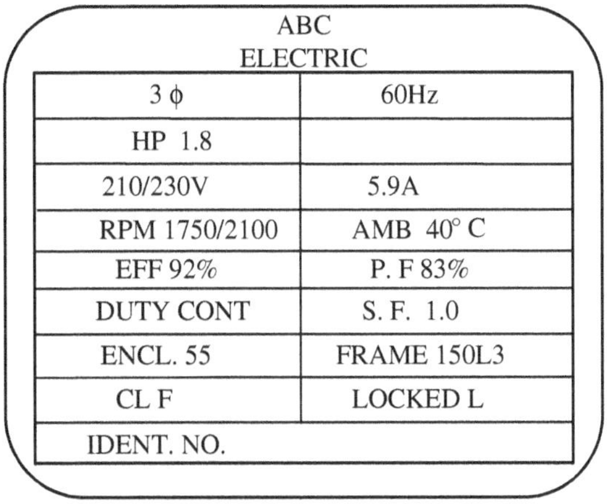

```
                    ABC
                  ELECTRIC
    ┌─────────────────────┬─────────────────────┐
    │        3 φ          │        60Hz         │
    ├─────────────────────┼─────────────────────┤
    │      HP  1.8         │                     │
    ├─────────────────────┼─────────────────────┤
    │     210/230V        │        5.9A         │
    ├─────────────────────┼─────────────────────┤
    │   RPM 1750/2100     │    AMB  40° C       │
    ├─────────────────────┼─────────────────────┤
    │      EFF 92%        │     P. F 83%        │
    ├─────────────────────┼─────────────────────┤
    │    DUTY CONT        │     S. F.  1.0      │
    ├─────────────────────┼─────────────────────┤
    │     ENCL. 55        │   FRAME 150L3       │
    ├─────────────────────┼─────────────────────┤
    │       CL F          │    LOCKED L         │
    ├─────────────────────┴─────────────────────┤
    │    IDENT. NO.                              │
    └────────────────────────────────────────────┘
```

FIGURE 16.9   Motor nameplate.

16.4.1   Electrical Ratings

Normally, a motor is designed to operate at the voltage (±10 percent) given on the nameplate; when under full load the motor will stay under its maximum rated operating temperature. The current rating of a motor is the operating current drawn by the motor when it is being operated at its rated voltage, frequency, temperature, and loaded to deliver its rated horsepower. The starting or locked rotor current is normally many times larger than the operating current and many large motors require low-voltage starting to limit the startup current (see Fig. 16.11).

The horsepower rating of a motor is the power the motor is designed to supply to a load at its rated speed and should not be exceeded unless the motor has a conservative rating and its service factor (SF) is greater than 1.0 otherwise the motor can overheat.

Motors are typically designed to operate at a maximum ambient temperature of 40°C. Their internal working temperature is dependent on the insulating material. There are four temperature classifications for insulating material. These are given in Table 16.3.

These temperatures are the maximum internal operating temperatures of the motor. In cases where the ambient temperature is higher than 40°C the available horsepower from the motor will be reduced. The speed listed on the nameplate is the speed of the motor at

Class of Insulation	Maximum Operating Temperature °C
A	105
B	130
F	155
H	180

TABLE 16.3   Temperature Rating Classification

its rated working load. Under lighter loads the speed of the motor will be higher. However, if the loading or torque (see Chap. 7) is higher than the rating of the motor, it will slow down. The torque is related to the horsepower of the motor as follows:

$$HP = T \times S \div 5252 \qquad (16.1)$$

where HP = horsepower
  T = torque
  S = speed

When the stall torque on a motor is reached due to heavy loading the motor will stall and try to restart. The current drawn from the supply will be the motor's starting current.

In applications where a motor is used intermittently, it has a duty cycle rating for the specified horsepower to prevent the motor overheating. This is normally expressed as a percentage; i.e., 4 minutes operation followed by 12 minutes off would give a duty cycle of 4/16 or 25 percent.

Motors are designed for maximum efficiency at the rated horsepower. In large motors, this can be high but due to power losses from winding resistance, eddy current or core losses, and friction the efficiency is about 95 percent. The efficiency is given by the power factor (PF) which is the rated horsepower divided by the input power at the rated horsepower and is expressed as a percentage.

Example 16.1   What is the efficiency of a 220 V motor with a PF of 75 percent that delivers 2.3 hp when the supply current is 15 A? Given 1 hp = 746 W.

Solution

$$\text{Power in} = 15 \text{ A} \times 220 \text{ V} \times 75 \div 100 = 2475 \text{ W}$$

$$\text{Power out} = 2.3 \text{ hp} \times 746 \text{ W} = 1715.8 \text{ W}$$

$$\text{Percent eff.} = (1715.8 \div 2475) \times 100$$
$$= 69.3 \text{ percent}$$

16.4.2   Control Equipment Ratings

Motor control equipment must meet the standards set by the National Electrical Code (NEC). Article 430 describes the code as it applies to motors and controllers in industrial control and Article 500 gives the code for hazardous locations. The codes give the operating currents for the various types of motors which depend on their operating voltage to establish relay contact size and rating, wiring size, protection and overload rating, and so on. The parameters for motor ratings and standards established by NEMA will assist users in the proper selection of control and ancillary equipment. A number of terms used in motor operation are defined by NEMA as follows:

Jogging (inching) is the rapid and repeated closure of a circuit to start a motor from rest for the purpose of obtaining small movements in a position.

Plugging is motor breaking by reversing the motor connections to develop a reverse torque for rapid stopping and motion reversal.

Antiplugging protection is obtained when a device prevents a counter torque until the speed of the motor is reduced to an acceptable level.

16.4.3 Enclosure Standards

Motor enclosure standards have been developed by NEMA. The dimensions of the enclosures (length, height, mounting, and so on) are set for given horsepower ratings. The dimensions are indicated by a frame number. However, the size of motor for a given horsepower can have large variations due to different insulation materials and different core types.

Motor enclosures can be classified as *open motors* or *fully enclosed motors*. The open motor classification can be further divided to allow for various operating environments. Some examples of these are as follows:

Drip-proof motor openings are designed to prevent droplets falling on the motor at an angle less than 15° from the vertical from entering the motor.

Splash-proof motor openings are designed to prevent liquids splashed on to the motor from entering.

Lint-free motors have smooth apertures to prevent the buildup of dust and particulates.

Guarded motors have screens to prevent objects from entering the ventilation openings of the motor.

Some examples of fully enclosed motors specifications are as follows:

Nonvented motors have no external cooling fans.

Fan-cooled motors have external fans for cooling the outside of the motor.

Explosion-proof motor is totally enclosed and sealed to prevent combustible gases from entering the motor.

Waterproof motors are sealed to prevent liquids entering the motor.

Large motors may have cooling fins or use circulating water or air for cooling.

16.5   Motor Control Applications

There are numerous applications in industry where the use of a motor is required. It is only possible to scratch the surface in this text on motor starting, speed control, and so on.

16.5.1 Two- and Three-Wire Starting

Motor control circuits are often divided into two-wire and three-wire control. The two-wire control circuit is shown in Fig. 16.10a. It uses a simple ON/OFF switch that requires two leads from the control panel to energize the motor control relay, whereas in a three-wire system shown in Fig. 16.10b three leads are required from the control panel to operate the motor relay using ON and OFF push buttons. The relay (contactor) CR1 switches all phases in a three-phase motor.

The three-wire system is much more flexible than the two-wire system in that a motor can be controlled by multiple start and stop buttons using momentary make and break buttons, giving the ability to start and stop a motor from different locations, i.e., remote and local starting and stopping. Note, the Start buttons are wired in an OR configuration and the Stop buttons are wired in an AND configuration when the motor is controlled from more than one location.

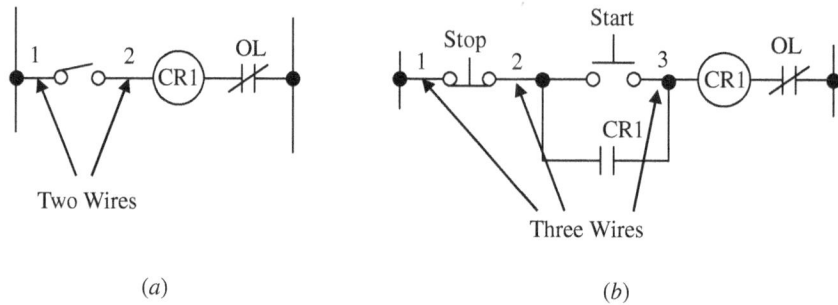

FIGURE 16.10   Illustration of (a) two-wire control and (b) three-wire control.

16.5.2   Startup Control

Slow acceleration on starting (soft start) is often required as opposed to starting a motor by connecting it directly to the supply (hard start) to protect machines and goods from sudden motion, reduce wear and tear, and to limit the inrush of starting current from the supply grid and prevent voltage surges. Soft start is normally achieved in ac motors by reducing the startup voltage using one of the following methods:

1. AC motor with wire wound rotor (see Fig. 16.12)
2. Variable frequency ac supply (see Fig. 16.13)
3. Connecting a resistance in series with the motor on startup (see Fig. 16.11)
4. Autotransformer to reduce the starting voltage
5. Connecting impedance in series with the motor on startup
6. In a three-phase motor the windings can be wye connected for starting and delta when running
7. Magnetic clutch
8. Use a brushless dc motor with bridge rectifiers

An example of a soft start using series resistors is shown in Fig. 16.11. High-wattage low-value resistors are connected in series with each input on startup by relay R1 which also starts a timer T1, that after several seconds operates relay R2 which switches out a third of each resistor and starts a second timer T2, that after several seconds operates relay R3 which switches out another third of each resistor and starts a third timer T3, that after several more seconds connects the motor directly to the three-phase supply via relay M. This type of switching network can also be used for motor sequencing. Relays R1, R2, and R3 could be used to drive motors instead of switching out resistors.

16.5.3   Wound Rotor Motor

The three-phase induction motor with a wound rotor has good starting torque, low starting current, and speed control making the motor suitable for applications where a speed controlled high startup horsepower is required without stressing the power grid. The motor is used in such applications as driving conveyer belt, hoists, pumps, mixers,

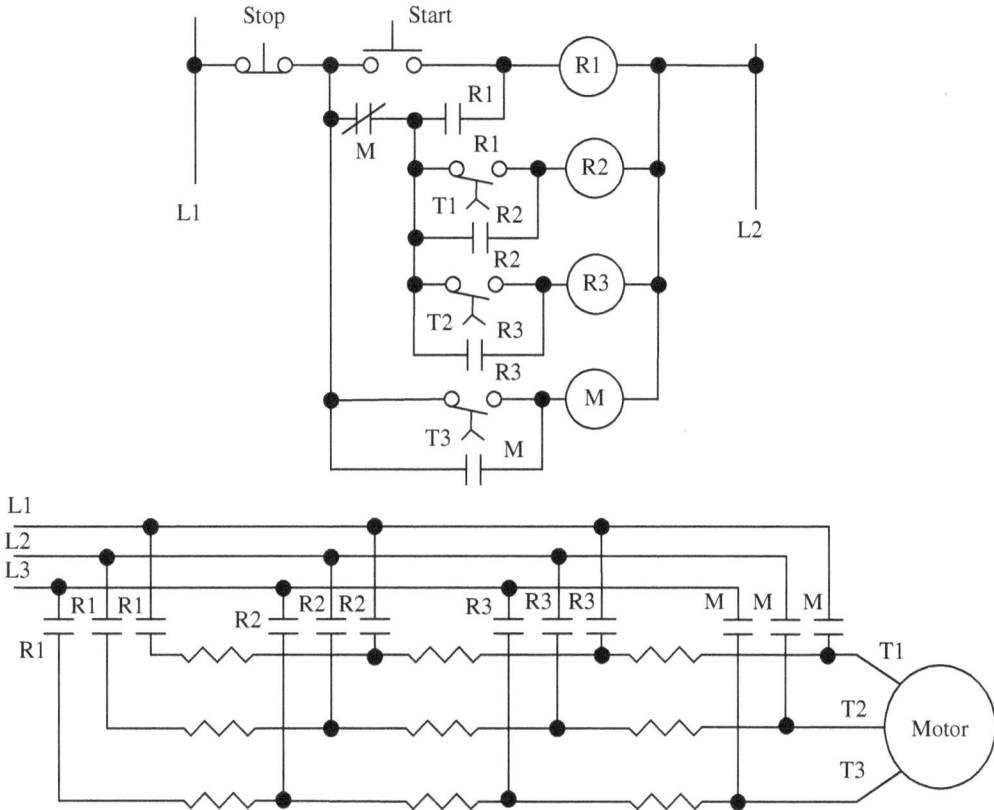

FIGURE 16.11   Four point starter.

and similar types of equipment. Figure 16.12 shows the control circuit for the wound rotor motor. The three-phase windings in the rotor are brought out via slip rings and carbon brushes. The motor is started with the resistors in series with the windings as shown to limit the startup current and increase the torque current ratio. After startup, the resistors can be adjusted for speed or shorted for running by the controller.

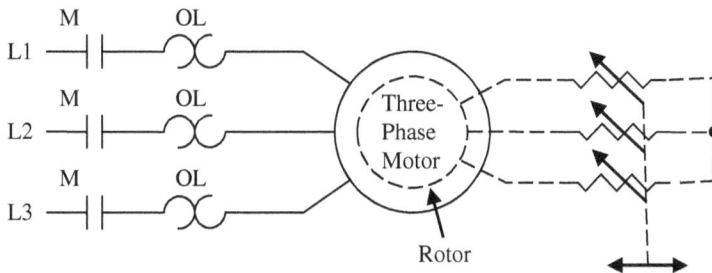

FIGURE 16.12   Wound rotor.

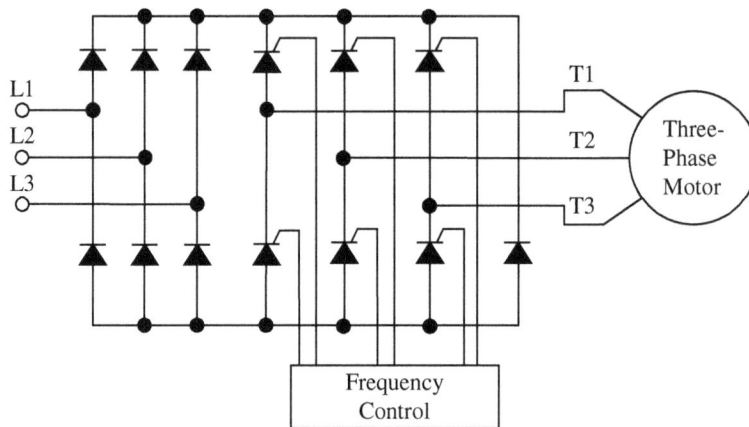

FIGURE **16.13**   Three-phase variable frequency control.

16.5.4   Speed Control

The speed of a three-phase induction motor is dependent on the supply frequency so that by varying the supply frequency the motor's speed can be controlled. Care must be taken at low frequencies not to damage the stator windings as the current flow is limited by the inductive reactance which is low at low frequencies. Power devices are now available to rectify the 60-Hz supply; the dc voltage can then be converted back to ac using high current silicon controlled rectifiers (SCRs) that are controlled by a variable frequency generator. This gives the ac motors the low-frequency torque that is associated with dc motors, as well as current limiting, speed regulation, and acceleration control. These features enable the ac motor to replace the dc motor in many applications. Figure 16.13 shows rectification of the line voltage using diodes and SCRs for switching the current to the three-phase motor.

The synchronous speed of an induction motor is determined by the number of poles in the motor and the frequency (f) of the supply voltage. The synchronous speed is given by:

$$\text{RPM} = 120 \; f \div \text{number of poles} \tag{16.2}$$

In practice, the speed of an induction motor is less than its synchronous speed when it is loaded. The speed of an induction motor (rated speed) when fully loaded is about 95 percent of the synchronous speed. This speed difference is called *slip* which is normally expressed as a percentage. The percent slip is given by

$$\text{Percent slip} = (\text{ss} - \text{rs}) \times 100 \div \text{ss} \tag{16.3}$$

where ss = synchronous speed
       rs = rated speed

Example 16.2   What is the percent slip of a six-pole 60-Hz motor with a rated speed of 1150 rpm?

Solution

$$\text{Synchronous speed (ss)} = 120 \times 60 \div 6$$
$$= 1200 \text{ rpm}$$
$$\text{Percent slip} = (1200 \text{ rpm} - 1150 \text{ rpm}) \times 100 \div 1200 = 4 \text{ percent}$$

16.5.5 DC Motor

Because of the high starting torque and variable speed of the dc motor, it is used in many applications requiring starting under heavy loads conditions and for variable speed control. The brushless dc motor is low maintenance and is used for driving electric vehicles, hoists elevators, conveyer belts, and in steel rolling mills. Power for a dc motor can be obtained from the power grid. A fixed voltage is required for the field winding and a variable voltage for the armature. The ac voltage is rectified for the field winding but the variable voltage supply uses SCRs controlled from a phase-shift network that is used for voltage control. When the SCRs are triggered late in the positive half cycle the output voltage is low and increases the earlier the SCRs are triggered in the cycle giving the variable voltage necessary for speed control. The basic circuit is shown in Fig. 16.14. The field and rotor currents are both monitored to detect overload or motor failure. A tacometer can also be used to monitor motor acceleration on starting and to adjust the running speed to a preset value when running by varying the rotor voltage.

16.5.6 Actuator Control

Figure 16.15 shows a valve actuator circuit. The valve is controlled by a dc motor. It is shown in the open position which is set by a limit switch (LS) and the closed position is set by a torque sensor (TS). A green light indicates the valve is in the open position and the closed position is indicated by a red light. Note, the cross-coupling between the two control relays so that only one relay can be on at any one time to prevent shorting the dc supply. When the close button CR2 is pressed the motor moves the valve toward the closed position until the torque switch opens and the limit switch turn ON the red light. Pressing the open switch operates control relay CR1 that reverses the power to the field winding of the motor and the valve opens. The open limit is set by the limit switch (LS).

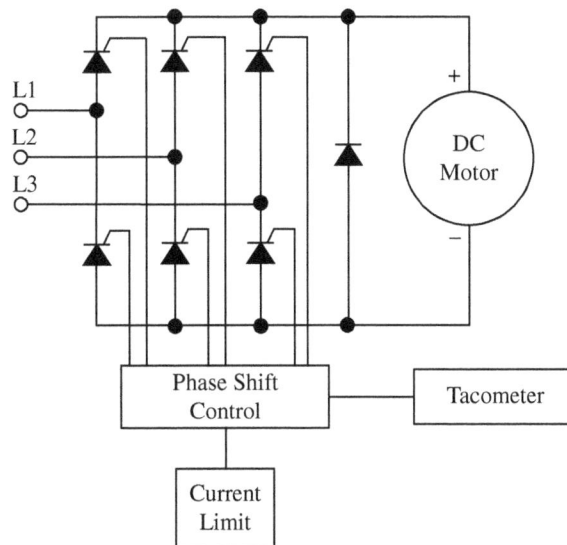

Figure 16.14   Variable voltage dc motor control.

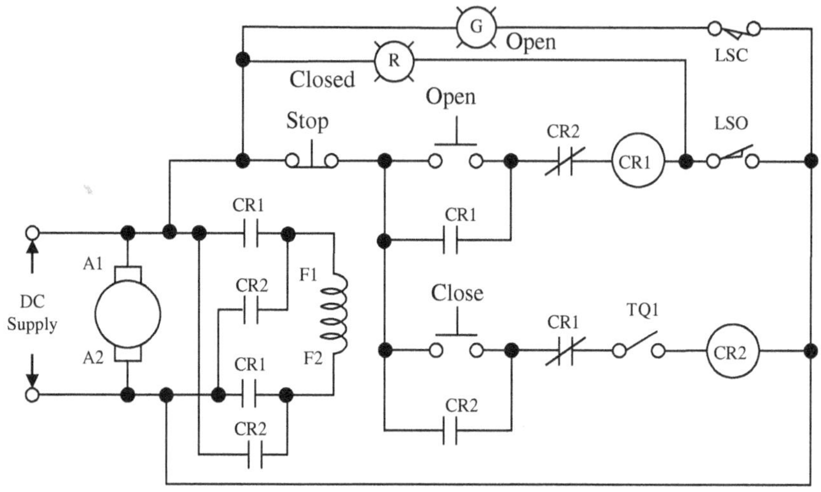

FIGURE **16.15** Control circuit of motor-operated valve.

16.5.7 Stepper Motor

Stepper motors have a wide range of applications in motion control positioning systems, and are normally digitally controlled for accurate indexing and positioning. Under computer control the motor can be operated in an open-loop system, step forward or reverse and has a wide range of operating speeds. The motor is used in the laser and optical fields for the precision positioning of linear actuators, linear stages, rotation stages, and mirror mounts. In the processing industry, the motor is used for accurate valve positioning for fluid control, and commercially they are used in floppy disc drives, flatbed scanners, printers, image scanners, compact disc drives, camera lens, intelligent lighting, and robotics.

16.6 Motor Protection

Motors are a critical part of all process control systems and as such a motor failure will shut down a process and production will cease, while basically a reliable component failure can be caused by external problems such as overloading, loss of a power phase, failure of control devices, dirt or moisture, wrong choice of motor type for an application, as well as internal failure which can be a motor protection open or short or in some cases a bad commutator or brushes due to poor maintenance. Thus proper motor protection is a necessity. Motors require supply current monitoring on all supply phases to detect excessive current or loss of current as well as over and under voltage detection. The types of protection used with ac motors are summarized as follows:

    Overload and excessive torque

    Open stator

    Loss of supply phase

    Reversed phase

Over travel

Over or under speed

Reversed current

Mechanical protection including protection from dirt and liquids

Short-circuit protection

Thermal protection from overloads

In some cases, it may be necessary to apply motor braking when a failure is detected

Summary

The main points described in this chapter were as follows:

1. The various types and sizes of dc and ac motors.
2. The variations between field and armature windings in a dc motor discussed and motor characteristics given.
3. AC motors and their operation are discussed together with the methods of producing a rotating magnetic field in a single-phase motor.
4. Three-phase wound rotors and commutator-less rotors compared in ac motors.
5. The application of stepper motors and synchro motors in process control is reviewed.
6. The NEMA operating standards for motors and motor sizes are introduced.
7. Applications and line power for dc and ac motors considered.

Problems

16.1 What are the three main motor classifications?

16.2 What are the four ways of connecting the armature and field windings in a dc motor?

16.3 How is the position of the magnet in a brushless dc motor sensed?

16.4 What is the difference between a dc and a universal motor?

16.5 What methods are used in a single-phase ac motor to produce an out-of-phase magnetic field?

16.6 Why is an out-of-phase magnetic field needed in a single-phase ac motor?

16.7 How is the magnetic field produced in the rotor of an induction motor?

16.8 Draw the supply line connections to the field windings in a delta motor and wye motors.

16.9 What is the difference between a stepper motor and an ac motor?

16.10 What is a synchro motor?

16.11 How many insulation classifications are there?

16.12 What is a guarded motor?

16.13   Determine the running torque of a 1.65 hp motor rated at 1873 rpm.

16.14   What current would a 2.1 hp motor draw from a 280 V supply if the PF was 74 percent and the efficiency was 68.3 percent?

16.15   What is the percent slip of an eight-pole motor operating from a 50 Hz supply with a rated speed of 625 rpm?

16.16   What type of dc motor is used when it is necessary to control its speed and how is the speed controlled?

16.17   Name three ways of starting up an ac motor.

16.18   How can the speed of an ac motor be controlled?

16.19   What is understood by the terms jogging and plugging?

16.20   What is a drip-proof motor?

APPENDIX A
Units

A standardized system of units is required for the measurement of physical properties. Over the years, two systems of measurements have been standardized. They are the English system of units, which is still in common use in the United States, and the SI (Systéme International d'Unités) system of units. However, efforts are being made to standardize on the SI system. The SI units are sometimes referred to as the centimeter-gram-second (CGS) units and are based on the metric system but it should be noted that not all of the metric units are used. The SI system of units is maintained by the Conférence Genérale des Poids et Measures. Because both systems are in common use, it is necessary to understand both system of units and to understand the relation between them. A large number of units (electrical) in use in the English system are SI units. Table A.1 gives the base units in both systems. Table A.2 gives SI units. Table A.3 gives English units. Table A.4 gives conversion between units. Table A.5 gives a list of some of the metric units that are not used in the SI system.

Quantity	English Units	English Symbol	SI Units	SI Symbol
Length	Foot	ft	meter	m
Mass	Pound (slug)	lb	kilogram	kg
Time	Second	s	second	s
Temperature	Rankine	R	Kelvin	K
Liquid measure	Gallon (US)	gal	Liter	L
Electric current	Ampere	A	Ampere	A
Luminous intensity	Candle	c	Lumen	lm
Angle	Degree	°	Radian	rad

TABLE A.1 Base Units

Quantity	Name	Symbol	Units	Base Unit
Frequency	Hertz	Hz	s^{-1}	s^{-1}
Energy	Joule	J	Nm	m^2 kg s^{-2}
Force	Newton	N	m kg/s^2	m kg s^{-2}
Pressure	Pascal	Pa	Nm^2	m^{-1} kg s^{-2}
Power	Watt	W	J/s	m^2 kg s^{-3}
Wavelength	Meter	m	m	m
Charge	Coulomb	C	s A	s A
Electromotive force	Volt	V	AΩ or W/A	m^2 kg s^{-2} A^{-1}
Resistance	Ohm	Ω	V/A	m^2 kg s^{-3} A^{-2}
Conductance	Siemen	S	A/V	m^{-2} kg^{-1} s^2 A^2
Capacitance	Farad	F	As	m^{-2} kg^{-1} s^4 A^2
Inductance	Henry	H	Wb/A	m^2 kg s^{-2} A^{-2}
Magnetic flux	Weber	Wb	Vs	m^2 kg s^{-2} A^{-1}
Flux density	Tesla	T	Wbm^2	kg s^{-2} A^{-1}
Illuminance	Lux	lx	lm/m^2	m^{-2} cd sr
Luminous flux	Lumen	lm	cd sr	cd sr
Capacity	Liter	L	dm^3	dm^3

TABLE A.2 SI Units Derived From Base Units

Quantity	Name	Symbol	Unit
Energy	Foot-pound	ft-lb	lb ft^2 s^{-2}
Force	Pound	lb	lb ft s^{-2}
Pressure	Pound per in^2	psi	lb in^{-2}
Power	Horsepower	hp	lb ft^2 s^{-3}
Specific heat	British thermal units	BTU (Btu)	ft^2 s^{-2} $°F^{-1}$
Volume	Gallon	gal	0.1337 ft^3

TABLE A.3 English Units Derived From Base Units

Quantity	English Units	SI Units
Length	1 ft	0.305 m
Mass	1 lb (slug)	14.59 kg
Weight	1 lb	0.454 kg
Volume	1 gal	3.78 L (l)
Force	1 lb	4.448 N
Angle	1°	$2\pi/360$ rad
Temperature	1°F	5/9°C
Energy	1 ft lb	1.356 J
Pressure	1 psi	6.897 kPa
Power	1 hp	746 W
Heat	1 BTU	252 cal = 1055 J
Conduction	1 BTU/h ft °F	1.37 W/m K
Expansion	$1\ \alpha/°F$	$1.8\ \alpha/°C$
Specific weight	$1\ lb/ft^3$	$0.157\ kN/m^3$
Density	$1\ slug/ft^3$	$0.516\ kg/m^3$
Dynamic viscosity	$1\ lb\ s/ft^2$	49.7 Pa s (4.97 P)
Kinematic viscosity	$1\ ft^2/s$	$9.29 \times 10^{-2}\ m^2/s$ (929 St)
Torque	1 lb ft	1.357 N m

TABLE A.4   Conversion Between English and SI Units

Quantity	Name	Symbol	Equivalent
Length	Angstrom	Å	1 Å = 0.1 nm
Volume	Stere	st	$1\ st = 1\ m^3$
Force	Dyne	dyn	1 dyn = 10 μN
Pressure	Torr	torr	1 torr = 133 Pa
Energy	Calorie	cal	1 cal = 4.1868 J
	Erg	erg	1 erg = 0.1 μJ
Viscosity dynamic kinematic	Poise	P	1 P = 0.1 Pa s
	Stoke	St	$1\ St = 1\ cm^2/s$
Conductance	Mho	mho	1 mho = 1 S
Magnetic field strength	Oersted	Oe	1 Oe = 80 A/m
Magnetic flux	Maxwell	Mx	1 Mx = 0.01 μWb
Magnetic flux density	Gauss	Gs (G)	1 Gs = 0.1 mT

TABLE A.5   Metric Units Not Normally Used in the SI System

Thermocouple Tables

T he following give examples of the tables for J-, K-, S-, and T-type thermocouples. The thermocouple EMF is given in mV for 10°C degree temperature increments. The cold junction is held at 0°C. The output voltage for the different types of thermocouples may vary slightly between manufacturers.

Type J Iron—Constantan

	0	10	20	30	40	50	60	70	80	90
−100	−4.63	−5.03	−5.42	−5.80	−6.16	−6.50	−6.82	−7.12	−7.40	−7.66
−0	0.00	−0.50	−1.00	−1.48	−1.96	−2.43	−2.89	−3.34	−3.78	−4.21
+0	0.00	0.50	1.02	1.54	2.06	2.58	3.11	3.65	4.19	4.73
100	5.27	5.81	6.36	6.90	7.45	8.00	8.56	9.11	9.67	10.22
200	10.78	11.34	11.89	12.45	13.01	13.56	14.12	14.67	15.22	15.77
300	16.33	16.88	17.43	17.98	18.54	19.09	19.64	20.20	20.75	21.30
400	21.85	22.40	22.95	23.50	24.06	24.61	25.16	25.72	26.27	26.83
500	27.39	27.95	28.52	29.08	29.65	30.22	30.80	31.37	31.95	32.53
600	33.11	33.70	34.29	34.88	35.48	36.08	36.69	37.30	37.91	38.53
700	39.15	39.78	40.41	41.05	41.68	42.28	42.92			

Type K Chromel—Alumel

	0	10	20	30	40	50	60	70	80	90
−100	−3.49	−3.78	−4.06	−4.32	−4.58	−4.81	−5.03	−5.24	−5.43	−5.60
−0	0.00	−0.39	−0.77	−1.14	−1.50	−1.86	−2.20	−2.54	−2.87	−3.19
+0	0.00	0.40	0.80	1.20	1.61	2.02	2.43	2.85	3.36	3.68
100	4.10	4.51	4.92	5.33	5.73	6.13	6.53	6.93	7.33	7.73
200	8.13	8.54	8.94	9.34	9.75	10.16	10.57	10.98	11.39	11.80
300	12.21	12.63	13.04	13.46	13.88	14.29	14.71	15.13	15.55	15.98
400	16.40	16.82	17.24	17.67	18.09	18.51	18.94	19.36	19.79	20.22
500	20.65	21.07	21.50	21.92	22.35	22.78	23.20	23.63	24.06	24.49
600	24.91	25.34	25.76	26.19	26.61	27.03	27.45	27.87	28.29	28.72
700	19.14	29.56	29.97	30.39	30.81	31.23	31.65	32.06	32.48	32.89
800	33.30	33.71	34.12	34.53	34.93	35.34	35.75	36.15	36.55	39.96
900	37.36	37.76	38.16	38.56	38.95	39.35	39.75	40.14	40.53	40.92
1000	41.31	41.70	42.09	42.48	42.87	43.25	43.63	44.02	44.40	44.78
1100	45.16	45.54	45.92	46.29	46.67	47.04	47.41	47.78	48.15	48.52
1200	48.89	49.25	49.62	49.98	50.34	50.69	51.05	51.41	51.76	52.11
1300	52.46	52.81	53.16	53.51	53.85	54.20	54.54	54.88		

Type S Platinum (Rhodium 10%)—Platinum

	0	10	20	30	40	50	60	70	80	90
0	0.000	0.056	0.113	0.173	0.235	0.299	0.364	0.431	0.500	0.571
100	0.643	0.717	0.792	0.869	0.946	1.025	1.166	1.187	1.269	1.352
200	1.436	1.521	1.607	1.693	1.780	1.868	1.956	2.045	2.135	2.225
300	2.316	2.408	2.499	2.592	2.685	2.778	2.872	2.966	3.061	3.156
400	3.251	3.347	3.442	3.539	3.635	3.732	3.829	3.926	4.024	4.122
500	4.221	4.319	4.419	4.518	4.618	4.718	4.818	4.919	5.020	5.122
600	5.224	5.326	5.429	5.532	5.635	5.738	5.842	5.946	6.050	6.155
700	6.260	6.365	6.471	6.577	6.683	6.790	6.897	7.005	7.112	7.220
800	7.329	7.438	7.547	7.656	7.766	7.876	7.987	8.098	8.209	8.320
900	8.432	8.545	8.657	8.770	8.883	8.997	9.111	9.225	9.340	9.455
1000	9.570	9.686	9.802	9.918	10.035	10.152	10.269	10.387	10.505	10.623
1100	10.741	10.860	10.979	11.098	11.217	11.336	11.456	11.575	11.695	11.815
1200	11.935	12.055	12.175	12.296	12.416	12.536	12.657	12.777	12.897	13.018
1300	13.138	13.258	13.378	13.498	13.618	13.738	13.858	13.978	14.098	14.217
1400	14.337	14.457	14.576	14.696	14.815	14.935	15.054	15.173	15.292	15.411
1500	15.530	15.649	15.768	15.887	16.006	16.124	16.243	16.361	16.479	16.597
1600	16.716	16.834	16.952	17.069	17.187	17.305	17.422	17.539	17.657	17.774
1700	17.891	18.008	18.124	18.241	18.358	18.474	18.590			

Type T Copper—Constantan

	0	10	20	30	40	50	60	70	80	90
−100	−3.349	−3.624	−3.887	−4.138	−4.377	−4.603	−4.817	−5.018	−5.205	−5.379
−0	0.0000	−0.380	−0.751	−1.112	−1.463	−1.804	−2.135	−2.455	−2.764	−3.062
+0	0.0000	0.389	0.787	1.194	1.610	2.035	2.467	2.908	3.357	3.813
100	4.277	4.749	5.227	5.712	6.204	6.703	7.208	7.719	8.236	8.759
200	9.288	9.823	10.363	10.909	11.457	12.015	12.575	13.140	13.710	14.285
300	14.864	15.447	16.035	16.626	17.222	17.821	18.425	19.032	19.624	20.257

References and Information Resources

There are a large number of resources for additional reading on instrumentation and process control. The internet contains a large number of Web sites that can be used as resources for more information. A list of Web site references is given below; this list is by no means complete.

Magazines

1. *Control,* www.controlmagazine.com
2. *Instrument and Control Systems,* www.icsmagazine.com
3. *Instrument and Automation News,* www.ianmag.com
4. *Sensors,* www.sensors.com

Organizations

1. Institute of Electrical and Electronic Engineers, www.ieee.org
2. Instrumentation, Systems, and Automation Society, www.ias.org
3. National Institute of Standards and Technology, www.nist.gov
4. American National Standards Institute, www.ansi.org
5. National Electrical Manufactures Association, www.nema.org
6. Industrial Control and Plant Automation, www.xnet.com
7. Society of Automotive Engineers, www.sae.org/servlets/index

PLC Manufacturers

1. GE, www.geindustrial.com/cwc/gefanuc/index.html
2. Mitsubishi, www.mitsubishielectric.com/bu/automation/index

3. Rockwell, www.rockwellautomation.com

4. Siemens, www.simatic.com

5. Foxboro, www.foxboro.com

6. Honeywell, www.honeywell.com

Component Suppliers

1. Texas Instrument, www.ti.com

2. National Semiconductor, www.national.com

3. Design Info, www.designinfo.com

4. Valves: k Controls, www.k-controls.co.uk

5. Omega Engineering, www.omega.com

6. Burr-Brown, www.burr-brown.com

7. Analog Devices, www.analog.com

8. Alpha, www. alphatechnics.com

9. Micro Strain, www.microstrain.com

10. Entran, www.entran.com

11. Kavlico, www.kavlico.com

12. Flow Meters, www.flowmeters.com

13. Omron, www.omron.com

14. NXP Semiconductors, www.nxp.com

15. ON Semiconductor, www.onsemi.com

16. International Rectifier, www.irf.com

17. Siliconix, www.vishay.com/company/brands/siliconix/

18. GE, www.gesensing.com

19. Phillips, www.semiconductors.philips.com

20. Intersil Corporation, www.intersil.com

21. Heat Pipe Technology, Inc., www.heatpipe.com

Tutorial References

1. PLC Tutor, www.plcs.net

2. National Instruments, www.ni.com/white-paper/4045/en/

3. Cyber Research, www.cyberresearch.com/tech/DADesign.html

4. Temperature World, www.temperatureworld.com

5. BHL, www.bhl.com

6. Macrosensors, www.macrosensors.com/primer/primer.html

References

1. Charles, A.S., *Electronics Principles and Applications*, Glencoe McGraw-Hill, New York, 1999.
2. Rodger, L.T., *Digital Electronics*, Glencoe McGraw-Hill, New York, 2003.
3. Ljubisa, R., *Sensor Technology and Devices*, Artech House, Norwood, MA, 1994, pp. 377–456.
4. Cascetta, F. and Paolo, V., *Flowmeters: A Comprehensive Survey and Guide to Selection*, ISA, Research Triangle Park, NC, 1990.
5. Gillum, D.R., *Industrial Pressure, Level, and Density Measurement*, ISA, Research Triangle Park, NC, 1995.
6. McMillan, G.K., *pH Measurement and Control*, ISA, Research Triangle Park, NC, 1994.
7. Dunning, G., *Introduction to Programmable Logic Controllers*, 2nd ed., Delmar, Albany, NY, 2002.
8. Curtis, D.J., *Process Control Instrumentation Technology*, 7th ed., Prentice Hall, Upper Saddle River, NJ, 2003.
9. Rex, K., Jr., "Linearization of a Thermocouple," *Sensors Magazine*, Vol. 14, No. 12, Dec. 1997.
10. Davis, M., "Choosing and Using a Temperature Sensor," *Sensors Magazine*, Vol. 17, No. 1, Jan. 2000.

APPENDIX D
Abbreviations

Å	Angstrom
AC	Alternating current
ADC	Analog-to-digital converter
AM	Amplitude modulation
ANSI	American National Standards Institute
BCD	Binary coded decimal
BJT	Bipolar junction transistor
BTU	British thermal unit
C	Coulomb
CMOS	Complementary metal oxide semiconductor
CR	Control relay
DAC	Digital-to-analog converter
dB	Decibel
DIAC	Bidirectional trigger diode
EEPROM	Electrical erasable read only memory
EMI	Electromagnetic interference
F	Farad
FET	Field effect transistor
FM	Frequency modulation
FSD	Full-scale deflection
FSK	Frequency shift keying
GaAs	Gallium arsenide
GaAsP	Gallium arsenide phosphide
GaP	Gallium phosphide
H	Henry
HART	Highway addressable remote transducer
HF	High frequency
HVAC	Heating, ventilation, and air-conditioning
Hz	Hertz
IEEE	Institute of Electrical and Electronic Engineers
IGBT	Insulated gate bipolar transistor
I/O	Input/output
IR	Infrared
ISA	Instrument Society of America
J	Joule
K	Kelvin
LAN	Local area network

LED	Light-emitting diode
LSB	Least significant bit
LVDT	Linear velocity differential transformer
MCT	MOS controlled transistor
MFSK	Multiple frequency shift keying
MHz	Megahertz
MRE	Magnetoresistive element
MSB	Most significant bit
N	Newton
NEMA	National Electrical Manufacturers Association
NIST	National Institute of Standards and Technology
Pa	Pascal
P and ID	Pipe and identification diagram
PCM	Pulse code modulation
pF	Picofarad
PID	Proportional integral and derivative
PLA	Programmable logic array
PLC	Programmable logic controller
PPM	Pulse position modulation
PWM	Pulse width modulation
R	Rankin
RAM	Random access memory
RC	Resistance capacitance
RF	Radio frequency
RMS	Root mean square
ROM	Read only memory
RPM	Revolutions per minute
RTD	Resistance temperature device
SCR	Silicon controlled rectifier
SI	Systeme International D'Unites
SIS	Safety instrumented system
SPL	Sound pressure level
TC	Time constant
TCE	Temperature coefficient of expansion
TDM	Time division multiplex
TRIAC	Bidirectional AC switch
UPS	Uninterruptible power supply
W	Watt
WAN	Wide area network
Wb	Weber

Glossary

Absolute accuracy   The accuracy stated as a definite amount, i.e., not as a percentage.

Absolute position measurement   Position measured from a fixed point.

Absolute pressure   Pressure measured with reference to a perfect vacuum.

Accelerometer   A sensor for measuring acceleration or the rate of change of velocity.

Accuracy   A measure of the difference between the indicated value and the true value.

Actuator   A device that performs an action on one of the input variables of a process according to a signal received from the controller.

ADC   An analog-to-digital converter that converts an analog voltage or current into a digital signal.

Alarm   A warning that a variable has exceeded set limits.

Alternating current   Current that flows in one direction during one-half of a regular time period and the opposite direction during the other half.

Ammeter   An instrument for measuring electrical current or electron flow.

Ampere   The unit of current or electron flow.

Amplifier   An electrical circuit that increases the magnitude of a signal.

Analog   A continuously varying signal.

Aneroid barometer   A barometer which uses an evacuated capsule as a sensing element.

Anticipatory action   See **Derivative action**.

Aqueous solution   A solution containing water.

Atmospheric pressure   The pressure acting on objects on the earth's surface caused by the weight of the air in the earth's atmosphere, normally measured at sea level.

Barometer   An instrument used for measuring atmospheric pressure.

Bellows   A pressure sensor that converts pressure into linear displacement.

Bernoulli equation   A flow equation based on the conservation of energy which includes velocity, pressure, and elevation terms.

Beta ratio   The ratio of the diameter of a restriction to the diameter of the pipe containing the restriction.

Bimetallic   A thermometer with a sensing element made of two dissimilar metals with different thermal coefficients of expansion.

Binary   Two values, or a numbering system using the base 2.

Bit   A binary digit.

Bourdon tube   A pressure sensor that converts pressure to movement. The device is a coiled metallic tube that straightens when pressure is applied.

Bridge   A network of passive components arranged so that small changes in one of the components can be easily measured.

British thermal unit   A measure of heat energy, i.e., the amount of heat required to raise 1 lb of water 1°F at 68°F and atmospheric pressure.

Buffer amplifier   A circuit for matching the output impedance of one circuit to the input impedance of another.

Buoyancy   The upward force on an object floating or immersed in a fluid caused by the difference in pressure above and below the object.

Byte   Eight bits of binary information.

Calorie   A measure of heat energy, i.e., the amount of heat required to raise the temperature of 1 g of water 1°C.

Capacitance   A measure of a device's ability to store electrical charge.

Capacitance probe   An instrument using the capacitance between two metal plates for measuring fluid level.

Capacitor   A device that can store electrical charge.

Cell   A simple power source that provides EMF, usually by means of a chemical reaction.

Celsius   One of the commonly used temperature scales.

Coefficient of heat transfer   A term used in the calculation of heat transfer by convection.

Coefficient of thermal expansion   A term used to determine the amount of linear expansion due to heating or cooling.

Comparator   A device which compares two signals and outputs the difference.

Concentric plate   A plate with a hole located at its center (orifice plate) used to measure flow by measuring the differential pressures either side of the plate.

Conduction   The movement of heat energy in a material by the transfer of energy from one molecule to another.

Conductivity probe   An instrument using two electrodes to measure fluid level.

Continuity equation   A flow equation which states that, if the overall flow rate is not changing with time, the flow rate past any section of the system must be constant.

Continuous level measurement   A level measurement that is continuously updated.

Controlled variable   The variable measured to indicate the condition of the process output.

Controller   The element in a process-control loop that evaluates any error of the measured variable and initiates corrective action by changing the manipulated variable.

Convection   The movement of heat by the motion of warm or hot material.

Converter   A device that changes the format of a signal but not the type of energy used as the signal carrier, i.e., voltage to current.

Correction signal   The signal to the manipulated variable.

DAC   A device that converts a digital signal into an analog voltage or current.

Dead weight tester   A device for calibrating pressure-measuring devices, which uses weights to provide the forces.

Decibel (dB)   A unit used to compare amplitude or power levels.

Density   The amount of mass in a unit volume.

Derivative action   Action that is proportional to the rate at which the measured variable is changing.

Dew point   The temperature at which the water vapor in a mixture of water vapor and gas becomes saturated and condensation starts.

Dielectric constant   The factor by which the capacitance between two plates changes when a material fills the space between the plates.

Differential amplifier   An amplifier that amplifies the difference between two inputs.

Digital   Signals having two discrete levels.

Dry-bulb temperature   The temperature indicated by a thermometer whose sensing element is dry.

Dynamic pressure   That part of the total pressure in a moving fluid caused by the fluid motion.

Dynamometer   An instrument used for measuring torque or power.

Eccentric plate   An orifice plate with a hole located below its center to allow for the passage of suspended solids.

Effective value   The dc voltage or dc current that would produce the same power in a load as the ac voltage or ac current being measured.

Electromagnetic flow meter   A flow-measuring device which senses the change in a magnetic field between two electrodes as a fluid flows between them.

Electromagnetism   The relationship between magnetic fields and electric current.

Error signal   The difference in value between a measured signal and a set point.

Fahrenheit   One of the commonly used temperature scales.

Farad   The unit of capacitance.

Feedback   (1) The voltage fed from the output of an amplifier to the input in order to control the characteristics of the amplifier. (2) The measured variable signal fed to the controller in a closed-loop system, so that the controller can adjust the manipulated variable to keep the measured variable within set limits.

Fiber optics   The transmission of information through optical cables using light signals.

Flow nozzle   A device placed in a flow line to provide a pressure drop that can be related to flow rate.

Flow rate   The amount of fluid passing a given point in a given interval of time.

Flume   An open-channel flow-measuring device.

Form drag   The force acting on an object due to the impact of fluid.

Foundation Fieldbus   Process-control bus used in the United States.

Free convection   Movement of heat as a result of density differences.

Free surface   The surface of the liquid in an open-channel flow that is in contact with the atmosphere.

Frequency   The number of cycles completed in 1 second.

Gage pressure   Is the measured pressure above atmospheric pressure.

Gas thermometer   A temperature sensor that converts temperature to pressure in a constant volume system.

Hall-effect sensor   A transducer that converts a changing magnetic field into a proportional voltage.

HART   Computer bus protocol.

Head   Sometimes used to indicate pressure, i.e., 1 ft of "head" for water is the pressure under a column of water 1-ft high.

Heat   A form of energy related to the motion of atoms or molecules.

Heat transfer   The study of heat energy movement.

Henry (H)   The unit of inductance.

Hertz (Hz)   A measure of frequency in cycles/second.

Hot-wire anemometry   A velocity-measuring device for gas or liquid flow that senses temperature changes, due to the cooling effect of gas or liquid moving over a hot element.

Humidity   A term to indicate the amount of water vapor present in the air or a gas.

Humidity ratio   The mass of water vapor in a gas divided by the mass of dry gas in the mixture.

Hydrometer   An instrument for measuring liquid density.

Hydrostatic paradox   The fact that pressure varies with depth in a static fluid, but is the same throughout the liquid at any given depth.

Hydrostatic pressure   The pressure caused by the weight of static fluid.

Hygrometer   A relative humidity-measuring device.

Hygroscopic   A material that absorbs water and whose conductivity changes with moisture content.

Hysteresis   The nonreproducibility in an instrument caused by approaching a measurement from opposite directions, i.e., going from low up to the value, or high down to the value.

Impact pressure   The sum of the static and dynamic pressure in a moving fluid.

Impedance   An opposition to ac current or electron flow caused by inductance, and/or capacitance.

Incremental position measurement   An incremental position measurement from one point to another, absolute position is not recorded, and position is lost if the power fails.

Indirect level-measuring device   Extrapolates the level from the measurement of another variable, i.e., liquid level from a pressure measurement.

Inductance   An electrical component that opposes a change in current or electron flow.

Inductor   A device that exhibits inductance.

Instrument   A device used to measure a physical variable.

Integral action   The action designed to correct for long-term loads.

Jogging   Incremental turning of motor for position adjustment.

Kelvin   The absolute temperature scale associated with the Celsius scale.

Ladder logic   The programmable logic used in PLCs to control automated industrial processes.

Lag time   The time required for a control system to return a measured variable to its set point.

Laminar flow   A smooth flow in which the fluid tends to move in layers.

LED   Light-emitting diode.

Linearity   A measure of the direct proportionality between actual value of the variable being measured and the value of the output of the instrument to a straight line.

Load   The process load is a term used to denote the nominal values of all variables in a process that affect the controlled variable.

Load cell   A device for measuring force.

Logic gates   Computer building blocks.

Loudness   A subjective quantity used to measure relative sound strength.

LVDT   A linear variable transformer that measures displacement by conversion to a linearly proportional voltage.

Magnetorestrictive element (MRE)   A magnetic field sensor that converts a changing magnetic field into a proportional resistance.

Manipulated variable   The variable controlled by an actuator to correct for changes in the measured variable.

Measured variable   The variable measured to indicate the condition of the process output.

Meniscus   The convex or concave surface of a column of liquid in a tube.

Moment   The effect of a force acting at a given perpendicular distance from a point.

Natural convection   The movement of heat as a result of density differences.

Newtonian fluid   A fluid in which the velocity varies linearly across the flow section between parallel plates.

Nibble   Four binary bits.

Node   A junction of three or more conductors.

Noise   The term usually used to indicate unwanted or undesirable sounds.

Nutating disk meter   A flow-measuring device using a disk that rotates and wobbles in response to the flow.

Offset   The nonzero output of a circuit when the input is zero.

Ohmmeter   An instrument used to measure resistance.

ON/OFF control   A system in which a process actuator has only two positions, i.e., ON and OFF.

Open-channel flow   The flow in an open conduit (e.g., as in a ditch).

Operational amplifier   A circuit used to amplify electronic signals.

Orifice plate   A plate containing a hole which when placed in a pipe causes a pressure drop which can be related to flow rate.

Over pressure   The term used to describe the maximum amount of pressure a gage can withstand without damage or loss of accuracy.

Overshoot   The overcorrection of the measured variable in a control loop.

Parabolic velocity distribution   Occurs in laminar flow when the velocity across the cross section takes on the shape of a parabola.

Parallel transmission   Simultaneous transmission of a number of binary bits.

Pascal   Pressure reading units (SI), i.e., Newtons per square meter.

Pascal's law   The pressure applied to an enclosed fluid is transmitted to every part of the fluid.

Percent of reading   The accuracy given in terms of the percentage of the reading.

Percentage full-scale accuracy   The accuracy determined by dividing the accuracy of an instrument by its full-scale output taken as a percentage.

Period   A fixed amount of time during which alternating current is completing one full cycle and is the inverse of the frequency in Hertz.

pH A term used to indicate the activity of the hydrogen ions in a solution. It helps to describe the acidity or alkalinity of the solution.

Phase A term used to describe the state of matter, i.e., solid, liquid, or gas.

Phons A unit for describing the difference in loudness levels.

Photodiode A sensor used to measure light intensity by measuring the leakage across a pn junction.

PID Proportional control with derivative and integral action.

P and ID Stands for piping and instrument diagrams.

Piezoelectric effect The electrical voltage developed across certain crystalline materials when a force or pressure is applied to the material.

Pitot-static tube A device used to measure the flow rate using the difference between dynamic and static pressures.

PLC Programmable logic controller.

Plugging A motor breaking by reversing motor's supply.

Pneumatic System that employs gas for control or signal transmission.

Poise The measurement unit of dynamic or absolute viscosity.

Potentiometer (Pot) An adjustable resistance device.

Precision The smallest division that can be read on an instrument.

Pressure The magnitude of a force divided by the area over which it acts, i.e., psi or Pa.

Pressure differential The difference in pressure amplitudes at two locations.

Process A sequence of operations carried out to achieve a desired end result.

Process control The automatic control of certain process variables to hold them within given limits.

Processor A digital electronic computing system that can be used as a control system.

Profibus Process-control bus used in Europe.

Proportional action A controller action in which the controller output is directly proportional to the measured variable error.

Psychrometric chart A chart dealing with moisture content in the atmosphere.

Pyrometer An instrument for measuring temperature by sensing the radiant energy from a hot body.

Radiation The emission of energy from a body in the form of electromagnetic waves.

Range The lowest to the highest readings that can be made by a sensing device.

Rankine The absolute temperature scale associated with the Fahrenheit scale.

Rate action See **Derivative action**.

Reactance   The opposition to an ac current or electron flow caused by a capacitor or an inductor.

Relative humidity   The amount of water vapor present in a given volume of a gas, expressed as a percentage of the amount that would be present in the same volume of gas under saturated conditions at the same pressure and temperature.

Reluctance   The opposition in a material to carrying magnetic flux. It is the magnetic equivalence to resistance.

Repeatability   A measure of the closeness between several consecutive readings of a value.

Reproducibility   The ability of an instrument to produce the same reading of a variable with repeated readings.

Reset action   See **Integral action**.

Resistance   A measure of the opposition to electron or current flow in a material.

Resistance thermometer (RTD)   A temperature sensor that provides temperature readings by measuring the resistance of a metal wire (usually platinum).

Resistivity   A temperature-dependent "constant" that reflects a material's resistance to electron flow.

Resistor   A component that exhibits resistance.

Resolution   The minimum detectable change of a variable in a measurement.

Reynolds Number   A dimensionless number indicating whether the flow is laminar or turbulent.

Rotameter   A flow-measuring device in which a float moves in a vertical tapered tube.

Saturated   The condition when the maximum amount of a material is dissolved in another material at the given pressure and temperature conditions, i.e., water vapor in a gas.

Sealing fluid   An inert fluid used in a manometer to separate the fluid whose pressure is being measured from the manometer fluid.

Segmented plate   An orifice plate with a hole located so as to allow suspended solids to pass through.

Sensitivity   The ratio of the change in output to input magnitudes.

Sensor   A device that can convert a physical variable into a measurable quantity.

Serial transmission   A sequential transmission of digital bits.

Set point   The reference value for a controlled variable in a process-control loop.

Signal conditioning   The conversion of a signal to a format that can be used for transmission.

Single-point level measurement   Indicates when a particular level has been reached.

Sling psychrometer   A device for measuring relative humidity.

Smart sensor   Integration of a processor directly into the sensor assembly to give direct control of the actuator and digital communication to a central controller.

Sone   A unit for measuring loudness.

Sound pressure level   The difference between the maximum air pressure at a point and the average air pressure at that point.

Span   The difference between the lowest and the highest reading for an instrument.

Specific gravity   The ratio of the specific weight of a solid or liquid material and the specific weight of water, or for a gas, the ratio of the specific weight of the gas and the specific weight of air under the same conditions.

Specific heat   The amount of heat required to raise a definite amount of a substance by $1°$, i.e., 1 lb $1°F$ or 1 g $1°C$.

Specific humidity   The mass of water vapor in a mixture divided by the mass of dry air or gas in the mixture.

Specific weight   The weight of a unit volume of a material.

Static pressure   The part of the total pressure in a moving fluid not caused by the fluid motion.

Stoke   The measurement unit of kinematic viscosity.

Strain gauge   A sensor that converts information about the deformation of solid objects when they are acted upon by a force into a change of resistance.

Sublimation   Passing directly from solid to vapor or vapor to solid.

Telemetry   The electrical transmission of information over long distances usually by radio frequencies.

Temperature   The term used to describe the hotness or coldness of an object.

Thermal conductivity   A measure of the ability of a material to conduct heat.

Thermal expansion   The expansion of a material as a result of its being heated.

Thermal time constant   The time required for a body to heat or cool by 63.2% of the difference between the initial temperature and the aiming temperature.

Thermistor   A temperature sensing element made from a metal oxide that usually has a negative temperature coefficient.

Thermocouple   A temperature sensing device that uses dissimilar metal junctions to generate a voltage proportional to the differential temperature between the metal junctions.

Thermometer   An instrument used to measure temperature.

Thermopile   A number of thermocouples connected in series.

Time constant (electrical)   The amount of time needed for a capacitance C, to discharge or charge through a resistance R, by 62.3% of the difference between the initial voltage and the aiming voltage; the product of RC gives the time constant in seconds.

Torque   The name given to a force moment that tends to create a twisting action.

Torr   The pressure caused by the weight of a column of mercury 1 mm high.

Total flow   The amount of flow past a given point over some length of time.

Total pressure   The sum of the static and dynamic pressures in a moving fluid.

Transducer   A device that changes energy from one form to another.

Transfer function   An equation that describes the relationship between the input and output of the function.

Transmission   The transferring of information from one point to another.

Transmitter   A device that conditions the signal received from a transducer so that it is suitable for sending to another location with minimal loss of information.

Turbine flow meter   A flow-measuring device utilizing a turbine wheel.

Turbulent flow   An agitated flow in which there are random velocity fluctuations on top of the average flow.

U-tube manometer   A glass tube in the shape of the letter U that is used to measure pressure or pressure differences.

Ultrasonic probe   An instrument using high-frequency sound waves to measure fluid levels.

Vacuum (pressure)   The amount that the measured pressure is below atmospheric pressure.

Velocity   Is a measure of speed, and in a flow is the average speed across the flow and the direction of movement of a liquid.

Vena contracta   The narrowing down of the fluid flow stream as it passes through an obstruction.

Venturi tube   A specially shaped restriction in a section of pipe that provides a pressure drop which can be related to flow rate.

Viscometer (viscosimeter)   An instrument for measuring viscosity.

Viscosity   The term describing the resistance to flow of a fluid.

Volt   The unit of electromotive force.

Voltage   An electromotive force that causes electrons or a current to flow.

Voltage drop   The difference in voltage between two points.

Vortex   Swirling or rotating fluid motion.

Wavelength   The time for an alternating source to complete a full cycle.

Weir   An open-channel flow-measuring device.

Wet-bulb temperature   The temperature indicated by a thermometer whose sensing element is kept moist.

Wheatstone bridge   The most common electrical bridge circuit used to measure small changes in the value of an element.

Answers to Odd-Numbered Questions

Chapter 1: Introduction

1.1 The controlled variable is the monitored or measured output variable from a process that must be controlled to within set limits. The manipulated variable is the input variable to a process that is controlled by a signal from a controller to an actuator. By controlling the input variable, the output variable is held to within its set limits.

1.3 $1 \text{ lb} = 0.454 \text{ kg}$
$63 \text{ kg} = (63/0.454) \text{ lb} = 138.77 \text{ lb}$

1.5 $1 \text{ psi} = 6.897 \text{ kPa}$
$38.2 \text{ kPa} = (38.2/6.897) \text{ psi} = 5.54 \text{ psi}$

1.7 $1 \text{ lb} = 4.448 \text{ N}$
$385 \text{ N} = (385/4.448) \text{ lb} = 86.55 \text{ lb}$

1.9 $1 \text{ ft·lb} = 1.356 \text{ J}$
$27 \text{ ft·lb} = (1.356 \times 27) \text{ J} = 36.6 \text{ J}$

1.11 % FSD accuracy $= \pm (3 \times 100/120)\% = \pm 2.5 \text{ percent}$

1.13 % FSD accuracy $= \pm (2 \times 100/125)\% = \pm 1.6 \text{ percent}$
% Span accuracy $= \pm (2 \times 100/95)\% = \pm 2.1 \text{ percent}$

1.15 Span is the total range of the scale and full scale is the maximum reading.

1.17 The environmental concerns are spills, leaks of hazardous gases, emissions, and dumping of waste.

1.19 See Figure A1.1.
Accuracy = 10 percent Reading
Accuracy = 10 percent of FSD

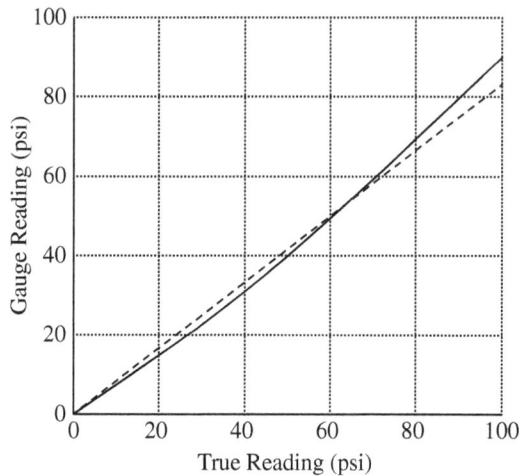

Graphs for Problem 1.19.

Chapter 2: Pressure

2.1 $p = \gamma h$

$h = \dfrac{17.63 \times 12 \times 12}{62.4}$ ft $= 40.66$ ft

2.3 1 psf $= 0.048$ kPa

1038 psf $= 1038 \times 0.048$ kPa $= 49.8$ kPa

2.5 Volume $= 2.2 \times 3.1 \times 1.79 = 12.2078$ ft³

$SW = \dfrac{1003 - 173}{12.2}$ lb/ft³ $= 67.98$ lb/ft³

$SG = 67.98/64.2 = 1.09$

2.7 Pressure $=$ Force/area $= \dfrac{763 \times 4}{3.14 \times 3.2 \times 3.2} = \dfrac{27 \times 4}{3.14 \times r^2}$

$r = \sqrt{\dfrac{27 \times 3.2^2}{763}}$ ft $= \sqrt{0.36}$ ft $= 0.6$ ft $= 7.2$ in

2.9 Buoyancy force $= (15.5 - 8.7) \times 9.8$ N $= 66.64$ N

$V = \dfrac{66.64}{9.8 \times 770}$ m³ $= 0.0088$ m³

$SW = \dfrac{15.5 \times 9.8}{0.0088}$ N/m³ $= 17.26$ kN/m³ $= 1761$ kg/m³

2.11 $SW = 7.38 \times 62.43$ lb/ft³ $= 460.7$ lb/ft³

$SW = 7.38 \times 1000$ kg/m³ $= 7380$ kg/m³

2.13 Force $= 2.9 \times 1.7 \times 14.3 \times 12 \times 12$ lb $= 10{,}151.86$ lb

2.15 1.9 m $= \dfrac{1.9}{0.305}$ ft $= 6.23$ ft

10.3 cm $= \dfrac{10.3}{100 \times 0.305}$ ft $= 0.34$ ft

$$\text{Pressure} = 6.23 \times 62.43 + 0.34 \times 62.43 \times 13.55 + 14.7 \times 12 \times 12 \text{ psfa}$$
$$= 389 + 287.5 + 2116.8 \text{ psfa} = 2793.3 \text{ psfa}$$

2.17 $\text{Force} = \left(\dfrac{8.7 \times 0.305}{12}\right)^2 \times \dfrac{3.7 \times 3.14}{4} \text{ N} = 0.14 \text{ N}$

2.19 $\text{Pressure} = 270 \times 0.019 \text{ psig} = 5.13 + 14.7 \text{ psia} = 19.83 \text{ psia}$

Chapter 3: Level

3.1 $p = \gamma h$

$\gamma = \dfrac{4.7 \times 144}{17} \text{ lb/ft}^3 = 39.8 \text{ lb/ft}^3$

3.3 $15 \text{ lb} = 15 \times 4.448 \text{ N} = 66.72 \text{ N} = 6.81 \text{ kg}$
$V = 6.81/785 \text{ m}^3 = 0.00867 \text{ m}^3$

3.5 $\text{Buoyancy} = 17 - 3 \text{ lb} = 14 \text{ lb}$

$V = \dfrac{14}{62.4} \text{ ft}^3 = 0.22 \text{ ft}^3$

$\gamma = \dfrac{17}{0.22} \text{ lb/ft}^3 = 77.27 \text{ lb/ft}^3$

3.7 $\text{Weight of liquid} = 533 - 52 \text{ lb} = 481 \text{ lb}$

$L = \dfrac{4W}{\gamma \pi d^2} = \dfrac{4 \times 481}{63 \times 3.14 \times 4.5^2} \text{ ft} = 0.48 \text{ ft} = 5.7 \text{ in}$

3.9 $d = \dfrac{(Cd - Ca)r}{\mu Ca}$

$\mu = \dfrac{(Cd - Ca)r}{d \times Ca} = \dfrac{(283 - 25)13 \times 12}{4 \times 31 \times 25} = 13$

3.11 $\text{Weight} = p \times A = \dfrac{32 \times 3.14 \times 3.2^2}{4} \text{ N} = 257.4 \text{ N} = 26 \text{ kg}$

3.13 $d = p/\gamma = \dfrac{28}{560} \text{ m} = 0.05 \text{ m} = 5 \text{ cm}$

3.15 $d = \sqrt{\dfrac{\delta F}{\gamma \pi h}} = \sqrt{\dfrac{3.2 \times 12}{33 \times 3.14 \times 45}} \text{ ft} = 0.0907 \text{ ft} = 1.09 \text{ in}$

3.17 $d = (Cd - Ca) \, r/\mu Ca = \dfrac{(7400 - 157) \times 2.7}{79 \times 157} \text{ m} = 1.58 \text{ m}$

3.19 $t = d/\text{Vel} = \dfrac{2 \times 10.5 \times 0.305}{340} \text{ s} = 0.019 \text{ s} = 19 \text{ ms}$

Chapter 4: Flow

4.1 $Q = VA$

$V = Q/A = \dfrac{3.2 \times 4 \times 12 \times 12}{3.14 \times 7 \times 7} \text{ ft/s} = 11.98 \text{ ft/s}$

4.3 $d = \sqrt{\dfrac{239 \times 0.1337 \times 4}{60 \times 27 \times 3.14}}$ ft $= 0.158$ ft $= 1.9$ in

4.5 $Q = \dfrac{0.73 \times 3.14 \times 23 \times 23}{4 \times 100 \times 100}$ m³/s $= 0.03$ m³/s $= 30$ L/s

$d = \sqrt{\dfrac{0.03 \times 4}{3.14 \times 1.66}}$ m $= 0.152$ m $= 15.2$ cm

4.7 $Q = \dfrac{3.14 \times 5.5 \times 5.5 \times 97 \times 0.1337}{4 \times 12 \times 12}$ ft³/s $= 2.14$ ft³/s $= 16$ gal/s

$Q\,(3.2) = \dfrac{2.4 \times 3.2 \times 3.2}{3.2 \times 3.2 + 1.8 \times 1.8}$ ft³/s $= 1.63$ ft³/s $= 12.19$ gal/s

$Q\,(1.8) = \dfrac{2.4 \times 1.8 \times 1.8}{3.2 \times 3.2 + 1.8 \times 1.8}$ ft³/s $= 0.514$ ft³/s $= 3.85$ gal/s

4.9 $m \times h = mV^2/2g$
$V = \sqrt{(2 \times 32.2 \times 273)}$ ft/s $= 132.6$ ft/s

4.11 $\dfrac{p_1}{\gamma} + \dfrac{V_1^{\,2}}{2g} + Z_1 = \dfrac{p_2}{\gamma} + \dfrac{V_2^{\,2}}{2g} + Z_2$
$0 + 0 + 17 = 0 + V_2^2/2g + 1.5$
$V_2 = \sqrt{15.5 \times 2 \times 32.2} = 31.6$ ft/s

4.13 $h_L = \dfrac{fLV^2}{2Dg}$

$h_L = \dfrac{0.027 \times 118 \times 17 \times 17 \times 12}{7 \times 2 \times 32.2}$ ft $= 24.5$ ft

4.15 $A_2 = \dfrac{3.14 \times 4.1 \times 4.1}{4 \times 12 \times 12}$ ft² $= 0.092$ ft²

$A_1 = \dfrac{3.14 \times 9.7 \times 9.7}{4 \times 12 \times 12}$ ft² $= 0.51$ ft²

$V_2 = \dfrac{28200 \times 0.1337}{60 \times 60 \times 0.092}$ ft/s $= 11.38$ ft/s

$V_1 = \dfrac{11.38 \times 0.092}{0.51}$ ft/s $= 2.05$ ft/s

$\dfrac{p_1}{\gamma} + \dfrac{V_1^{\,2}}{2g} + Z_1 = \dfrac{p_2}{\gamma} + \dfrac{V_2^{\,2}}{2g} + Z_2$

$\dfrac{p_1}{62.4} + \dfrac{2.05 \times 2.05}{2 \times 32.2} + 0 = \dfrac{(65 + 14.7)144}{62.4} + \dfrac{11.38 \times 11.38}{2 \times 32.2} + 0$

$p_1 = (183.92 + 2.01 - 0.065)62.4 = 11\,597.97$ psfa $= 80.54$ psia $= 65.84$ psig

4.17 $F = \dfrac{C_D \gamma A V^2}{2g}$

$V = \sqrt{\dfrac{4.8 \times 2 \times 32.2 \times 4 \times 12 \times 12}{0.35 \times 0.79 \times 6.3 \times 6.3 \times 3.14 \times 62.4}}$ ft/s $= 9.1$ ft/s

4.19 $8 \times 32/\text{rev} \times 570$ rev $= 145\,920$ in³/h
Flow rate $= 145\,920/231$ gal/h $= 631.7$ gal/h $= 631.7/60$ gpm $= 10.53$ gpm

Chapter 5: Temperature and Heat

5.1 $°F = (°C \times 9/5) + 32 = (115 \times 9/5) + 32 = 239°F$
$°F = (456 - 273)9/5 + 32 = 361.4°F$
$°F = -460 + 423 = -37°F$

5.3 $°C = (115 - 32) \times 5/9 = 46.1°C$
$°C = 356 - 273 = 83°C$
$°C = (533 \times 0.555) - 273 = 22.81°C$

5.5 $\text{Heat} = 3 \text{ ft}^3 \times 62.43 \text{ lb} \times 15 \text{ BTU} = 2809.35 \text{ BTU} \times 252 = 708 \text{ kcal}$

5.7 $T \text{ increase} = \dfrac{4.3 \times 0.092 \times 50 \times 13.2 \times 17 \times 60}{1055} °F = 258.2°F$

5.9 $\text{Heat} = \dfrac{220 \times 3.14 \times 7 \times 7 \times 36.4 \times 12}{4 \times 12 \times 12 \times 9} \text{ BTU/h} = 2852.11 \text{ BTU/h}$

5.11 $\dfrac{220 \times 3.14 \times 7 \times 7 \times 12(59.4 - t)}{24 \times 24 \times 9} = 30 \times 0.22(t - 23)$
$78.35 \,(59.4 - t) = 6.6 \,(t - 23)$
$11.87 \times 59.4 - 11.87t = t - 23$
$t = 705/12.87 = 54.8°F$

5.13 $Q = 15 \times 19 \times 0.19 \times 10^{-8}\{(125 + 460)^4 - (74 + 460)^4\} = 0.5415 \times 10^{-8} (11.712{-}8.131) \, 10^{10}$
$Q = 194 \text{ BTU/h}$

5.15 $\dfrac{2.5}{12} = 115 \times 156 \times \alpha$

$\alpha = \dfrac{2.5}{12 \times 115 \times 156} = 1.15 \times 10^{-5}/°F = 11.5 \times 10^{-6}/°F$

5.17 $R_2 = R_1 \,(1 + \delta R \,\{342 \times 5/9\})$

$3074/2246 = 1 + 190\delta R$
$\delta R = 0.3686/190 = 1.94 \times 10^{-3} \, \Omega/°C$

5.19 $V_{\text{out}} = (1773 - 67) \times 40 \times 10^6 \times 5/9 = 0.0379 \text{ V} = 37.9 \text{ mV}$

Chapter 6: Humidity, Density, and pH

6.1   Relative humidity = (a) 33 percent, (b) 20 percent, (c) 12 percent

6.3   Relative humidity = 64 percent

6.5   Relative humidity = 18 percent

Absolute humidity = 0.01 lb/lb (70 grains/lb)

6.7   25 percent: 0.0114 lb/lb (80 grains/lb), 95 percent: 0.0343 lb/lb (240 grains/lb)

Water required = 0.0343 − 0.0114 lb/lb (240 − 80 grains/lb)

$= 0.0229$ lb/lb (160 grains/lb)

6.9   Water = 0.027 lb/lb (190 grains/lb)

6.11   Space = 14 ft^3/lb × 4.7 lb = 65.8 ft^3

6.13 $SW = 32.2 \times 1.234$
$p = 32.2 \times 1.234 \times 54 \text{ psf} = 2145.7 \text{ psf} = 14.9 \text{ psi}$

6.15 $F = \dfrac{\mu A V}{y} = \dfrac{7.3 \times 2 \times 1.2 \times 1.2 \times 14.7 \times 12}{0.11 \times 10^5} \text{ lb} = 0.337 \text{ lb}$

6.17 $pH = \log 1/0.0006 = 3.22$

6.19 $13.2 = \log 1/c$
$c = \dfrac{1}{10^{13.2}} = \dfrac{1}{1.58 \times 10^{13}} \text{ g/L} = 6.33 \times 10^{-14} \text{ g/L}$

Chapter 7: Position, Motion, and Force

7.1 $F = ma = 17 \times 21 \text{ lb} = 357 \text{ lb}$

7.3 $\text{Torque} = Fd = 33 \times 13 \text{ lb·ft} = 429 \text{ lb·ft}$

7.5 $\text{Couple} = Fd$
$d = 53/15 \text{ m} = 3.5 \text{ m}$

7.7 $w_1 \times d_1 = w_2 \times d_2$
$d_2 = \dfrac{10 \times 0.5}{16} \text{ m} = 0.31 \text{ m} = 31 \text{ cm}$

7.9 $w = 3 \times 2.7 \text{ lb} = 8.1 \text{ lb}$
$\text{weight of basket} = 8.1 - 6 \text{ lb} = 2.1 \text{ lb}$

7.11 $p = f/A = \dfrac{10 \times 10^4}{75 \times 75 \times 3.14} \text{ Pa} = 5.66 \text{ Pa}$

7.13 $\lambda = v/f = \dfrac{340}{13 \times 10^3} \text{ m} = 0.026 \text{ m} = 2.6 \text{ cm}$

7.15 $\text{Difference} = 10 \log \dfrac{375}{125} \text{ dB} = 4.77 \text{ dB}$

7.17 $3.83 = 10 \log d/20$
$d = 10^{0.383} \times 20 = 2.415 \times 20 \text{ ft} = 48.3 \text{ ft}$

7.19 $\text{Angular sensitivity} = \dfrac{360}{115 \times 16} = 0.2°$

Chapter 8: Safety and Alarm

8.1   The first trip system is the process logic controller.

8.3   Fire sensors are ionization chambers, photoelectric sensors, and heat sensors.

8.5   Oxygen sensors are needed to ensure that there is an adequate oxygen supply for personnel and that it has not been replaced by cryogenic substances.

8.7   The natural radiation from the earth is from 0.2 rem to 1 rem per day.

8.9   A SIS system is used for handling critical trips and is the second line of defense to the PLC.

8.11   The estimated logic failure rates in a SIS are 8 percent.

8.13 An accelerometer is used to check the acceleration rate of equipment such as conveyer belts to ensure the rate of starting and stopping is not excessive.

8.15 Environmental hazards are emissions, ground contamination, water pollution, ordinance, fires, industrial waste, and radiation.

8.17 To protect personnel safety barriers, limited access, motor kill switches, protective clothing, oxygen masks, fire alarms, gas detectors, oxygen sensors, and radiation detectors.

8.19 The types of heat sensors are fixed temperature detectors and rate of temperature rise detector.

Chapter 9: Electrical Instruments and Conditioning

9.1 Hysteresis is the difference in readings obtained when an instrument reads a signal when going from zero and the reading of the same signal when going from full-scale deflection.

9.3 % FSD accuracy $= \pm (0.5/129.9)\% = \pm 0.385$ percent
Absolute accuracy $= \pm 0.05°C$
Resolution $= \pm 0.1°C$

9.5 % FSD accuracy $= \pm (5 \times 100/150)\% = \pm 3.3$ percent
% Span accuracy $= \pm (5 \times 100/100)\% = \pm 5$ percent

9.7 $I_{out}/E_{in} = 8.5 = 100/3.5 \, R_3$
$R_3 = 100/8.5 \times 3.5 \, k\Omega = 3.36 \, k\Omega$

9.9 The two magnetic field sensors most commonly used are the Hall effect device and the magneto resistive element.

9.11 To generate a 0.21 V reference voltage from 10 V (see Figure A9.1)
$0.21(10 + R_1) = 10 \times R_1$
$R_1 = \dfrac{2.1}{9.79} \, k\Omega = 0.2145 \, k\Omega = 214.5 \, \Omega$
Feedback resistor $(R_2) = \dfrac{5 \times 10}{0.56 - 0.21} \, k\Omega = 142.86 \, k\Omega$

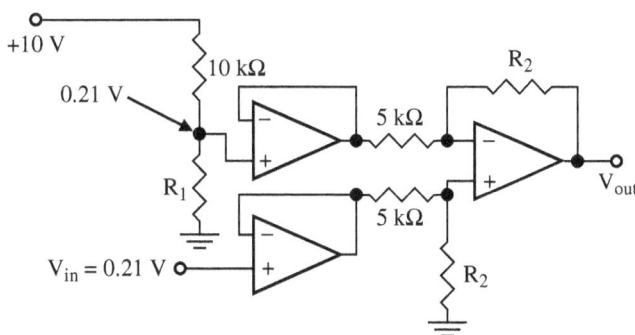

FIGURE A9.1 Circuit for use with Problem 9.11.

9.13 Strain gauges are normally mounted in pairs at right angles, so that only one gauge is under strain and the other gauge is used for temperature correction in a resistive bridge circuit.

9.15 Amplifiers with nonlinear elements in their feedback can be used for linearization, such as logarithm and antilogarithm amplifiers.

9.17 The pressure ranges used in pneumatic equipment are 3 to 15 psi or 6 to 30 psi. Zero is not used because of the difficulty of transmitting low pressures, and it can be used as a failure mode.

9.19 A differential capacitive measurement can be made in an ac bridge or using digital techniques. An external capacitor can be used with a single capacitance to obtain a differential capacitance.

Chapter 10: Regulators, Valves, and Actuators

10.1 Regulators are self-compensating pressure reducers. The regulators can have internal or external feedback and can use spring, weight, or external pressure for a reference.

10.3 An instrument pilot-operated pressure regulator is a pressure regulator that is an externally compensated regulator; it uses an external air supply to obtain feedback amplification to enhance regulation and range.

10.5 Regulators can be spring, weight, or pressure loaded.

10.7 A positioner is used to compare the operating position of a valve in a smart control loop to the set point.

10.9 A smart controller is used to monitor the flow rate in a line and adjust the operating position of a valve to give the flow rate set by the controller.

10.11 The current range used in control is from 4 to 20 mA.

10.13 The common types of plugs are quick opening, linear, and equal percentage.

10.15 There are five valve families in common use, they are globe, butterfly, diaphragm, ball, and rotary plug.

10.17 $C_V = Q \times \sqrt{(SG/P_d)}$

$$P_d = \left(\frac{Q}{C_V}\right)^2 \times SG = \left(\frac{1.8 \times 60}{88}\right)^2 \times \frac{78}{62.4} \text{ psi} = 1.51 \times 1.25 \text{ psi} = 1.9 \text{ psi}$$

10.19 A butterfly valve is used to control flow rate with a disc attached to a rotating shaft.

Chapter 11: Process Control

11.1 ON/OFF action is the simplest form of control. The output variable from a process is compared to a reference, turning the control signal to the input variable to the process "ON" or "OFF" depending on whichever is the greater.

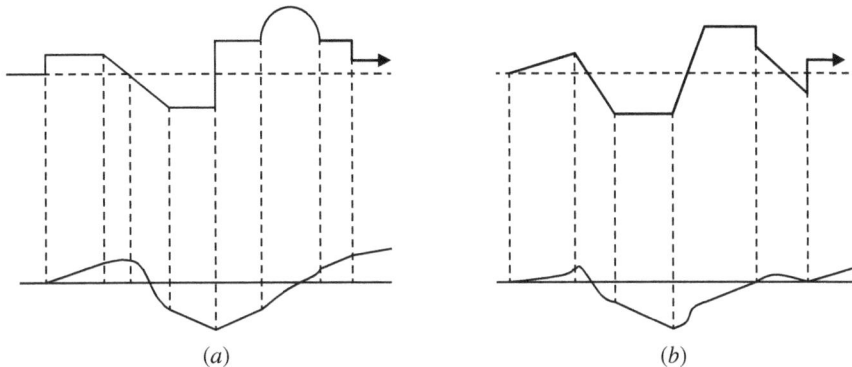

(a)   (b)

FIGURE A11.1   Answers for Problems 11.7 and 11.9.

11.3   In proportional action, the amplitude of the output variable is compared to a reference, giving an output error signal with an amplitude proportional to the amount of the deviation of the variable signal from the reference signal. The error signal is then used to control the input variable by an amount proportional to the amplitude of the error signal.

11.5   Derivative, rate, or anticipatory action is used to reduce the correction time that occurs with proportional action alone. Derivative action senses the rate of change of the measured variable, and applies a correction signal that is proportional to the rate of change of the measured variable only.

11.7   See Figure A11.1a.

11.9   See Figure A11.1b.

11.11 "ON/OFF" sensing can be used for level sensing, positioning sensing, limit sensing, HVAC, etc.

11.13 The measured variable is the amplitude of the signal being measured. The error signal is the difference between the measured variable and the set point.

11.15 The error signal is the difference between the measured variable and the set point. The offset is that fraction of the error signal, which when amplified produces a correction signal for a change in the measured variable.

11.17 Dead-band is a set hysteresis between the turn "ON" level and turn "OFF" levels in a system to prevent rapid switching between the "ON" and "OFF" points.

11.19 Derivative action is not normally used for pressure control, level control, or flow control.

Chapter 12: Documentation and Symbol Standards

12.1   A hydraulic supply line.

12.3   Discrete and inaccessible to operator voltage indicator.

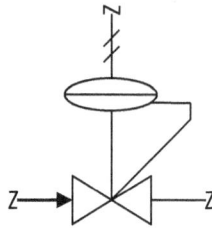

FIGURE **A12.1** Pressure-loaded regulator.

12.5 A converter used to change a 3 to 15 psi pressure measurement to a 4 to 20 mA current measurement from a flow sensor.

12.7 Two-way digitally operated valve, which is closed in the power-fail mode.

12.9 Conveyer belt with a weight measurement sensor and transmitter.

12.11 Documentation must be kept up-to-date to prevent time lost in maintenance, repair, and modifications, as well as to prevent catastrophic errors.

12.13 P and IDs are normally developed by an engineering team made up of engineers from process engineering and control engineering.

12.15 Information in a PLC documentation should be:

System overview and description of control process.

Block diagram of units in the system.

List of all inputs and outputs, destination, and number.

Wiring diagram of I/O modules, address of I/O points, and rack location.

Rung description, number, and function.

12.17 The SIS is an alarm and trip system to alert operators or maintenance of a malfunction, to shut down a system in an orderly fashion when a malfunction occurs, or to switch failed units over to standby units.

12.19 See Figure A12.1.

Chapter 13: Signal Transmission

13.1 Data can be transmitted as analog signals using voltage or current levels, or as a digital transmission over hard wired connections, digital transmission can be used over fiber-optic cables or as RF signals.

13.3 RTDs use two-wire, three-wire, or four-wire connections. The two-wire system is the least expensive with the four-wire system the most expensive but most accurate. The three-wire system uses compensation to correct for errors introduced in the wiring so that it approaches the accuracy of the four-wire system at medium cost.

13.5 Two techniques are normally used to convert digital-to-analog signals. A resistor network can be used to convert the signals or pulse width modulation can be used.

13.7 There are several digital transmission standards, the two most common were the IEEE-488 ("1" > 2 V and "0" < 0.8 V) and the RS-232 ("1" +3 to +25 V and "0" −3 to −25 V) but in many cases are being replaced by other standards.

13.9 Digital signals transfer data faster and more accurate than analog signals, are unaffected by noise, can be isolated if the ground voltage levels are different, can be transmitted over very long distances without loss of accuracy, and data can be stored.

13.11 Foundation fieldbus has a transmission speed of 31.25 kb/s for the H1 and 100 Mb/s for the HSE.

13.13 Analog signals are dc signals whereas FSK signals are low-level ac signals which can be separated from the analog signals using filtering techniques.

13.15 Amplitude modulation uses less power than frequency modulation conserving on battery power.

13.17 Pneumatic signals are used in place of electrical signals for safety reasons, such as when there is a chance that a spark from an electrical signal could ignite combustible material, or cause an explosion in a volatile atmosphere.

13.19 There are 2^{12} steps—1 for zero when using a 12-bit DAC, or 4095 steps, the percentage resolution is 0.024 percent.

Chapter 14: Logic Gates

14.1

A	B	C	Y	$\overline{Y}$
0	0	0	0	1
0	0	1	0	1
0	1	0	0	1
0	1	1	0	1
1	0	0	0	1
1	0	1	0	1
1	1	0	0	1
1	1	1	1	0

TABLE **A14.1**   Truth Table for Problem 14.1

14.3

A	B	C	Y	$\overline{Y}$
0	0	0	0	1
0	0	1	1	0
0	1	0	1	0
0	1	1	0	1
1	0	0	1	0
1	0	1	0	1
1	1	0	0	1
1	1	1	1	0

TABLE A14.2   Truth Table for Problem 14.3

14.5

A	B	A + B	$\overline{A + B}$
0	0	0	1
0	1	1	0
1	0	1	0
1	1	1	0

A	B	$\overline{A}$	$\overline{B}$	$\overline{A} \cdot \overline{B}$
0	0	1	1	1
0	1	1	0	0
1	0	0	1	0
1	1	0	0	0

TABLE A14.3   Truth Table for Problem 14.5

14.7

A	B	$\overline{A}$	$\overline{B}$	$\overline{A} + \overline{B}$	$\overline{\overline{A} + \overline{B}}$
0	0	1	1	1	0
0	1	1	0	1	0
1	0	0	1	1	0
1	1	0	0	0	1

A	B	$A \cdot B$
0	0	0
0	1	0
1	0	0
1	1	1

TABLE A14.4   Truth Table for Problem 14.7

14.9 $0037 = 100101$

14.11 $111000111010 = 1110\text{-}0011\text{-}1010 = \text{E-3-A}$

14.13

$Y = \overline{\overline{A} \cdot \overline{B} \cdot \overline{C}}$

TABLE A14.5   Equation for Problem 14.13

14.15

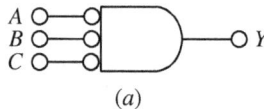

(a)

FIGURE A14.1   (a) Gate for Problem 14.15.

14.17

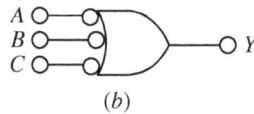

FIGURE **A14.1** (b) Gate for Problem 14.17.

14.19

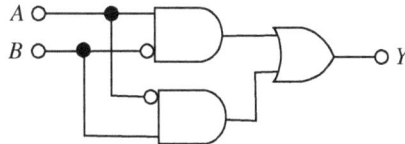

FIGURE **A14.2** Gates for Problem 14.19.

Chapter 15: Programmable Logic Controllers

15.1 A PLC has two modes of operation—the I/O scan mode where the inputs are scanned for new data, and the execution mode where the PLC evaluates the data received and sends out the necessary corrective action signals.

15.3 A ladder diagram is a method of programming a PLC, by representing the interconnections between the elements by symbols, the interconnections look similar to the rungs in a ladder.

15.5 The HART protocol is a hybrid protocol that can combine a single analog signal with low-level digitally encoded FM signals.

15.7 Electrical isolation is obtained using optical isolators.

15.9 The resolution of the input analog to digital converter is 1.2207 µA.

15.11 Programming languages are ladder instruction list, Boolean flowchart, functional blocks, sequential function charts, and high-level language (ANSI, C, structured text).

15.13 The number of drivers in an output module are 4 or 8.

15.15 The types of memory used in a PLC are ROM, RAM, and EEPROM.

15.17 The main blocks in an ac output module are optical isolator, a D to A converter, amplifier, and driver.

15.19 Intelligent modules are serial and network, communication, computer coprocessor, closed loop, position and motion, process specific, artificial intelligence, and HART protocol.

Chapter 16: Motor Control

16.1   The main motor classifications are dc, ac, and universal.

16.3   The position of the magnet is sensed with a magnetic field sensor such as a Hall effect device or magneto resistive element.

16.5   An out of phase supply can be generated with an inductive element such as an inductor or capacitor.

16.7   The magnetic field in the rotor is induced by the magnetic field in the stator windings.

16.9   The difference between an ac motor and stepper motor is an ac motor is used as a continuous power source where as a stepper is low power, synchronized to the ac power and can be used to step at specified increments.

16.11   There are four insulation classifications.

16.13   Power = torque × speed ÷ 5252

Therefore torque = 5252 × 1.65 ÷ 1873
                              = 4.627 lb·ft

16.15   Sync speed = 120 × 50 ÷ 8
                            = 750

Percent slip = sync speed – rated speed × 100 ÷ sync speed
                      = 750 – 625 × 100 ÷ 750
                      = 16.6 percent

16.17   An ac motor is normally started with a reduced supply voltage using series resistors, auto transformer, or impedance in series with windings, a variable frequency supply is used, or with a wound rotor.

16.19   Jogging is powering the motor in small steps to obtain small incremental changes in its angular position, and plugging is braking of the motor by reversing its connections.

Index

www.ingramcontent.com/pod-product-compliance
Lightning Source LLC
Chambersburg PA
CBHW080923220326
41598CB00034B/5659